NATURAL and ENGINEERED SOLUTIONS for DRINKING WATER SUPPLIES

Lessons from the Northeastern United States and Directions for Global Watershed Management

Edited by
Emily Alcott • Mark S. Ashton
Bradford S. Gentry

CRC Press is an imprint of the
Taylor & Francis Group, an informa business

CRC Press
Taylor & Francis Group
6000 Broken Sound Parkway NW, Suite 300
Boca Raton, FL 33487-2742

Printed in the United States of America on acid-free paper
Version Date: 20121030

International Standard Book Number: 978-1-4665-5164-0 (Hardback)

Library of Congress Cataloging-in-Publication Data

Natural and engineered solutions for drinking water supplies : lessons from the Northeastern United States and directions for global watershed management / editors, Emily Alcott, Mark S. Ashton, Bradford S. Gentry.
p. cm.
Includes bibliographical references and index.
ISBN 978-1-4665-5164-0 (hardback)
1. Watershed management--Northeastern States. 2. Drinking water--Purification--Cost effectiveness. 3. Wellhead protection--Northeastern States. I. Alcott, Emily. II. Ashton, Mark S. III. Gentry, Bradford S.

TC423.1.N38 2013
628.1'10974--dc23 2012037918

Visit the Taylor & Francis Web site at
http://www.taylorandfrancis.com

and the CRC Press Web site at
http://www.crcpress.com

Contents

Preface

Almost every large city in a forested biome in the northeastern United States is dependent upon an upland watershed for some or all of its drinking water supply. Some cities have protected most of their watershed and manage their forests to act as a natural filter from water pollutants. Other cities have developed watersheds, therefore relying upon engineering of filtration plants to provide drinking water that meets applicable quality standards. Still others have adopted a blend of built and natural infrastructure, combining watershed protection and filtration/treatment.

In this book we evaluate how, when, and where six cities in the northeastern United States have made environmental, economic, and social decisions to protect and manage upland forests to produce drinking water as a downstream service. This book analyzes a series of city watersheds that have been managed under different state regulations, planning and development incentives, biophysical constraints, social histories, and ownerships.

Some of the overarching questions that this book addresses relate to how managers should optimize the investments in their drinking water systems. What is the balance between the use of concrete/steel treatment plants and forested/grassland/wetland areas to protect surface water quality? We seek to provide insights through the use of case studies comparing how engineered and/or natural systems are employed to protect water quality. To this end, our book seeks to (a) understand the decision processes shaping how municipal water systems choose to deliver safe drinking water to residents from surface water sources; and (b) illuminate opportunities to develop a more integrated approach to system design, particularly in an increasingly energy- and carbon-constrained world.

This book is a product of a seminar for graduate professional students at the Yale School of Forestry and Environmental Studies that focused on payments for ecosystem services. The seminar was led jointly by Bradford S. Gentry, professor in the practice of sustainable investments and director of the Yale Center for Business and the Environment, and Mark S. Ashton, Morris K. Jessup professor of silviculture and forest ecology. The seminar was part of the professional master's degree curriculum and took place in spring 2010.

The seminar started with a thorough review of the current literature on the topics being covered, followed by in-depth class discussion. Leaders in the field were invited to give seminal talks, followed by lengthy discussion and debate with the class, to help set the stage for the students' review and analysis. The class also spent a week in the Seattle area, speaking with and observing the range of efforts underway to use watershed management to provide high-quality drinking water to the cities of Seattle and Tacoma. The resulting papers were written by the graduate students in the seminar, with input from faculty at Yale and elsewhere, followed by external and internal reviews primarily by the original workshop participants and Yale faculty.

Overall, this book supplies analysis on alternative scientific, management, and policy approaches to protect the quality of drinking water supplied from surface watersheds. It contains recommendations for managers and policymakers that reflect the scientific realities of how forests and engineering can be integrated and under what circumstances. We welcome comments and feedback. This is a work in progress amid an evolving scientific understanding of a complex topic and an equally complex dialogue on the role of forests and engineered systems in the future management of our surface drinking watersheds.

Acknowledgments

The Yale School of Forestry and Environmental Studies (YF&ES) funded the seminar, workshops, and field visits, as well as the initial costs for the publication of this book. We are very grateful to former dean, Gus Speth, who enthusiastically supported the idea and provided the funding.

We also wish to acknowledge our colleagues on the YF&ES faculty (Shimon Anisfeld, Gaboury Benoit, Mathew Kotchen, Julie Newman, and Julie Zimmerman) and graduate students who co-led the seminars, hosted guest speakers, and reviewed many drafts of the papers. We especially thank Erin Derrington for helping to edit the case study chapters.

We benefited tremendously from our guest speakers and other collaborators in this effort, not only from their presentations, but especially from the discussions that followed and their willingness to continue the engagement through outside reviews and input into the manuscripts as they were developed. They enthusiastically engaged with the students. We thank those who took time out of their busy schedules to travel to New Haven, Connecticut, as well as all of those who shared their knowledge with the students during our field trip to Seattle and Tacoma, Washington. We also thank David Tobias at NYC Department of Environmental Protection and Mark Hunt for the use of their photographs on the front cover of this book. In particular, we thank the following individuals

Joel Anderson—Chief operator, Sebago Lake Water Treatment Facility, Portland Water District, Maine
Al Appleton—Former commissioner, NYC Department of Environmental Protection, New York
Paul Barten—Professor of Water Resource Management, University of Massachusetts
Claire Bennitt—Former chair, South Central Regional Water Authority, Connecticut
J. Paul Blake—Director, Communications, Seattle Public Utilities, Washington
Jesse Bloomfield—Forester, Olympic Resources Management, Washington
Maria Calvi—Restoration ecologist, Tualip Tribes, Washington
Thomas Chaplik—VP water quality, South Central Regional Water Authority, Connecticut
Craig Chatburn—Green stormwater infrastructure specialist, Seattle Public Utilities, Washington
Alex Chen—Senior water quality engineer, Seattle Public Utilities, Washington
Chuck Clarke—CEO, Cascade Water Alliance, Washington
Paul Fleming—Manager, Climate and Sustainability, Seattle Public Utilities, Washington
Jerry Franklin—Professor of ecosystem analysis, University of Washington
Ray Fryberg—Director, Fish and Wildlife, Tualip Tribes, Washington
Peter Galant—Consultant, Tighebond, Bridgeport, Connecticut
Diana Gale—Former director, Seattle Public Utilities, Washington
Todd Gardner—Director of Ecosystem Services, World Resources Institute, Washington, DC

John Gunn—Senior program leader, Manomet Center for Conservation Sciences, Maine
Fred Gliesing—Watershed forester, Croton, NYC Department of Environmental Protection, New York
Ray Hoffman—Acting director, Seattle Public Utilities, Washington
Cyndy Holtz—Major watersheds business area manager, Seattle Public Utilities, Washington
Abby Hook—Hydrologist, Tualip Tribes, Washington
John Hudak—Environmental planning manager, South Central Regional Water Authority, Connecticut
Paul Thomas Hunt—Environmental manager, Portland Water District, Maine
Chris Johnson—Distribution engineer, Tacoma Public Utilities, Washington
Bryan King—Watershed manager, Tacoma Public Utilities, Washington
Thom Kyker-Snowman—Natural resource specialist, Department of Conservation and Recreation, Massachusetts
Amy LaBarge—Senior forest ecologist, Seattle Public Utilities, Washington
Rand Little—Senior fish biologist, Seattle Public Utilities, Washington
Hilary Lorenz—Treatment facility manager, Tacoma Public Utilities, Washington
Robert Mack—Deputy director for Government and Community Affairs, Tacoma Public Utilities, Washington
Derek Marks—Manager, Timber, Fish, and Wildlife, Tualip Tribes, Washington
Brad Marten—Marten Law, Washington
Linda McCrea—Water superintendent, Tacoma Public Utilities, Washington
Ralph Naess—Public and cultural programs manager, Seattle Public Utilities, Washington
Jim Nilson—Senior water quality engineer, Washington
Valerie O'Donnell—Watershed forester, South Central Regional Water Authority, Connecticut
William Pula—Chief engineer, Quabbin Reservoir, Department of Conservation and Recreation, Massachusetts
Mark Quehrn—Bellevue's office managing partner, Perkins Coie, Washington
J. Kevin Reilly—U.S. Environmental Protection Agency, Region I, Massachusetts
Martha Smith—Former director of the Center for Coastal and Watershed Systems, Yale University, Connecticut
Tracy Stanton—Senior program manager, Earth Economies, Washington
Dan Stonington—Policy director, Cascade Land Conservancy, Washington
Andy Tolman—Water resources team leader, Department of Health and Human Services, Maine
Alison van Gorp—Senior advisor at City of Seattle, Office of the Mayor, Washington
Greg Volkhardt—Environmental programs manager, Tacoma Public Utilities, Washington
David Warne—Assistant commissioner, NYC Department of Environmental Protection, New York
Carol Youell—Natural resources administrator, Metropolitan District Commission, Hartford, Connecticut

Editors

Emily Alcott, MES, is a fluvial geomorphologist and an ecologist who works for Inter-Fluve Inc. of Hood River, Oregon. She has received a master of environmental science from Yale University and a bachelor of science in biology from Hobart and William Smith Colleges. Her area of expertise is in the management of water resources and the restoration and rehabilitation of cold water stream systems. Her work has primarily been in the northeast and Pacific northwest of the United States.

Mark S. Ashton, PhD, is the Morris K. Jessup professor of silviculture and forest ecology at the School of Forestry and Environmental Studies, Yale University. He received his bachelor of science from the University of Maine, College of Forest Resources, and his master of forestry and PhD from Yale University. Professor Ashton conducts research on the biological and physical processes governing the regeneration of natural forests. The results of his research have been applied to the development and testing of silvicultural techniques for restoration of degraded lands and for the management of natural forests for a variety of values. In particular his work is focused at land rehabilitation for watershed management and water quality. His field sites include tropical forests in Sri Lanka and Panama, temperate forests in India and New England, and boreal forests in Saskatchewan, Canada. He has authored or edited more than 10 books and monographs and more than 100 peer-reviewed papers relating to forest regeneration and natural forest management.

Bradford S. Gentry is the director of the Center for Business and the Environment as well as a senior lecturer and research scholar at the Yale School of Forestry and Environmental Studies. He received his BA from Swarthmore College and his JD from Harvard University. Trained as a biologist and a lawyer, his work focuses on strengthening the links between private investment and improved environmental performance. His area of expertise is on the business of water resource management for urban environments. He is also an advisor to GE, Baker & McKenzie, Suez Environment, and the UN Climate Secretariat, as well as a member of Working Lands Investment Partners and board chair for the Cary Institute of Ecosystem Studies.

Contributors

Caitlin Alcott
Consultant
Madrona Environmental Consulting
Portland, Oregon

Emily Alcott
Geomorphologist/ecologist
Inter-Fluve, Inc.
Hood River, Oregon
and
School of Forestry and Environmental Studies
Yale University
New Haven, Connecticut

Mark S. Ashton
Forest manager
School of Forestry and Environmental Studies
Yale University
New Haven, Connecticut

Alex Barrett
School of Forestry and Environmental Studies
Yale University
New Haven, Connecticut

Michael Blazewicz
Geomorphologist/hydrologist
Round River Design
Salida, Colorado
and
School of Forestry and Environmental Studies
Yale University
New Haven, Connecticut

Peter Caligiuri
Forester
The Nature Conservancy
Bend, Oregon
and
School of Forestry and Environmental Studies
Yale University
New Haven, Connecticut

Erin Derrington
Consultant
Eco-Management and Design
Seattle, Washington
and
School of Forestry and Environmental Studies
Yale University
New Haven, Connecticut

Justin Freiberg
Co-founder
Encendia Biochar
and
School of Forestry and Environmental Studies
Yale University
New Haven, Connecticut

Bradford S. Gentry
School of Forestry and Environmental Studies
Yale University
New Haven, Connecticut

Xiaoting Hou
Program manager
The Forest Dialogue
and
School of Forestry and Environmental Studies
Yale University
New Haven, Connecticut

Jennifer Hoyle
PhD student
Hydrology and Water Resources
and
School of Forestry and Environmental Studies
Yale University
New Haven, Connecticut

Lisa Hummon
Legislative aide
United States Senate
Washington, DC
and
School of Forestry and Environmental Studies
Yale University
New Haven, Connecticut

Claire Jahns
Project director
The Nature Conservancy
San Francisco, California
and
School of Forestry and Environmental Studies
Yale University
New Haven, Connecticut

Nathan Karres
Water resource consultant
Enviroissues
Seattle, Washington
and
School of Forestry and Environmental Studies
Yale University
New Haven, Connecticut

Jason Nerenberg
Forester
Department of Forests, Parks, and Recreation
Vermont
and
School of Forestry and Environmental Studies
Yale University
New Haven, Connecticut

Fauna Samuel
Water resource engineer
Cambridge, Massachusetts
and
School of Forestry and Environmental Studies
Yale University
New Haven, Connecticut

Tien Shiao
Senior asssociate
World Resources Institute
Washington, DC

CHAPTER 1

Gray to Green

An Introduction to Four Case Studies on Drinking Water Supply in the Northeastern United States

Caitlin Alcott, Emily Alcott, Mark S. Ashton, and Bradford S. Gentry

CONTENTS

1.0 EXECUTIVE SUMMARY

Restoring or protecting drinking water watersheds can result in reduced turbidity and significantly reduce the cost of water filtration plant upgrades. In addition, upland restoration projects can create ancillary benefits such as recreation opportunities, carbon sequestration, habitat conservation, and production of nontimber forest products (e.g., maple syrup). Nationwide, protection of surface water generated from rain and snowmelt in these upland areas is an underutilized and overlooked cost-effective management tool.

Often, upland watershed management has been underemphasized. However, with the improvement of targeting technologies (e.g., GIS) that can help select priority areas for restoration and leveraged implementation methodologies, upland watershed and stream restoration has become a pragmatic strategy for reducing the costs and improving the safety and quality of drinking water delivery systems. Upland watershed management is now considered a critical strategy in a "multibarrier" approach for the delivery of clean, safe drinking water.

This book outlines four case studies in separate chapters, Chapters 2 through 5, that describe the efforts of six different water providers across four states to provide high-quality drinking water at the lowest possible cost. Each of these water providers utilized surface water as their primary water supply, and each provider was faced with unique attributes of place. These attributes include, but are not limited to, unique biophysical attributes (e.g., underlying geology, land cover type), land ownership, political climate, and policy drivers. These case studies are then followed by Chapter 6, which synthesizes lessons learned from each of these four states. Lastly, Chapter 7 presents opportunities for countries and regions with developing infrastructure and water supplies to apply lessons from New England abroad.

1.1 INTRODUCTION: DEFINING THE ISSUE

Turn on the tap and water comes out clean and ready to use. In the United States rarely do we wonder if one day it might not be there, or whether or not it is safe to drink. At an average price of $2.00 per 1,000 gallons (EPA, 2004), drinking water remains an undervalued resource. Huge effort and expense are required to supply this resource and, as drinking water prices slowly climb and new contaminants continue to emerge (e.g., Ernst et al., 2004), protecting the quality of and finding efficiencies for delivery of clean water is a focus of water managers everywhere.

Throughout this book we focus on the processes and techniques for bringing "raw," untreated water up to safe drinking standards. Dating back to as early as 4000 B.C., drinking water has been treated to improve its taste and appearance. Later, with the development of sand filtration and chlorination, treatment was used to improve its safety and quality (U.S. EPA, 2000). Today, new engineering technologies for providing clean, safe drinking water continue to emerge although often at increased cost. A variety of technologies currently treat drinking water for different attributes that range from color and taste to particulates, pathogens, and minerals. Treatments include filtration, chlorination, ultraviolet radiation, ozonation, and coagulation. Engineered solutions for water treatment or "gray infrastructure," enable drinking water suppliers to ensure that across the United States when people turn on the tap, clean, safe drinking water comes out.

While water treatment infrastructure is costly, there are alternative methods for ensuring safe drinking water supplies. In addition to gray infrastructure, the management and conservation of forested watersheds, or "green infrastructure," has become increasingly recognized as a pragmatic strategy for reducing the costs and improving the safety and quality of drinking water delivery systems (e.g., Wickham, 2011; Vilsack, 2009). This strategy works because forests infiltrate water more slowly than developed areas and act as a natural filter. As water moves through the vegetation and soil, contaminants are removed in much the same way as in engineered solutions for cleaning drinking water.

The effectiveness of natural filters depends on the particular biophysical attributes of the watershed including vegetation type (e.g., age, composition, and structure), soil type, geology, topographic relief, and climate (e.g., seasonality, quality and quantity of precipitation falling on that watershed). Although this natural filtration process works more slowly and is seemingly less controlled than gray infrastructure, it delivers cleaner raw water to the treatment plant than water that has runoff from more developed areas (Wickham et al., 2005; Gilliom et al., 2006).

It is a delicate combination of these two water treatment strategies—green and gray infrastructure—known as a "multibarrier approach" that has been described as the most resilient, pragmatic approach for drinking water delivery. These multiple barriers are a series of protective measures that work together to provide multiple defenses against drinking water contamination. These barriers are (1) source watershed protection and management, (2) engineered drinking water treatment, (3) distribution, and (4) monitoring (Barnes et al., 2009). Where on the spectrum of green to gray infrastructure use a system falls relies on the biophysical attributes of a watershed and the social attributes of place (e.g., regulations, landownership, management choices, and land use within a watershed).

In addition to providing higher quality raw water, upland watershed management provides a number of ancillary benefits (e.g., wildlife habitat, carbon sequestration) and presents a number of unique collaboration opportunities for the restoration and conservation communities to work with drinking water suppliers and private landowners. Americans already pay for clean drinking water—and in many surface water systems portions of ratepayer dollars are being used to manage upland watersheds. Along with improvements to safe drinking water, upland watershed management provides improved ecological resilience and function, as well as opportunities for biodiversity conservation, increased water yield, and carbon sequestration.

There have been multiple attempts (e.g., Ernst et al., 2004; Gray et al., 2011) to quantify exactly how much money can be saved in the water treatment process by preserving upland watersheds. Ernst et al. (2004) demonstrated an incremental decreasing treatment cost with increasing forest cover. Variability between watersheds and a lack of upland watershed monitoring data make a widespread, generalizable, and statistically sound relationship (e.g., 1 acre of protected watershed = X savings in treatment costs) nearly impossible (Ernst, 2004). However, despite the lack of this widespread relationship, upland watershed protection and management presents opportunities for long-term treatment cost savings, recreational opportunities, improved ecological function and resilience, and safer, cleaner drinking water.

It is therefore a very timely opportunity to assess what we know and do not know about relationships between watershed protection and management and cost of water treatment. This book is directed at helping managers and decision makers weigh the costs and benefits of engineered treatment solutions (gray infrastructure) and watershed protection (green infrastructure) and help determine where on the spectrum of green to gray they should place their utility. Through analysis of four case studies of cities in the northeastern United States (Chapters 2, 3, 4, and 5), this book illustrates how the existing market for drinking water can be utilized for upland management, to protect our drinking water supplies while achieving significant ancillary benefits such as recreation, ecological resilience, and habitat.

This introduction provides a rationale for the book by briefly describing the resource issues around drinking water supply from surface waters nationally and more specifically in the northeastern United States, and then introducing the case studies. The book ends with a synthesis (Chapters 6 and 7) that describes the relevance of the book in relation to drinking water issues elsewhere.

1.2 NATIONAL TRENDS

Each day, two-thirds of Americans depend on drinking water that has come from surface water sources (Levin et al., 2002; EPA, 2009). Surface source water is defined as "water from rivers or lakes that is used to provide public drinking water" (EPA, 2012). Seventy-eight percent of the land contained within the lower 48 states lies within a drinking water watershed (Wickham et al., 2010). Large municipal areas are often particularly dependent on surface water, including New York, New York; Boston, Massachusetts; Portland, Maine; Portland, Oregon; San Francisco, California; Denver, Colorado; Seattle, Washington; Albuquerque, New Mexico; and Atlanta, Georgia. These cities and others dependent upon surface watersheds and reservoirs can have some of the best drinking water qualities in the nation particularly when source watersheds are proactively protected. The top ten U.S. cities with best and worst drinking water are listed with notes about the water source and kinds of programs the utility engages in (see Tables 1.1 and 1.2).

Because surface water can become contaminated from storm water runoff, pesticide application, sedimentation and erosion, hazardous material spills, wildlife, and other sources, water managers must ensure that the water is protected and treated before consumption either through natural filtration by the ecosystem or through technology. In a few locations, surface source water is not filtered in a treatment plant and the source watersheds are highly protected and allowed to "treat" the water naturally instead. In these cases, the Environmental Protection Agency (EPA) grants water managers filtration waivers. Several well-known examples exist including Boston; New York City; Seattle; Portland, Oregon; San Francisco, and Portland, Maine (Table 1.3).

However, as drinking water standards become more stringent, adding a new drinking water system to this list may not be feasible. Regardless, most water managers agree that it is cheaper to treat high-quality raw water (Ernst et al., 2004).

Table 1.1 Results of Analysis of Best U.S. Cities for Drinking Water

City	Population	Utility	Source of Water	Notes
Arlington, TX	365,438	Arlington Water Utilities	Surface water (Tarrant Regional Water District)	Flood management program; Wetlands managed for water filtration; Outreach for landscaping and conservation
Providence, RI	178, 042	Providence Water	Surface water reservoir (Scituate Reservoir)	Surface water reservoir protected and closed to public
Fort Worth, TX	741,206	Fort Worth Water Department	Surface water (Tarrant Regional Water District)	Flood management program; Wetlands managed for water filtration; Outreach for landscaping and conservation
Charleston, SC	120,083	Charleston Water System	Surface water (Edisto River and the Bushy Park Reservoir)	Treatment plants included filtration and disinfection
Boston, MA	617,594	Massachusetts Water Resources Authority	Surface water (Quabbin, Ware, Wachusett reservoirs)	Surface watershed managed for water supply protection and closed to the public
Honolulu, HI	374,658	Board of Water Supply	Groundwater	Watershed management plans in development; Underlying volcanic geology helps filter water supply
Austin, TX	790,390	Austin Water Utility	Surface water (Colorado River, stored in Lake Austin)	Filtered and treated at two treatment plants; Outreach programs included landscaping
Fairfax County, VA	1,086,743	Fairfax Water	Surface water (Occoquan and Potomac Watersheds)	Watersheds 42% and 31% forested
St. Louis, MO	319,294	City of St. Louis Water Division	Surface water (Missouri River and Mississippi River)	Two treatment plants
Minneapolis, MN	382,578	City of Minneapolis Water Department	Surface water (Mississippi River)	

Sources: EWG, 2009; population data from US Census 2010.

Table 1.2 Results of Analysis of Worst U.S. Cities for Drinking Water

City	Population	Utility	Source of Water	Notes
Pensacola, FL	51,932	Emerald Coast Water Utility	Groundwater	45 pollutants detected
Riverside, CA	300,000	City of Riverside Public Utilities	Groundwater	Agricultural pollutants
Las Vegas, NV	583,756	Las Vegas Valley Water District	Surface water (Colorado River)	12 pollutants that exceed EPA guidelines including radium-226, radium-28, arsenic, lead
Riverside County, CA	2,203,332	Eastern Municipal Water District	75% surface water (imported from Colorado River); groundwater	
Reno, NV	225,221	Truckee Meadows Water Authority	Surface water (Truckee River)	
Houston, TX	2,099,451	City of Houston Public Works	Surface water (Trinity River)	46 pollutants detected
Omaha, NE	408, 958	Metropolitan Utilities District	Surface water (Missouri & Platte Rivers)	Agricultural pollutants
North Las Vegas, NV	216,961	City of North Las Vegas Utilities Department	Groundwater & Surface water (Colorado River)	
San Diego, CA	1,307,402	San Diego Water Department	Surface water (imported through aqueducts from Colorado River and northern CA)	Excessive trihalomethanes, manganese
Jacksonville, FL	821,784	JEA	Groundwater	

Sources: EWG, 2009; population data from US Census 2010.

Ultimately, the question of how a utility delivers clean drinking water is not whether to use only engineered filtration technology or only watershed protection but rather how much of each method is appropriate for the particular biophysical and social circumstance of the watershed and consumers.

Multiple challenges face water managers in the United States from environmental, social and economic sources. Major environmental threats to clean and abundant drinking water supplies come from pathogens, chemicals, excess sediment, nutrients, and unpredictable high or low flows. Each of these environmental threats can be addressed through watershed protection and/or treatment processes. Social or

Table 1.3 Partial List of Drinking Water Systems with a Waiver from Filtration (as of spring 2010)

City	State
Auburn	ME
Bangor	ME
Bar Harbor	ME
Bethel	ME
Boston	MA
Brewer	ME
Camden	ME
Concord	MA
East Northfield	MA
Falmouth	MA
Great Salt Bay	ME
Hancock	NH
Holyoke	MA
Lewiston	ME
Mount Desert Island	ME
New York	NY
Newbury	VT
Portland	ME
Portland	OR
San Francisco	CA
Seattle	WA
Syracuse	NY
Tacoma	WA
Vinal Haven	ME
Wilmington	VT

economic challenges to drinking water suppliers include (1) increasing water demand from population growth; (2) decreasing revenues as water users conserve resources; (3) lack of capital for infrastructure improvements; (4) increased occurrence of pollutants in drinking water watersheds (agricultural chemicals and fertilizers, runoff from urban areas, pharmaceutical use); and (5) increasingly stringent water quality regulations. Addressing social and economic challenges requires building political capital, forward-thinking land use planning, sustainable funding, and effective public outreach.

1.3 THE NORTHEASTERN UNITED STATES

There is a particular difference between most of the filtration waivers in the western United States and those in the northeastern United States. Specifically, those in the northeast are sourcing surface water from watersheds that have developed land and human access. In the western United States, Seattle, Portland, and San Francisco's source watersheds are almost completely protected from human access. Rarely now will an untouched watershed become available to provide drinking water, and so, in many ways, the phenomenon of having a completely protected source water watershed is a historic relic.

Looking forward, municipalities or other water managers hoping to protect source watersheds will need to manage watersheds with at least some human impact. Therefore, the juggling acts that the northeastern waived systems are conducting are impressive and allow for careful examination of when a system might benefit from upland protection, ways to interact with current uses and users of the watershed, and circumstances that merit engineering and technology. The region's history serves to illustrate a trend for many regions and developing nations.

In the northeastern United States, after colonization, forestlands were cleared largely for subsistence agriculture in the 1700s and the land was farmed until the mid-1800s (Whitney, 1994). After this period the region served as the initial heart of the industrial revolution in North America (1850–1920). During that period, agricultural lands were abandoned for better land further west and for better jobs with higher salaries in the cities. The land subsequently grew back to forest, and cities within the region, with their rapidly growing populations, sought to purchase much of this land cheaply for protection of surface drinking waters and newly constructed reservoirs. The region has since deindustrialized (1920–1960) and is now (1970 to present) largely fueled by an economy based on technology (computers, biotechnology, medicine), service (insurance, market investments), and education (universities and colleges) (Cronon, 1983). As the ability to travel to work has become relatively cheap and easy, the land surrounding cities that was not acquired or protected is now under current threat of conversion from second growth forest into cleared development from expansion of cities into growing suburbs (Cronon, 1983; Whitney, 1994). USFS analysis has shown that the Northeast has strong development and water source pressures (see Figures 1.1 and 1.2).

Suburbanization and expansion into second growth forest that was once agricultural land begs a series of questions. Are the developmental pressures that affect drinking water quality in the Northeast relevant elsewhere? How are cities in the Northeast making decisions between engineered infrastructure and watershed protection? Does the population and political establishment have the ability to make long-term economic decisions to maintain the quality of drinking water? How will future trends in consumption and regulations influence future planning for drinking water?

We believe the questions being asked now of the Northeast are relevant to other forested or formerly forested regions of the United States and abroad where population centers are growing along with development pressures and customer numbers. Southern New England especially is likely facing problems and potential solutions to drinking water issues that other regions eventually must face. Understanding the northeastern United States, and southern New England in particular, will help watershed managers and policymakers make better-informed decisions elsewhere.

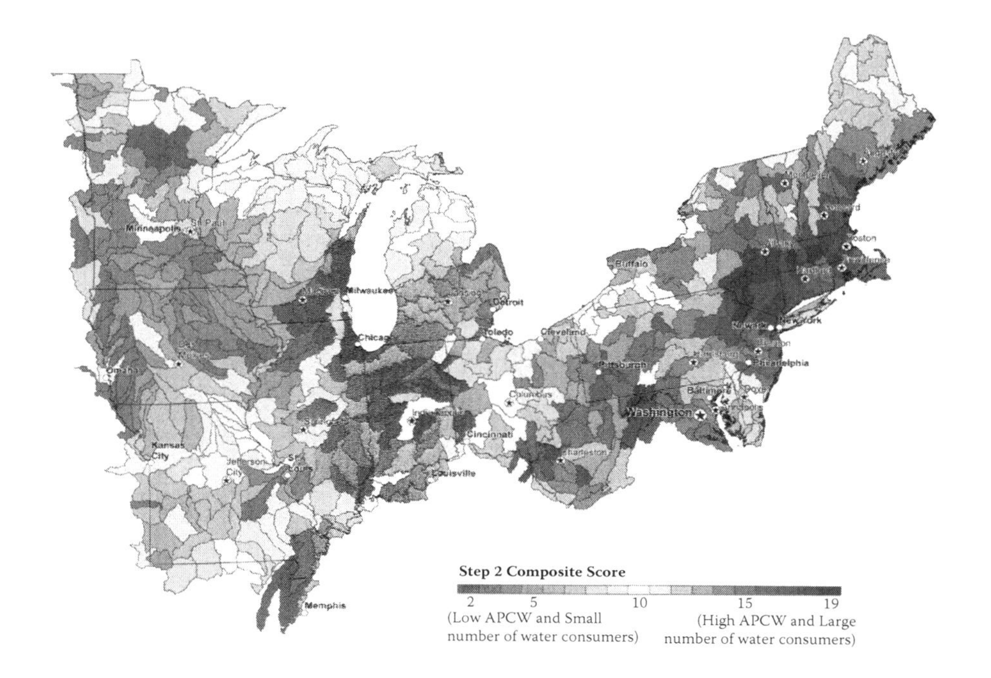

Figure 1.1 A map illustrating the importance of watersheds for drinking water supplies for each of the 540 watersheds in the Northeast and Midwest. It highlights those areas that provide surface drinking water to the greatest number of consumers. The higher a watershed's ability to provide drinking water, the darker brown it appears on the map and the higher its Ability to Produce Clean Water (APCW) score (From USFS, 2009). (See color insert.)

1.4 DESCRIPTION OF THE CASE STUDIES

This book is directed at helping decision makers weigh the costs and benefits of engineered treatment solutions (gray infrastructure) and watershed protection (green infrastructure). It has evolved from a seminar on this subject held in the spring of 2010 at the Yale School of Forestry and Environmental Studies. The purpose was to evaluate how, when, and where it makes environmental, economic, and social sense to protect and manage upland forests to produce water as a downstream service—primarily drinking water to large cities. Graduate students in the seminar analyzed a series of city watersheds in the northeastern United States that have been managed under different state regulations, planning and development incentives, biophysical constraints, social histories, and ownerships.

This work is the culmination of a series of seminars in prior years around ecosystem services and water that complemented and added information on watershed issues for the development of this book and that in some cases was published previously (e.g., *Emerging Markets for Ecosystem Services: The Case of the Panama Canal Watershed*, Gentry et al., 2007).

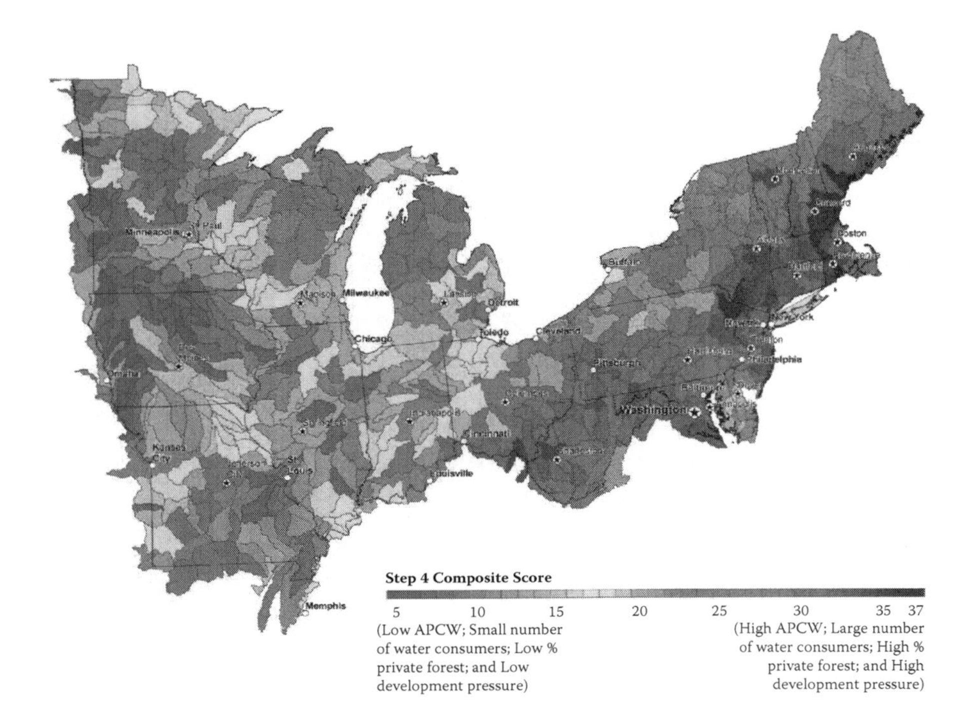

Figure 1.2 A map showing the development pressure on forests and drinking water supplies. The map combines data on the ability to produce clean water, surface drinking water consumers served, percent private forest land, and housing conversion pressure, to highlight important water supply protection areas that are at the highest risk for future development. The greater a watershed's development pressure, the more blue it appears on the map, and the higher its Ability to Produce Clean Water (APCW) score (From USFS, 2009). (See color insert.)

1.4.1 Research Methods

The seminar focused on case studies from six water systems in four states. These systems included New York, New York; Boston and Worcester, Massachusetts; New Haven and Bridgeport, Connecticut; and Portland, Maine. These case studies represented a wide array of water supply protection scenarios—from the unfiltered (via a filtration avoidance determination from the EPA) in New York City where landowner partnerships for watershed management keep water quality high—to states with required filtration, like Connecticut, where water suppliers still utilize land acquisition and land management programs to keep raw water quality high (Table 1.4).

The seminar grouped graduate student researchers into three teams. Each team comprised students with specializations in engineering, forestry, policy, and business. Each team represented one of three state-level analyses (New York, Massachusetts, and Connecticut). The fourth analysis on Maine was done by a graduate student for a graduate project requirement.

Table 1.4 Status of Filtration Requirements in New England Drinking Water Systems

City	Filtration Determination
New York, NY	Filtration avoidance waiver
New Haven, CT	Filtration required
Bridgeport, CT	Filtration required
Boston, MA	Filtration avoidance waiver
Worcester, MA	Filtration required
Portland, ME	Filtration avoidance waiver

Experts from all four states were invited to make separate presentations and then to participate in a panel discussion. Panel discussions were conducted at intervals over a six-week period. The four panels represented each of the student specializations—water quality engineers, watershed foresters, policymakers, and regulators and business managers.

Student teams conducted interviews of experts and key informants for their respective watershed assignments. Reports, historical archival materials, gray literature, and peer-reviewed papers were reviewed, compiled, and synthesized. Together with the presentations and panel discussions, all information from both interviews and written documents were used in the writing of each state chapter. For Massachusetts and Connecticut, comparative case studies were performed to examine how two water systems differed in their management within the same state but with different biophysical and social settings and the same state- and federal-level policies and regulations.

1.4.2 The Case Studies

The case studies are reported in separate chapters that describe the efforts of six different water providers across the four states to provide high-quality drinking water at the lowest possible cost. Each system relied upon source water from a watershed as its primary raw water supply. The strategies undertaken by each respective system are driven by uniqueness of place. Each system has differing biophysical circumstances, varying landownership patterns, and differing land use policies and employs unique strategies to meet these varying sets of challenges. However, the goal was to identify similarities upon which to generalize to develop better management and policy.

1.4.2.1 Chapter 2: Connecticut

Connecticut provides a comparative analysis of two drinking water utilities—the South Central Connecticut Regional Water Authority (SCCRWA) and the Aquarion Water Company—one public and one private, and how each approach drinking water delivery. The parallel stories of these two utilities highlight how the general public's interest in watershed lands as open space can influence the ability of these utilities to make decisions regarding their businesses.

1.4.2.2 Chapter 3: Massachusetts

The Commonwealth of Massachusetts example compares two drinking water suppliers: Boston's Massachusetts Water Resources Authority (MWRA) and Worcester's Department of Public Works. While MWRA's Quabbin reservoir provides the luxury of protected watershed lands with little development, Worcester is faced with challenges of mixed landownership and a smaller budget. Here we see two suppliers faced with two very different circumstances but find that each system thrives in their unique place, while employing similar strategies to prioritize and utilize their green assets.

1.4.2.3 Chapter 4: New York

One of the most storied drinking water systems in the United States, the New York City case outlines ongoing involvement and investment by the city into watersheds and communities in upstate New York. Land stewardship and incentive programs have allowed New York City to attain and maintain a filtration avoidance waiver saving anywhere from $4 to $8 billion dollars. While this case outlines, if not defines, *upstream/downstream* or *urban/rural* tensions, New York City's Department of Environmental Protection continues to invest in upstream landowners and abides by the adage "an ounce of prevention is worth a pound of cure."

1.4.2.4 Chapter 5: Maine

Portland is the largest and fastest-growing urban center in Maine. Sebago Lake and the Crooked River Watershed not only serve as Portland's water supply but are also recreational destinations and face significant development pressures. With a progressive and emerging watershed management division, this chapter draws on lessons learned from the previous three states and provides recommendations for the Portland Water District (PWD) moving forward.

1.4.3 Synthesis and Conclusions

Following the case studies, two additional chapters focus on regional comparisons that may clarify common and different approaches, and on global relevance of lessons from the northeastern United States.

1.4.3.1 Chapter 6: A Synthesis: Comparing Drinking Water Systems in the Northeastern United States

Each system discussed faces similar challenges: tightening water quality regulations, a changing climate, development pressures, and decreasing revenues. Though each system faces somewhat unique circumstances and has employed its own strategies to provide clean, safe drinking water, a number of themes emerge from each example. Further, though our case studies draw on experiences within the Northeast, the strategies and lessons learned across these four states are valuable across the country. This chapter outlines emerging themes from each state's experiences and outlines considerations and strategies for watershed managers in the United States.

1.4.3.2 Chapter 7: Global Relevance of Lessons from the Northeastern United States

As nations around the globe work to provide clean, safe drinking water to their expanding populations, we draw on lessons learned from established drinking water systems within the Northeast. Here we outline questions to consider as drinking water systems take shape and face similar conflicting resource-use issues as in the northeastern United States. Driven by uniqueness of place, we offer guiding questions and important considerations with the aim to take full advantage of green infrastructure for drinking water delivery.

1.5 CONCLUSIONS

Everyone needs clean, safe drinking water. Drinking water suppliers and rate payers agree that they want the best-quality product at the lowest possible delivery cost. As water utility managers face decisions of how to manage for the highest quality water at the lowest possible cost, they are faced with a multitude of demands to meet, but with a strong arsenal of management alternatives. In addition, watershed managers, engineers, policy makers, and utility managers agree: protecting and managing upland watersheds will provide you with higher raw quality water, save treatment costs, and limit operating and maintenance costs now and into the future. Though a generalizable, statistically sound relationship alludes us, we know that protecting source water areas is saving rate payers and utilities money, and it is providing higher quality, safer raw water.

The chapters in this book will demonstrate that a balance is necessary between social and biophysical management. This balance is achieved by understanding the various factors at play and that are discussed in the chapters: watershed ownership, biophysical attributes, political climate, financial resources, and risk (natural and anthropogenic). This book presents concrete management options for watershed managers, which will not only save their utility money over the long term, but provide for a safe, resilient water supply.

REFERENCES

Barnes, M. C., A. H. Todd, R. W. Lilja, and P. K. Barten. (2009). *Forests, water and people: Drinking water supply and forest lands in the Northeast and Midwest United States*. Available online at http://na.fs.fed.us/pubs/misc/watersupply/forests_water_people_watersupply.pdf.

Barten, P. K. (2007). The conservation of forests and water in New England ... again. *New England Forests* (Spring): 1–4.

Barten, P. K., and C. Ernst. (2004). Land conservation and watershed management for source protection. *Journal of the American Water Works Association* 96(4): 121–135.

Barten, P. K., T. Kyker-Snowman, P. J. Lyons, T. Mahlstedt, R. O'Connor, and B. A. Spencer. (1998). Managing a watershed protection forest. *Journal of Forestry* 96(8): 10–15.

Congressional Budget Office. (2002). *Future investment in drinking water & wastewater infrastructure: A CBO study.* http://www.cbo.gov/publication/14205. (Accessed December 6, 2012.)

Cronon, W., (1983). *Change in the land: Indians, colonists, and the ecology of New England.* New York: Hill and Wang. 237 p.

Dudley, N., and S. Stolton. (2003). *Running pure: The importance of forest protected area to drinking water.* Prepared for the World Bank/WWF Alliance for Forest Conservation and Sustainable Use.

Ernst, C. (2004). *Protecting the source: Land conservation and the future of America's drinking water.* Water Protection Series. San Francisco: Trust for Public Land. 56 p.

Ernst, C. (2006). Land conservation: A permanent solution for drinking water source protection. *On Tap* Spring: 18–40.

Ernst, C. (2010) Interview. In partial requirement for *Emerging Market for Ecosystem Services: Optimizing "Natural" and "Engineered" Systems for Protecting the Quality of Surface Drinking Waters.* Yale School of Forestry & Environmental Studies. March 12, 2010.

Ernst, C., and K. Hart. (2005). *Path to protection: Ten strategies for successful source water protection.* Water Protection Series. San Francisco: Trust for Public Land. 28 p.

Ernst, C., R. Gullick, and K. Nixon. (2004) Protecting the source: Conserving forests to protect water. *Opflow* 30:1–7

EPA (Environmental Protection Agency). (2000). *The History of Drinking Water Treatment.* EPA-816-F-00-006. US EPA Office of Water, Washington DC.

EPA. (2001). *Drinking Water Infrastructure Needs Survey: Second Report to Congress.* EPA 816-R-01-004. U.S. EPA Office of Water, Washington, D.C.

EPA. (2002). *The clean water and drinking water infrastructure gap analysis.* EPA-816-R-02-020. U.S. EPA Office of Water, Washington, D.C.

EPA. (2004). *Drinking water costs & federal funding.* EPA 816-F-04-038. U.S. EPA Office of Water, Washington, D.C.

EPA. (2007). *Drinking water infrastructure need survey and assessment summary.* Available online at http://water.epa.gov/infrastructure/drinkingwater/dwns/upload/2009_03_26_needssurvey_2007_fs_needssurvey_2007.pdf.

EPA. (2009). *Factoids: Drinking water and ground water statistics for 2009.* EPA 816-K-09-004. U.S. EPA Office of Water, Washington, D.C.

EPA. (2009). *Geographic information systems analysis of the surface drinking water provided by intermittent, ephemeral and headwater streams in the U.S.* http://water.epa.gov/lawsregs/guidance/wetlands/surface_drinking_water_index.cfm (Accessed December 6, 2012).

EPA. (2012). *Water: Source water protection.* Available online at http://water.epa.gov/infrastructure/drinkingwater/sourcewater/protection/basicinformation.cfm.

EWG (Environmental Working Group). (2005). *Cities with the best and worst tap water quality.* Reprinted by Sustainlane.com. Available online at http://www.sustainlane.com/us-city-rankings/categories/tap-water-quality.

EWG (Environmental Working Group). (2009). *Cities with the best and worst tap water quality.* Available online at http://www.ewg.org/tap-water/methodology.

Gentry, B., S. Anisfeld, Q. Newcomer, and M. Fotos (Eds.). (2007). *Emerging Markets for Ecosystem Services: A Case Study of the Panama Canal Watershed.* New York: Haworth Press.

Gilliom R. J, J. E. Barbash, C. G. Crawford, P. A. Hamilton, J. D. Martin, N. Nakagaki, L. H. Nowell, J. C. Scott, P. E. Stackelburg, G. P. Thelin, and D. M. Wolock. (2006). *The quality of our nation's waters— pesticides in the nations streams and groundwater, 1992–2001.* U.S. Geological Survey, Circular 1291, Reston, VA. http://pubs.usgs.gov/circ/2005/1291/. Accessed August 2011.

Gray, E., J. Talberth, L. Yonavjak, and T. Gartner. (2011). *Green vs. gray infrastructure options for the Sebago Lake watershed: Preliminary analysis and scoping study.* Washington, DC: World Resources Institute. Project report, unpublished.

Hanson, C., J. Talbert. and L. YonavJak. (2011). *Forests for water: Exploring payments for watershed services in the U.S. South.* World Resources Institute. Available online at http://www.seesouthernforests.org/files/sff/wri_forests_for_water.pdf.

Levin, R. B., P. R. Epstein, T. E. Ford, W. Harrington, E. Olson, and E. G. Richard. (2002). U.S. drinking water challenges for the twenty-first century. *Environmental Health Perspectives* 110 (Suppl 1): 43–52.

McIntyre, D. A. (2011). *Cities with the worst water.* 24/7 Wall St. Available online at http://www.dailyfinance.com/2011/01/31/ten-american-cities-with-worst-drinking-water/

Natural Resources Defense Council. (2003). *What's on tap? Grading drinking water in U.S. cities.* Available online at http://www.nrdc.org/water/drinking/uscities/pdf/whatsontap.pdf.

USFS (U.S. Department of Agriculture Forest Service). (2009). Northeastern Area State and Private Forestry, *Forests, water and people: Drinking water supply and forest lands in the Northeast and Midwest United States. Report No. NA-FR-01-08.* Available online at: www.na.fs.fed.us.

Vilsack, T. (2009). First public speech as Secretary of Agriculture. Seattle, WA. August 14. http://www.fs.fed.us/video/tidwell/vilsack.pdf. Accessed August 2011.

Whitney, G. G. (1994). *From coastal wilderness to fruited plain: A history of environmental change in temperate North America from 1500 to the present.* Cambridge, UK: Cambridge University Press.

Wickham, J. D., Y. G. Wade, and K. H. Ritters. (2010). An environmental assessment of United States drinking water watersheds. *Landscape Ecology.* 26: 605–616.

CHAPTER 2

An Assessment of Drinking Water Systems in Connecticut

Optimizing Natural and Engineered Systems for Protecting the Quality of Surface Drinking Waters

Michael Blazewicz, Lisa Hummon, Claire Jahns, and Tien Shiao

CONTENTS

2.0 EXECUTIVE SUMMARY

Two drinking water utilities—the South Central Connecticut Regional Water Authority (SCCRWA) and the Aquarion Water Company (Aquarion)—manage their watershed land and engineered treatment assets to optimize investments in drinking water quality. Since 1974, utility management decisions have been shaped by the U.S. Safe Drinking Water Act, several state-level regulations governing drinking water utilities and their landholdings, and public values. The parallel stories of SCCRWA and Aquarion illustrate how the general public's interest in watershed lands as open space can influence the ability of utilities to make decisions regarding their natural, or *green*, and built, or *gray*, water filtration assets.

Utilities must provide high-quality water at the lowest possible cost, while accommodating public cultural, environmental, and economic values that are also significant drivers of water system optimization efforts. The lessons learned over the past 35 years suggest a number of recommendations utilities may wish to consider to optimize their green and gray assets going forward. First, utilities should pursue monitoring that more clearly establishes the linkages between forestland conservation, engineered filtration requirements and costs, and drinking water quality. Second, utilities should base operational and asset management decisions on long-term cost effectiveness. This should be measured in terms of dollars spent per unit of water quality improvement, reflecting long-term cost savings by considering avoided capital, operating, and maintenance costs. Third, utilities should also seek non-land-acquisition protection measures to achieve water quality improvements and maintain water quality. Such measures may include more intensive outreach programs with residents of the watershed and the purchase of conservation easements and funding of best management practices on currently unprotected open space, agricultural, and low-density watershed lands.

In addition to fiscal and water quality considerations, utilities should acknowledge and attempt to quantify the additional benefits of maintaining watershed land as open space. These benefits include but are not limited to public access to open space, biodiversity conservation, temperature regulation, and nontimber forest products (e.g., fuelwood, maple syrup). Stakeholders who may have a financial interest in these additional benefits should be identified and consulted on land matters to avoid conflict and explore their willingness to partner on management, monitoring, and land acquisition costs. In Connecticut, the State Department of Energy and Environmental Protection (DEEP), and the State Department of Health and Human Services, municipalities, and a number of conservation groups have exhibited a clear willingness to fund watershed land conservation and cooperative planning and management to support shared land use objectives. Land values should be combined to evaluate potential land acquisitions given their highest and best uses as contributors to improved water quality and lower engineered treatment costs as well as cultural, recreational, and ecological assets.

2.1 INTRODUCTION

Drinking water suppliers are faced with a number of management alternatives when working to ensure the delivery of safe drinking water. Until recently, these approaches have largely been focused on the downstream end of drinking water delivery—treatment plants. Significant time and money have gone into the development of engineered drinking water treatment facilities and technologies. These technologies have enabled the delivery of large quantities of drinking water to continually expanding population centers. Though engineered treatments play a critical role in the delivery of safe drinking water, management of source watersheds at the upstream side of the system is often underemphasized. The 1996 amendment to the Safe Drinking Water Act (SDWA, 1974) recognized the linkage between watershed conservation and clean surface waters downstream.

In particular, forests serve as a natural water filter, creating high-quality raw, or untreated, water (Barten, 2006). Water quality engineers agree that it is more cost effective and less resource intensive to treat high quality raw water than it is to treat low quality raw water. Improving the quality of raw water in source watersheds requires careful land use planning, management, and an understanding of both biophysical and social factors within the source watershed. Moreover, the conservation and stewardship of source watersheds provides multiple co-benefits such as preserving open space and biodiversity, as well as carbon sequestration. With a clear understanding of these principles, and state and federal legislation to support them, two of Connecticut's largest water suppliers, SCCRWA and Aquarion, have embraced upland land conservation and management as a mechanism to optimize the delivery of safe, high-quality drinking water.

Connecticut provides a unique example of how public, regulatory, and financial demands can interact to drive watershed management for water quality. Despite federal funding supporting source water protection, state-level legislation requiring land conservation and management by drinking water suppliers, and significant pressure from the public to conserve open space, Connecticut utilities are in a constant struggle to maintain and justify their balance between upland watershed management and downstream-engineered treatments. High property taxes and management costs make land ownership a resource intensive investment for water suppliers. Further, a lack of statistically sound monitoring data to link the relationship between forest to land use and water quality, combined with the current economic downturn, makes it difficult to financially justify water company landholdings. Today, source water protection efforts in Connecticut are focused on the most essential aquifer protection areas but are not exhaustive (Figure 2.1) (Hopper and Pepper, 2003; CT DPH, 2005).

Given the difficulty of proving a strong and direct correlation between watershed protection acreage and raw water quality, and given the presence of third-party interests in maintaining watershed land as open space, Connecticut water utilities should continue to look beyond the immediate fiscal bottom line for upland watershed management. As the Office of Policy and Management for the State General Assembly's Energy and Technology and Appropriations Committee emphasized, sustainable water resources management must consider the interrelationship of social, economic, and environmental values (OPM, 2008).

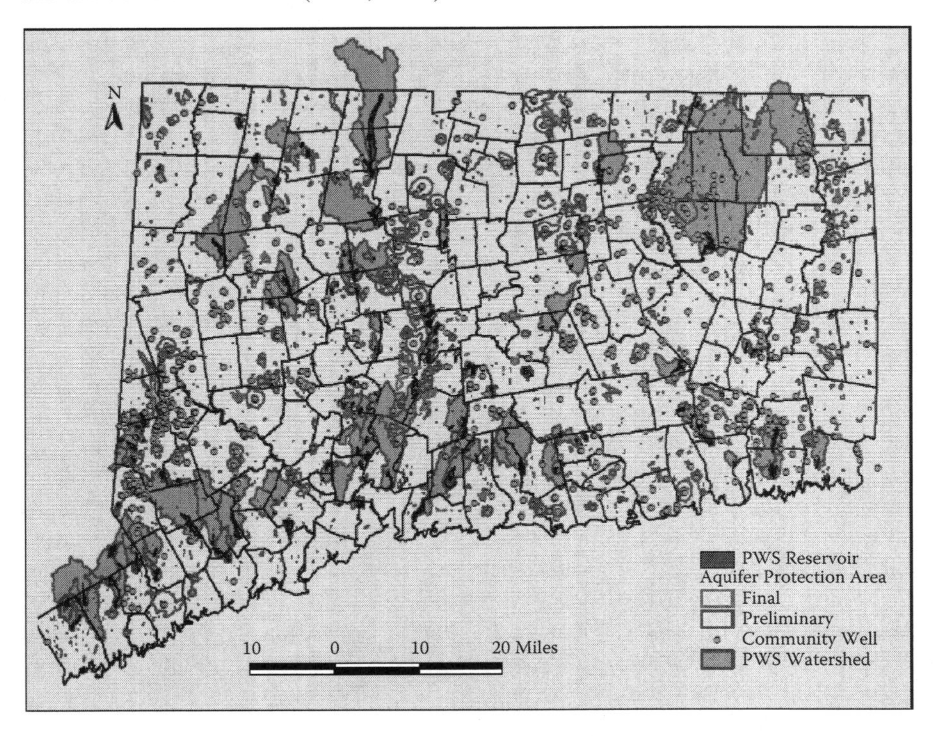

Figure 2.1 Map of Connecticut watersheds, illustrating surface watershed lands for cities and towns (green), community wells (gold) and preliminary (blue) and final (red) aquifer protection areas (From CT DPH, 2005). (See color insert.)

Upland watershed management not only improves raw water quality but also provides a number of other valuable ecological and social benefits, including improved ecological resilience, access to open space, carbon sequestration, and biodiversity conservation. This chapter will present a comparative analysis of how the publicly held SCCRWA and Bridgeport's privately held Aquarion have chosen to blend an array of management alternatives to deliver high-quality water while still meeting demand and being cost effective.

In this chapter we outline strategies for both SCCRWA and Aquarion, as well as other similarly situated utilities, to work cooperatively to quantify these third-party co-benefits of watershed land conservation and additional land acquisition, and work with the co-beneficiaries to plan and manage finances. Additionally, we detail how this strategy will help SCCRWA and Aquarion pre-empt pending water quality regulations as well as the proposed state minimum stream flow regulations. To avoid potential use conflicts and maintain safe water supplies, these and other emergent challenges will need to be factored into each utility's existing decision-making framework. We start with describing the history and status of current drinking water supply systems in Connecticut generally, and then provide more specific descriptions for SCCRWA and Aquarion. We then describe the legal aspects of water supply regulation and land management strategies taken, using Aquarion and SCCRWA as examples. We discuss future threats that water utilities in the region will face, again based on the examples of SCCRWA and Aquarion, and conclude with recommendations to support the mission of providing high-quality drinking water at low costs to consumers.

2.2 HISTORY AND CURRENT STATUS OF DRINKING WATER SUPPLY SYSTEMS IN CONNECTICUT

Early Connecticut settlers acquired water through shallow wells or by tapping into nearby springs or streams. During the mid-1800s, as the industrial revolution began to take hold and Connecticut towns began to grow rapidly, shallow groundwater and small stream systems soon were overextended or deteriorated. Additionally, as the use of steam-powered manufacturing and railroad transport increased, so did the need to have water available to fight fires. Numerous private water companies began to tap sources in the region to supply water for the growing communities and the industries that supported them. Companies such as the New Haven Water Company (NHWC) (chartered in 1849) and Bridgeport Water Company (chartered in 1853) began to develop reservoirs and acquire smaller community water systems.

Early land acquisition and reservoir development was not always a simple matter of transacting real estate. In the late 1930s, despite the protests of residents in that region, one Bridgeport Hydraulic Company project flooded the Saugatuck River Valley, displacing the once-thriving industrial village of Valley Forge. Similar conflicts transpired with the creation of Hartford's Barkhamstead Reservoir and New Haven's Lake Gaillard, a project that flooded 22 home sites and farms in 1933 (McCluskey and Bennitt, 1996; Lomuscio, 2005).

The importance of stewardship of source watersheds to improve water quality was recognized as early as the early 1900s. The New Haven Water Company reported efforts to increase timber growth on watershed lands between 1900 and 1941. These forestry initiatives were tied to an understanding of the importance of planned forest management to managing soil erosion and runoff. The company planted between 100,000 and 150,000 trees annually during the first 40 years of this program under the supervision and management of the new Yale Forest School, one of the nation's first forestry schools (NHWC, 1941).

The early NHWC lands were mostly located within the Eli Whitney Forest. The Yale Forest School used the forest as an outdoor classroom and laboratory until 1950, when the NHWC assumed full responsibility for land management. At this time, the Eli Whitney Forest comprised 22,000 acres of woodland spread out over 10 tracts. In addition to protecting water quality and accelerating timber growth, the company also articulated its aim to manage forests for aesthetic value.

Physical assessment and reservoir protection was also practiced in the early years. From as early as 1941 the NHWC, along with the State Board of Health and local health officials, made a general inspection of the watersheds once a year. The inspections inventoried buildings, livestock, and human activity in the area, particularly near drainages into reservoirs. In addition to this annual inspection, sanitary engineers for the water company were responsible for investigating cases of careless disposal of waste and furnishing advice for proper sanitary installations (NHWC, 1941).

2.3 THE SOUTH CENTRAL CONNECTICUT REGIONAL WATER AUTHORITY

2.3.1 History of the SCCRWA

In 1849, the growing city of New Haven decided it could no longer rely upon wells for its drinking water and chartered the New Haven Water Company (NHWC, 1941; McKluskey and Bennitt, 1996) to provide for the region's water needs. Ten years later, Eli Whitney II built a dam to impound two and a half miles of the Mill River and form Lake Whitney (NHWC, 1941). Two waterwheels, 30 feet in diameter, pushed water from the dam up into a reservoir atop Sachem's Hill (now Prospect Street). From the reservoir, the water was gravity fed through 18 miles of pipe to the city. On January 1, 1862, the privately owned New Haven Water Company began distributing this water to the residents of the city (NHWC, 1941).

Limited water resources and substantial financial requirements made it increasingly difficult to manage small community water systems in the surrounding towns. Recognizing these challenges, in 1875, the New Haven Water Company began acquiring neighboring water companies. These acquisitions added extensive infrastructure to NHWC's water supply system including reservoirs, well fields, water mains, and pumping stations. The New Haven Water Company's service area quickly expanded to include the towns of Bethany, Branford, Cheshire, East Haven, Hamden, Milford,

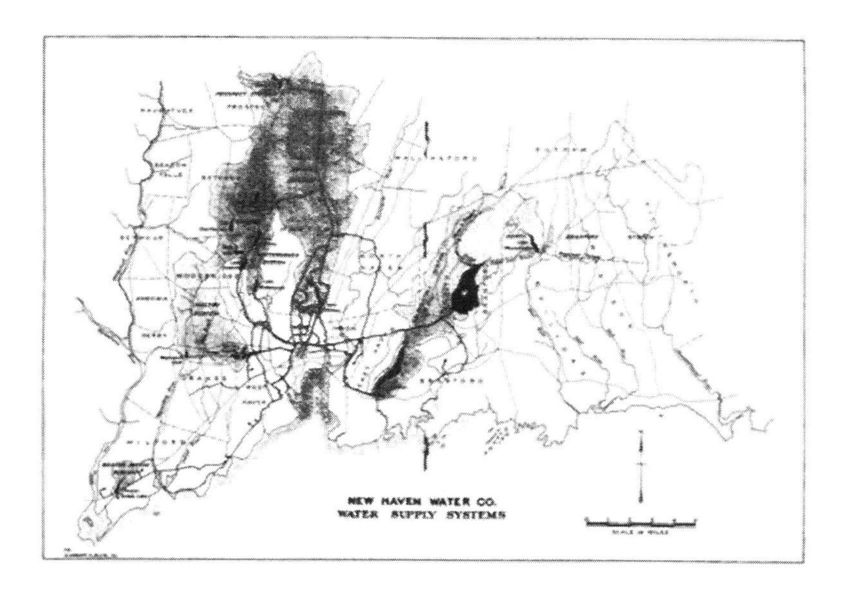

Figure 2.2 1941 Map of the New Haven Water Company's supply system (From NHWC, 1941).

North Branford, North Haven, Orange, West Haven, and Woodbridge (Figure 2.2) (NHWC, 1941).

2.3.1.1 Treatment History

By 1904 the watershed of Lake Whitney had become thickly populated and the New Haven Water Company deemed water filtration necessary to ensure the safety of its drinking water. Sand filtration treatment had begun to appear in the 1890s in other American cities, and in 1906 the NHWC built "twelve filter beds, filled with fine, clean sand and covering four acres" near the Whitney Dam (NHWC, 1941). Lake Whitney water would slowly seep through three feet of sand at a rate of three million gallons per acre per day. The living biofilm that formed on the top of the sand filters helped to eliminate bacteria, algae, and other organisms as well as to reduce unwanted color and odor. Periodically this upper layer would be scraped off and washed thoroughly, in a scheduled rotation so that 11 filters would be operating at any given time (NHWC, 1941). With maintenance, this original treatment plant operated until 1991 (SCCRWA, 2011a).

2.3.1.2 Ownership History

The 1974 Safe Drinking Water Act, a federal response to nationwide water degradation concerns, as well as subsequent amendments and regulations, presented new filtration requirements for the City of New Haven (SDWA, 1974, 1986, 1996;

SDWAR, 1989). These requirements presented significant financial challenges to the New Haven Water Company. In an attempt to comply with new requirements and meet budget shortfalls, the NHWC announced plans to sell more than 16,000 acres, or 61 percent of its total 26,000-acre landholdings, throughout 17 towns at that time (McCluskey and Bennitt, 1996). This decision generated public outcry from local residents, state health officials, the Connecticut Department of Energy and Environmental Protection (DEEP), and others. This public opposition stemmed from concerns over increased development, loss of open space, and development burdens on town budgets.

In response to the public outcry surrounding the NHWC's proposal, legislators imposed a moratorium on the land sale and suggested public ownership of the water company (CGA, 1979). New Haven's Mayor Frank Logue announced that the city planned to exercise a 1902 purchase option, while neighboring suburban towns advocated for a regional water supply (McCluskey and Bennitt, 1997). After feasibility studies and continued political turmoil, the Connecticut General Assembly authorized the transition of the company from the privately held NHWC to the public SCCRWA in 1977 (McCluskey and Bennitt, 1997). The SCCRWA was directed to both provide high-quality drinking water and to advance the compatible uses of Authority lands, including conservation and recreation (CGA, 1977).

As former Connecticut State Representative McCluskey explained, the regionalization and conversion of the New Haven Water Company to the publicly owned SCCRWA resulted from federal drinking water regulations as well as urban-suburban politics, tax policies, and rising water rates (McCluskey and Bennitt, 1996). The debate over private versus public ownership was a long-standing issue for water providers in the region. The City of New Haven proposed to purchase the private water works in 1881 and again in 1891 (McCluskey and Bennitt, 1996). In 1902, an agreeable contract was finally established that allowed for free water for fire and schools and fair rates for customers as well as a purchase option once every 25 years (McCluskey and Bennitt, 1996). Seventy years later, this purchase option provided the City of New Haven the legal leverage it needed to purchase the utility and establish a regionalized public water system to provide safe drinking water, watershed protection, and recreation opportunities to Connecticut residents.

2.3.2 The SCCRWA Today

Today, the SCCRWA provides drinking water to more than 400,000 people through 116,277 metered connections in 23 cities in south central Connecticut (Figure 2.3). The 1977 legislative act that enabled this agency defined its mission as "providing and assuring the provision of an adequate supply of pure water at reasonable cost ... and, to the degree consistent with the foregoing, of advancing the conservation and compatible recreational use of the land held by the authority" (CGA, 1977a, p. 1). Since the 1980s, the agency has invested over $17 million to purchase and protect nearly 5,000 acres of watershed land in the region to prevent water quality degradation and to minimize treatment expenses (SCCRWA, 2007a).

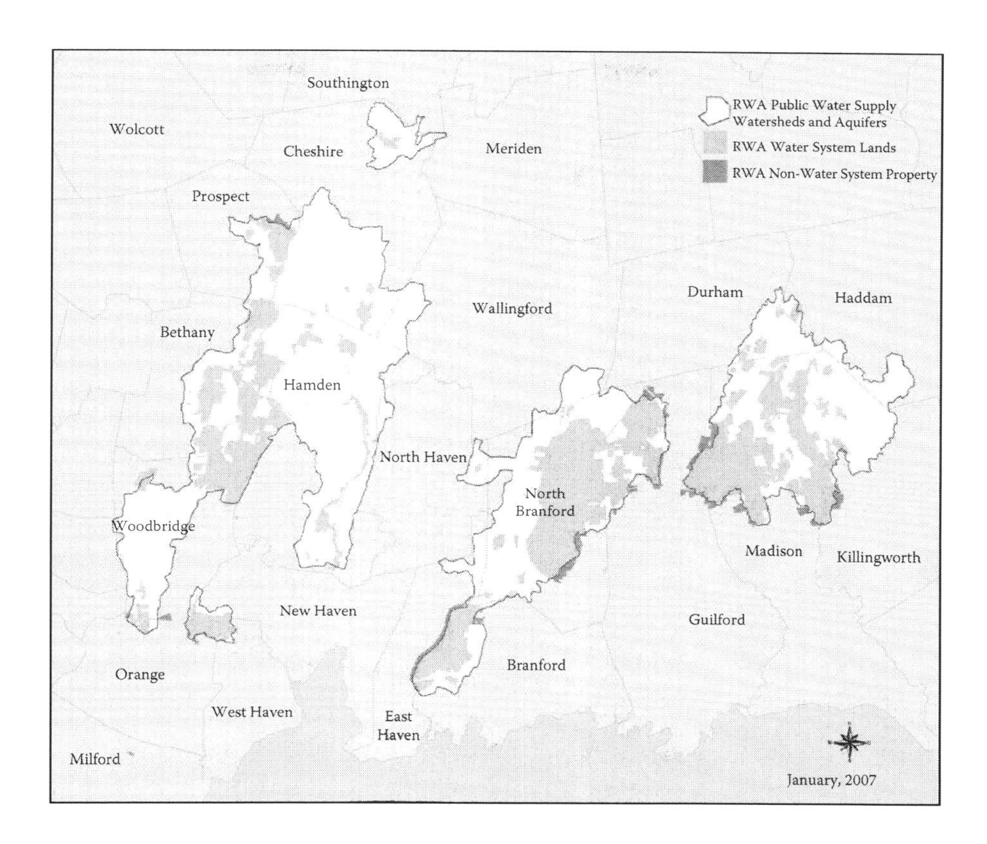

Figure 2.3 SCCRWA watershed lands. Green indicates land owned by SCCRWA. White indicates land within drinking water watersheds, about 3000 acres of which SCCRWA would like to purchase or protect. Orange indicates SCCRWA-owned land that is outside of their priority watershed protection areas (Class III), about 900 acres, that SCCRWA would like to sell (From SCCRWA, 2007b). (See color insert.)

SCCRWA presently owns more than 26,000 acres of land in 20 Connecticut towns (Figure 2.4) (SCCRWA, 2010b). These 26,000 acres equate to total ownership of 34 percent of the 77,000 acres of SCCRWA's source watersheds. Over 80 percent of the region's tap water comes from 10 different reservoirs located in the district towns of Bethany, Branford, East Haven, Guilford, Hamden, Killingworth, Madison, North Branford, and Woodbridge (SCCRWA, 2010b). These reservoirs are fed by both surface water and replenishment from the Mill River aquifer located in Hamden and Cheshire.

2.3.3 SCCRWA Land Stewardship Efforts

Land stewardship and management to protect and improve raw water quality are at the core of SCCRWA's operations. All land stewardship on SCCRWA properties

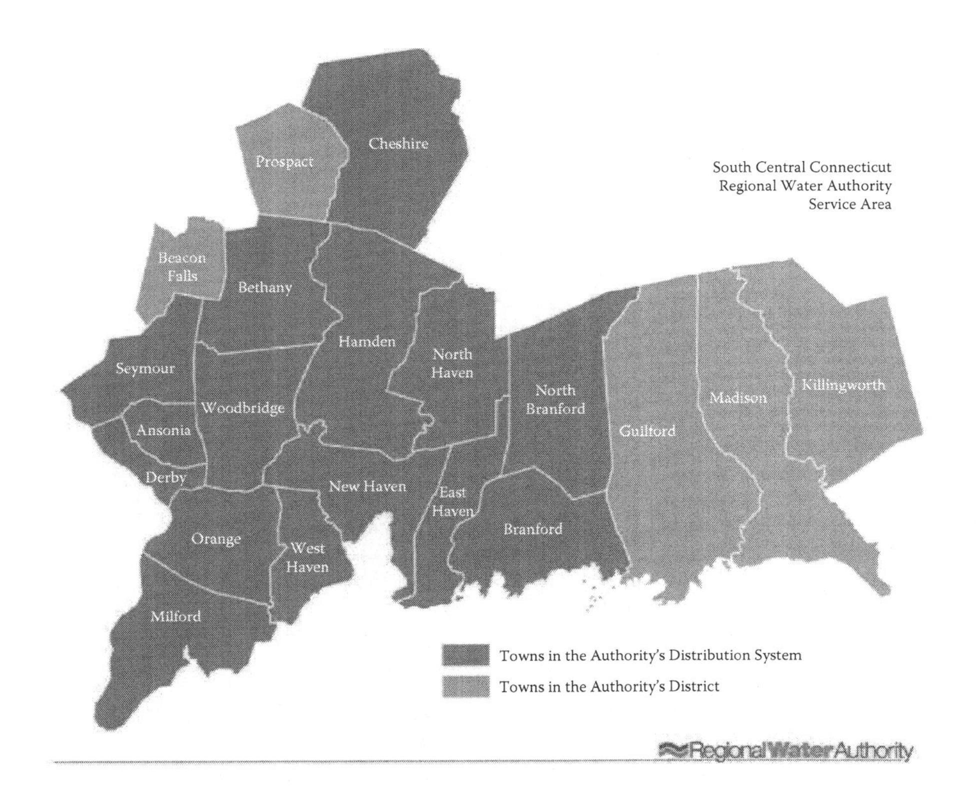

Figure 2.4 Connecticut towns in 2010 service area (From SCRWA, 2010c).

is guided by the Connecticut DEEP's Best Management Practices, a state-level guide to minimum actions required to protect water quality (CT DEP, 2007). SCCRWA's land stewardship is informed by SCCRWA's land use and forest management plans (SCCRWA, 1989). The Land Use Plan was approved in 1983 and updated in 1989 and 1996. The Forest Management Plan was completed in 1989 (SCCRWA, 1989). Forest management activities are carried out by SCCRWA's Forestry Department.

The SCCRWA's land stewardship goals include (1) watershed protection, (2) timber resource conservation, (3) wildlife resource protection, (4) open-space preservation, and (5) education and research. Each management goal is supported by a list of specific management objectives of the Land Use Plan and Forest Management Plan and are discussed in more detail in Section 2.5.

2.3.4 The SCCRWA Land Acquisition Program

Since the 1980s, SCCRWA has invested $17 million to protect 5,000 acres of watershed land in the region to ensure high water quality (SCCRWA, 2010d). Land has been purchased outright and has also been secured through the acquisition of conservation easements, option agreements, and right of first refusal purchase agreements (SCCRWA, 2010d). In 2010, SCCRWA held approximately 27,000 acres in fee-simple ownership, with an additional 900 acres protected through conservation easements (Tompkins, 2010).

2.3.4.1 Expanding Watershed Protection

In addition to SCCRWA's current landholdings, the utility has identified an additional 3,000 acres of privately owned land in the region that is desirable to purchase for watershed protection (SCCRWA, 2007b) (Figure 2.3). Given the high costs of land acquisition in Connecticut and SCCRWA's limited financial liquidity, the utility has developed a set of criteria to identify which of these lands will yield the greatest source water protection benefits for the cost (SCCRWA, 2010d). In general, these criteria for acquisition give priority to large tracts of undeveloped Class I– and II–type lands which are at high risk of development or are highly environmentally sensitive, are adjacent to or an inholding of existing SCCRWA property, and contain tributaries to SCCRWA reservoirs (CGA, 1996). They also weigh cost and opportunities for partnerships in the acquisition (SCCRWA, 2010d). SCCRWA's Land Committee regularly evaluates the established criteria and applies them to potential property procurement (Tompkins, 2010).

SCCRWA prefers to acquire additional watershed land through conservation easements and partnerships rather than fee-simple acquisition because these methods cost less and allow the utility to stretch its limited funds. The utility regularly partners with The Nature Conservancy (TNC), the Trust for Public Land, local land trusts, towns, and other conservation agencies to expand source water protection (Tompkins, 2010). However, in some instances fee-simple acquisition is necessary or even preferred, such as when land is being considered for major development or when landowners are only interested in the outright sale of their land. SCCRWA also regularly participates in the DEEP's open-space grant program, which provides funding to entities to accomplish land conservation using conservation easements that are managed by DEEP (Tompkins, 2010).

2.3.4.2 Selling Unnecessary Land

In addition to planning for land acquisition, SCCRWA continually reviews their holdings for unnecessary land that can be made available to sell. A recent review of their source water protection program has identified about 900 acres of non-water-system land (Class III) (SCCRWA, 2007b). These landholdings are not required for the protection of raw water, and selling them will aid SCCRWA to keep water rates low (SCCRWA 2010a).

State water utility regulations require that before selling off property, SCCRWA must obtain approval from their representative policy board or the chief executive officer of the town or city where the land is located (CGA, 2002). Upon receiving permission to sell, they must then send notification to local community officials and land and conservation agencies, the Connecticut Departments of Public Health and Environmental Protection, as well as the Connecticut Office of Policy and

Management (SCCRWA, 2007b). SCCRWA also conducts an in-depth land study and environmental evaluation of the parcel, and holds a public hearing prior to any sale of land (SCCRWA, 2007b). Because of this, SCCRWA anticipates that the proposed 900-acre land sale will take between 5 and 10 years (SCCRWA, 2007b). Since 1980, SCCRWA has received $24 million for the sale of 1,779 acres of "non-water-system land" and has used these funds to offset the cost of capital improvement programs, which includes land acquisitions (SCCRWA 2010a).

2.3.5 Management for Multiple Benefits

Source water protection is the primary use of SCCRWA land but it is not the only use. Residents of New Haven and its surrounding communities value SCCRWA lands as a vital cultural, environmental, and recreational resource. Former New Haven Mayor Frank Logue is credited with rallying community support for public ownership of water company land in the late 1970s, particularly highlighting recreational opportunities (McCluskey and Bennitt, 1996).

SCCRWA lands are managed for multiple public benefits including recreation, research, timber harvest, and nontimber products (e.g., maple syrup, hay). More than 4,500 resident permit holders use the utility's 54 miles of recreational trails and 17.1 stream miles of fishing access (SCCRWA, 1989). Despite a significant focus on providing recreational land, not all activities are welcome. For instance, ATVs, dogs, berry picking, mushroom gathering, trapping, and hunting (with the exception of a special-permit deer cull) are not allowed on SCCRWA land (SCCRWA, 2011b).

Though these benefits provide insight into SCCRWA's land management measures, SCCRWA Natural Resources Manager Tim Hawley reports that SCCRWA is not presently generating revenue from these co-benefits, and by the time the supervisory and support costs (rangers, toilets, boats, trail blazing, issuing of maps and permits) are factored in to the recreation program, no net revenue is generated (Hawley, 2010). Prior to the 2007 recession, when stumpage averaged $0.145 per board foot, timber and firewood harvesting was generating about $90,000 per year in net revenue for SCCRWA (Hawley, 2010). Not quantified in these revenue calculations are the many other ecosystem services that do not currently have an associated market value including air purification, water quantity and quality protection, carbon sequestration, scenic and spiritual resources, peak flow moderation, noise and light modification, soil formation and stabilization, and habitat biodiversity (Hawley, 2010).

2.3.6 Gray Infrastructure: Filtration and Purification Technology

Managers and water treatment plant operators at SCCRWA recognize that cleaner raw water is easier and cheaper to treat than more polluted raw water, a driving consideration of many source water protection efforts. Nevertheless, SCCRWA is required to filter its water, in order to comply with federal and state drinking water regulations (Ernst et. al., 2004; Freeman et. al., 2008; CGA, 1985; SDWA, 1986; SDWAR, 1989). After water leaves the SCCRWA reservoirs, it enters the Lake Whitney treatment plant facility. The current treatment plant processes include flash mixing, coagulation, flocculation, filtration, disinfection, fluoridation, and a pH adjustment (SCCRWA, 2011a).

2.3.6.1 SCCRWA Gray to Green and Green to Gray

Based on a mixture of regulatory drivers, citizen involvement, and regional setting, SCCRWA has adopted a progressive management approach to blend the use of land conservation and management with engineered treatments for water quality. SCCRWA recognizes that conservation requires both opportunity costs of not selling valuable land as well as direct expenses of land management activities (SCCRWA, 1989). Despite this, drinking water delivery for SCCRWA is not a choice between either source water protection and management or built infrastructure investment and maintenance. Delivering the highest quality drinking water at the lowest possible cost is a matter of finding the right balance between investing in both gray and green infrastructure. A similar approach to land management and infrastructure investment can be seen in the example of the privately held Connecticut water utility Aquarion.

2.4 THE HISTORY OF AQUARION WATER COMPANY OF CONNECTICUT

2.4.1 Ownership History

The City of Bridgeport's water supply can be traced back to 1818 when Reverend Elijah Waterman constructed a hollow log pipeline to carry water from Bridgeport's "Golden Hill" down to the city's waterfront. Entrepreneurs soon purchased this pipeline with a $10,000 grant from the Connecticut General Assembly and founded the Bridgeport Golden Hill Aqueduct Company. Three decades later the Great Bridgeport Fire of 1845 destroyed the log pipeline. This fire and subsequent service disruption highlighted the city's need for a safe and secure water supply. Consequently, an entrepreneur named Nathaniel Green invested $160,000 in capital to form the Bridgeport Water Company (BWC) in 1853. Green spearheaded the construction of two reservoirs, the Masonry Tank Reservoir and Lower Reservoir, and began laying water supply pipes throughout the city. An economic decline soon led to the decline and eventual foreclosure of BWC (Greenland, 2007).

Following the foreclosure of BWC, a special act passed by the Connecticut General Assembly formed the Bridgeport Hydraulic Company (BHC) in 1857. The BHC was formed to serve the residents of Bridgeport and its surrounding communities. The investor-owned company, which was run by circus tycoon and former Bridgeport mayor P. T. Barnum between 1877 and 1887, grew quickly to meet increasing demand (Lomuscio, 2005). Author James Lomuscio gives a poignant account of the land use battle that unfolded in the once-thriving industrial town of Valley Forge, where, between 1937 and 1938, residents fought to stop construction of BHC's largest reservoir, the Saugatuck Reservoir. The BHC had purchased land from residents outright, but residents alleged that the water company deceived them, paying below-market prices and hiding their true intentions. Construction began in 1939, despite public outcry, and the dam, which flooded and effectively obliterated the town, was completed in 1941 (Lomuscio, 2005).

BHC continued to expand and prosper until the late 1960s, when BHC opted to diversify its operations by both expanding its supply area and entering into business ventures outside of the water supply industry. To begin that transition, a holding company known as the Hydraulic Company was created in 1968. In 1969, the Hydraulic Company officially acquired BHC and subsequently acquired a number of smaller utilities. Throughout the 1970s and '80s the Hydraulic Company built new treatment plants and expanded watershed management activities to meet increasingly stringent federal drinking water quality requirements. In 1991, the Hydraulic Company changed its name to Aquarion (Greenland, 2007). The original BHC system, known as the Main System, included what are now Aquarion's four largest reservoirs—Saugatuck, Easton, Hemlocks, and Trap Fall—as well as four smaller reservoirs and two undeveloped reservoir sites (Roach, 2010; Galant, 2010).

In the 1990s Aquarion held a total of 17,000 acres of watershed lands (Roach, 2010). Aquarion purchased the Stamford Water Company in the 1980s and continued to acquire assets (Galant, 2010). Prior to the Safe Drinking Water Act (SDWA) no filtration was required for the surface water from its reservoirs. Instead, surface water was treated solely by the addition of chemicals—notably chlorine. After passage of the SDWA and further amendments to improve drinking water quality (CGA, 1985; SDWA, 1986; SDWAR, 1989), Aquarion took steps to meet filtration requirements and began to build filtration treatment plants starting in 1991 (Liberante, 2010).

Aquarion's steady growth led to its eventual purchase in 2000 by the United Kingdom–based Kelda Group for $596 million. Aquarion continued to expand its holdings in the 2000s as a subsidiary of Kelda Group, acquiring four other New England water companies—Connecticut-American Water, New York-American Water, Massachusetts-American Water, and Hampton Water—for $118 million in 2001. Connecticut-American Water and BHC combined to form the Water Company of Connecticut, Inc., which in 2002 purchased the Connecticut Water Company, acquiring major systems in Greenwich, Mystic, and Darien as well as small systems in East Hampton and Lebanon as a result (Galant, 2010). In 2006, the parent Aquarion Company was sold to Australia's Macquarie Bank for $860 million (Hoover's, 2010; Kelda Group, 2006). Today the company (renamed the Aquarion Water Company of Connecticut) is one of the top 10 largest investor-owned utilities in the United States (Hoover's, 2010).

2.4.2 The Aquarion Water Company of Connecticut Today

2.4.2.1 Current Ownership and Service

Aquarion of Connecticut (Aquarion) is the largest water utility in New England and serves approximately 580,000 people across Fairfield, New Haven, Hartford,

Litchfield, Middlesex, and New London counties (Aquarion, 2011a). It currently owns and operates 22 active reservoirs in Connecticut, as well as five stream diversions and nine treatment plants across 29 different water supply systems (Galant, 2010). The utility owns or manages approximately 19,546 acres (19 percent) of the more than 102,352 acres that make up the watershed that serves these reservoirs (Galant, 2010). It owns over 50 percent of the watershed for only four of its 22 reservoirs. The Saugatuck reservoir, the largest in the system at nearly 12 billion gallons capacity, is buffered by a watershed that is only 17 percent utility owned (Galant, 2010).

2.4.2.2 *Current Water Treatment Technology*

Aquarion operates nine surface water and several dozen groundwater treatment plants throughout Connecticut, which use either conventional settling or dissolved air flotation as their primary sedimentation technology (Table 2.1). This is followed by filtration and addition of disinfectant (chlorine), fluoridation, and pH adjustment. The treatment capacity of the system is 158.5 million gallons per day (Aquarion, 2011b). The main system is the largest, with a capacity of 95 million gallons per day (Roach, 2010).

2.4.3 National Fairways Conflict: Divisive Aquarion Land Sale Proposal

Aquarion's willingness to sell large tracts of land, subject to regulatory approval, became apparent when the utility entered a contract for sale of 730 acres of Class III forestland in 1997 with National Fairways, a golf course and real estate developer (Business Wire, 1998). The land, slated for golf course and luxury home development, was adjacent to 225 acres of open space owned by the Aspetuck Land Trust to the south and across the Saugatuck Reservoir from TNC's Devil's Den property to the west (Miner, 2009).

Table 2.1 Aquarion's Treatment Operations and Costs for Its Watersheds

Treatment Plant	Treatment Capacity (mpd)	Approx. Capital Cost ($M)	Technology
WT1	8	Unknown	Upflow clarifiers, filtration
WT2	4	Unknown	Sedimentation and filtration
WT3	20	Unknown	Sedimentation and filtration
WT4	30	16	DAF and filtration
WT5	50	47	DAF and filtration
WT6	0.75	4.1	DAF and filtration
WT7	20	27	High rate plate settler and filtration
WT8	25	18	High rate plate settlers and filtration
WT9	0.75	4.1	DAF and filtration Same as Lakeville

Source: Galant, 2010.

Residents in nearby Easton, Weston, Fairfield, and Westport joined to form the Coalition to Preserve Trout Brook Valley, as the property was called. Open-space activists and town leaders wrote letters and articles to local papers and *The New York Times* and met with the Aspetuck Land Trust, TNC and the DEEP to discuss acquisition for open space. The movement grew quickly and eventually attracted the attention of the Newman family—Paul, Nell, and Melissa—as well as Paul Newman's friend and conservation activist Robert Redford. Paul Newman donated $500,000 to the cause, and Redford made an undisclosed donation (Miner, 2009). The Aspetuck Land Trust, TNC, and the state bought National Fairways out of its contract with Aquarion for just over $12 million in late summer of 1999 (Patton, 2010). Aquarion donated an additional parcel to the deal, at an estimated value of $1.7 million (Business Wire, 1999). The land is now the Trout Brook Valley Preserve, managed by the Aspetuck Land Trust (Figure 2.5).

The Trout Brook Valley incident brought together citizens and organizations that shared a common interest in water company land as public open space—regardless of whether the water company was a public or private entity. This experience also put many Connecticut residents on edge. It was clear that the locally owned Aquarion/BHC was willing to sell off large tracts of land, and residents wondered how much land a privately held company without such local connections might sell.

2.4.4 Kelda Takeover of Aquarion

Fears of changing ownership and increased incentives to sell land were not unfounded—in 2000, the merger of UK-based Kelda Group and Aquarion/BHC was announced and approved by shareholders and state regulators (Kinsman, 2000; Staff Reports, 1999). At the time, Aquarion owned 16 active reservoirs, 3 undeveloped reservoir sites, and approximately 18,000 acres of Class I, II, and III lands and had just emerged from the high-profile Trout Brook Valley land transaction with National Fairways in 1999.

2.4.4.1 Kelda-Aquarion's Land and Easement Sale

The fight to stop National Fairways' 730-acre development project laid the groundwork for a public campaign to acquire more water company lands (Roach, 2010; Patton, 2010). The Connecticut Fund for the Environment and other conservation groups were already mobilizing to conserve Aquarion watershed land when Kelda's acquisition was approved in 2000. These organizations, as well as some state legislators, were concerned that when Aquarion was acquired by Kelda, the company would sell off all of its Class II and Class III landholdings to turn a quick profit (Fotos, 2010). These concerns were exacerbated by the fact that Kelda had reportedly taken on debt to pay more than double book value for Aquarion and the CEO of Kelda was dismissed in April 2000 (Moreau, 2000). Just days before he was fired, CEO Kevin Bond had been in Connecticut assuring state regulators that Kelda had no intentions to sell land (Moreau, 2000).

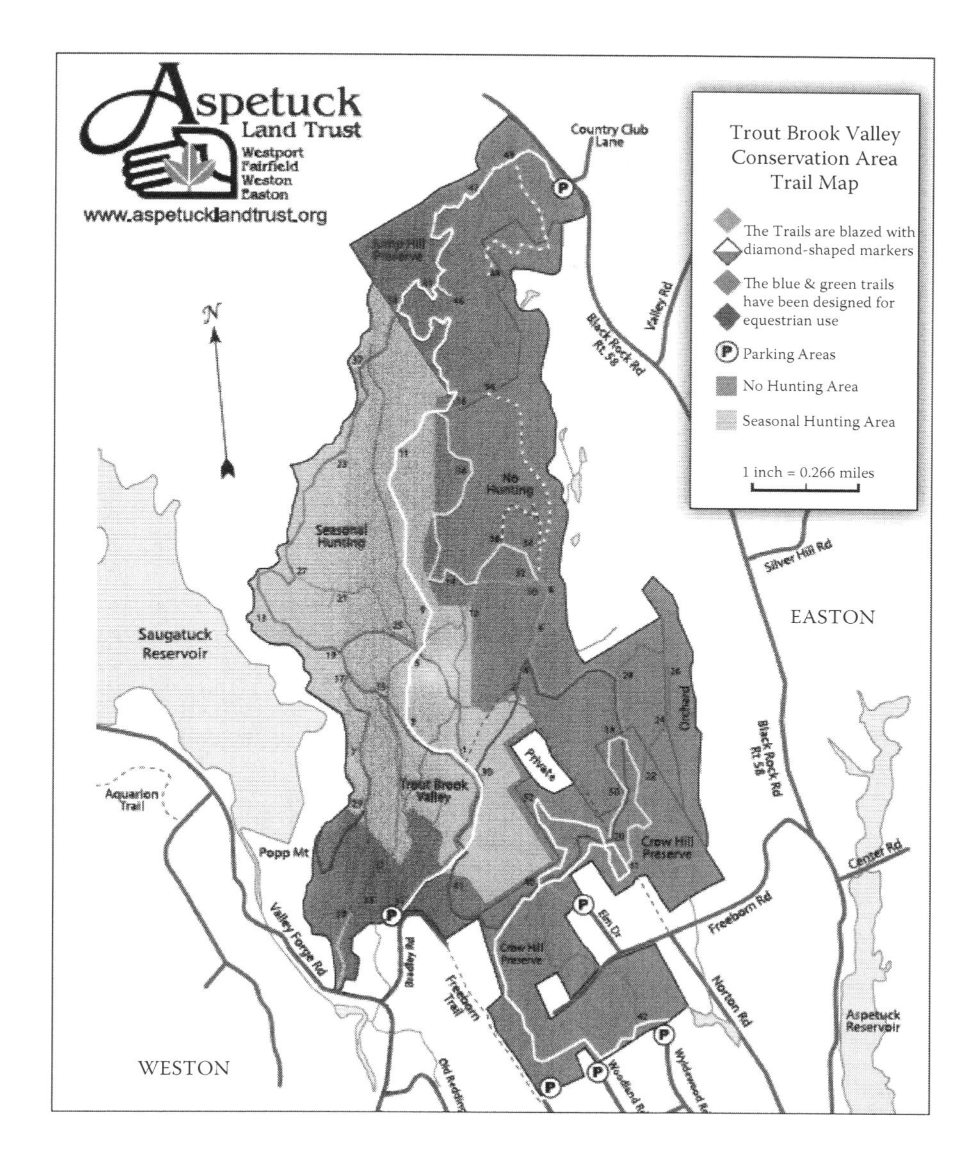

Figure 2.5 Trout Brook Valley Conservation Area trail map (From Aspetuck Land Trust, n.d.).

Despite Kelda's insistence that they had no plans to sell portions of the nearly 19,000 acres they acquired from the merger with Aquarion/BHC, Connecticut residents were wary and a number of proposals were introduced to wrest land control away from the company (Budoff, 2001). The Coalition for the Permanent Protection of Kelda Lands was formed in 2000, composed of 67 organizations and 47 elected officials as well as thousands of individuals from nearly every town and city in Connecticut—not just those adjacent to company lands or Aquarion customers

(Belaga et al., 2001). Fifteen of these towns donated $5,000 each for a feasibility study that would balance Kelda land protection with taxpayer and ratepayer interests (Belaga et al., 2001). The study identified four options: purchase a combination of fee interests and conservation easements on Kelda-Aquarion lands, an asset buy-down, and two methods of establishing a publicly controlled regional water authority. The latter options would involve government takings of Kelda watershed lands and was considered a wildcard option in the event Kelda refused to offer land or easement sales (Belaga et al., 2001).

The former Connecticut Attorney General Richard Blumenthal called on the legislature to impose a moratorium on all utility company land sales in November 1999 to block Aquarion from potentially selling its holdings (CAGO, 1999). Blumenthal also asked the legislature to consider enhancing condemnation rights and powers of local and state government. Instead, Kelda-Aquarion agreed to a voluntary three-year moratorium on sales, during which time the DEEP and conservation groups could work to develop a conservation proposal.

TNC began discussions with Aquarion and the Connecticut DEEP to arrange a conservation purchase of all Class II and III lands and a conservation easement for the Class I forestlands. Aquarion's holdings were large tracts of forestland in "a rapidly developing urban landscape" (Fotos, 2010). Both TNC and DEEP, as well as other conservation groups in the area, wanted to slow development to create open space and provide recreational opportunities to residents in the region (Fotos, 2010).

After months of negotiations, a deal was signed in February 2001. It was a bargain sale: Aquarion would receive $90 million and a significant tax benefit in exchange for fee sale of approximately 5,400 acres of Class II and III lands and the development rights on 9,025 acres of Class I lands. The "bargain" was a significant one: these lands and development rights had been appraised at $193 million, more than double the price paid (Patton, 2010). The state contributed $80 million, and TNC contributed the remaining $10 million, and after required approval and permitting from the Connecticut Department of Public Health, the deal was finalized in 2002 (Roach, 2010; Patton, 2010; NRMA, 2002). It was the largest open-space acquisition in state history (CT Executive Chambers, 2004).

Approximately 15,000 acres of the Kelda-Aquarion deal were dedicated to the creation of the Centennial Watershed State Forest, 90 percent of which lies in Fairfield County, with additional property in Litchfield, New Haven, and Hartford Counties (CT Executive Chambers, 2004). The Centennial State Forest has numerous hiking and cross-country trails and is open for fishing, hunting, and birding. These recreational opportunities were of considerable importance to DEEP when crafting the land and easement purchase and ongoing management plans; however, unlike other state forests, all activities in the Centennial Watershed State Forest require a permit (CT DEP, 2011a).

2.4.5 Conservation Lands Committee: Partnership for Land Management

Under the Natural Resources Management Agreement (NRMA), all of the land associated with the 2002 Centennial Forest deal is jointly managed by the Conservation Lands Committee (CLC), which is made up of representatives from Aquarion, DEEP, and TNC. Aquarion is the coordinating manager and is primarily responsible for forest management and maintenance. The land and easement sale has had very little impact on Aquarion's management policies and practices, particularly on the Class I lands that Aquarion still owns (Roach, 2010; Haines, 2010; Galant, 2010).

When TNC and DEEP purchased the land and easements, the parties agreed that the land would be managed uniformly, as though it had one owner (Patton, 2010). The CLC acts as a forest owner, and as such has assumed all the responsibilities of a forest manager. Aquarion pays for all management and improvements aside from kiosks related to recreational use and some research, which DEEP and TNC finance in part. The three parties divide proceeds from timber sales and other revenue-generating activities according to the following breakdown: Aquarion gets 65 percent, the State gets 30 percent, and TNC gets 10 percent. These interests are based on landownership and financial investments made in the sale (Haines, 2010). Aquarion is responsible for creating management plans, which CLC must then approve.

2.4.5.1 Aquarion/Bridgeport Hydraulic Company Land Stewardship

Aquarion/Bridgeport Hydraulic Company's (Aquarion) land management requirements are clearly outlined in CLC's Natural Resources Management Agreement (NRMA). Though Aquarion's management to protect reservoir water quality and flow quantity is guaranteed throughout the document, the forestland is not being managed for water quality alone. In fact, the plan gives a number of ecological factors precedence to management for water quality.

The CLC's approach to forest resources management is designed to promote the following goals:

1. Promote the growth and development of a healthy, diverse and resilient forest
2. Protect forests from fire (other than controlled fires permitted under the agreement), insects, disease, and other damaging agents
3. Protect and promote the recovery of threatened and endangered species regulated pursuant to Connecticut General Statutes Chapter 495
4. Encourage a continuing supply of forest products harvested in ways that sustain long-term site productivity and that consider aesthetic and ecological values
5. Encourage the safe conduct of forest practices in a manner that is in conformance with all applicable statutes and regulations
6. Afford protection to and improvement of air and water quality
7. Foster biological diversity
8. Allow for a variety of forest-resource-based, high-quality, environmentally responsible, public recreational opportunities

9. Foster and maintain significant tracts of naturally occurring, mature, diverse, and continuous forest cover
10. Support scientific research into the growth and development of a healthy, diverse, and resilient forest by accommodating the siting of research plots on Conservation Land (NRMA, 2010)

Though it can be argued that water quality and quantity targets will be achieved through goals 1 and 2, water quality protection is not explicitly stated until goal 6 and is bundled with air quality. Interestingly, the most recently completed Forest Management Plan for the Aspetuck and Hemlock Reservoirs, places water quality protection as the primary management objective (Ferrucci and Walicki, 2009). Such emphasis is consistent with the pursuit of healthy diverse forest stands and their associated values of watershed, wildlife, fisheries, recreation, and aesthetics.

2.4.6 Water Quality Monitoring

To meet the standards of the Safe Drinking Water Act, Aquarion has a comprehensive water quality monitoring program. Each year from May through November, the utility spends $140,000 on monitoring, testing for each of the regulated water quality contaminants (Aquarion, 2011b). These contaminants include turbidity, odor, iron, manganese, copper, ammonia, nitrate, total organic carbon, dissolved organic carbon, total trihalomethanes, and total haloacetic acids (Aquarion, 2008, 2009a, b). Aquarion samples in tributaries, at the reservoirs, across the treatment system, and throughout the distribution system. Its staff are divided into three teams for water sampling: Watershed, Water Quality, and Supply Operations. The Watershed team samples throughout the watershed, with a focus on the tributaries. The Water Quality team samples at every reservoir and distribution system. The Supply Operations team samples across the treatment plants. The Supply Operations team takes samples daily, while the other teams sample weekly and monthly (Hanover, 2010).

Water quality is sampled at different depths of each reservoir. The treatment plants can selectively withdraw water from different levels of the reservoir depending on where the best water quality is, as determined by the water quality sampling information (Aquarion, 2011b).

2.5 CURRENT LEGAL EFFORTS TO ADDRESS WATER SUPPLY CHALLENGES IN CONNECTICUT

2.5.1 Evolution of Connecticut Water Law

In response to increasingly strict standards in the SDWA, as well as public backlash against proposed sale of water company lands, the State of Connecticut adopted a series of regulations governing water companies and management of source water lands. In 1974 the Connecticut General Assembly enacted Public Act No. 74-303, requiring water companies to obtain approval from the Connecticut Department of

Public Health (DPH) before selling, leasing, or changing the use of any watershed land (CT CGA, 1974; McCarthy, 2001). DPH states that as of 1999, this regulation protected 20–30 percent of source water lands for surface water public drinking supply from being sold or leased (McCluskey and Bennitt, 1996).

In response to major proposals by SCCRWA to sell watershed lands, a second act from the Connecticut General Assembly came a year later with the passage of Public Act No. 75-405 (CT CGA, 1975). The act established the Council on Water Company Lands and directed the Council to inventory such lands, develop criteria to determine which of them were not needed for water supply, and recommend a comprehensive policy for disposing of these lands (McCarthy, 2001). In 1977 the recommendations of the Council were codified in Public Act No. 77-606 (CT CGA, 1977c). The Council's recommendations included the establishment of three classes of water-company-owned land as well as regulations regarding the sale, lease, and change of management of watershed lands. The regulations define land classes in terms of proximity to water sources:

Class I: Includes all land owned by a water company or acquired from a water company that is either within 250 feet of high water of a reservoir, 200 feet of groundwater wells, or 100 feet of agency-regulated watercourses, or composed of critical stream corridors or steep slopes without significant interception by wetlands (CGA, 1996). Water companies may not lease or assign Class I lands and can only sell these lands to the state, a municipality, or another water company, and the buyer must agree to use restrictions. Additionally, the owner cannot sell or change the use of these lands without a permit (McCarthy, 2001; Holstead, 2002).

Class II: Includes land owned by or acquired from a water company that is a public drinking supply watershed not included in Class I or within 150 feet of a distribution reservoir or its principal tributaries (CGA, 1996). Sale, lease, or changes in management of these lands requires a permit from DPH (McCarthy, 2001; Holstead, 2002).

Class III: Includes all land owned by or acquired from a water company that is unimproved land outside of drinking supply watersheds and beyond 150 feet from a distribution reservoir or its tributaries (CGA, 1996). The sale, lease, or change in management of these lands was not restricted by Public Act No. 77-606, but subsequently the state expanded regulations to require notification of the sale of these as well as any other water-company-owned lands (McCarthy, 2001).

In 1984 Public Act No. 84-554 was passed, requiring water companies to include future land sales in their water supply plans and submit them to DPH (CGA, 1984; McCarthy, 2001). This requirement was expanded in 1998 with the passage of Public Act No. 98-157, which requires water companies to notify land conservation organizations of both potential and actual land sales, as well as reclassification of land (CGA, 1998; McCarthy, 2001). In 2006, the State further improved communication regarding the sale of land within a public drinking water supply's watershed, placing regulations on developers. Public Act No. 06-53 requires developers to notify the DPH when they submit an application to a local zoning or planning commission for a project within the watershed of a water company. However, if the local planning or zoning commission determines the project will not adversely affect the public water supply, notification of DPH is not required (CGA, 2006).

2.5.2 Connecticut Source Water Assessment Program

To comply with the 1996 amendments to the SDWA, the State of Connecticut created a Source Water Assessment Program (SWAP) (CT DPH, 2007). The purpose of SWAP is to identify and inventory potential sources of contamination, land use, and other threats to the quality of public drinking water sources in Connecticut (see Figure 2.1). This assessment was intended to inform communities of these threats so that water systems, local governments, health officials, and others could plan and direct source water protection initiatives including zoning regulations, land acquisition, and implementation of best management practices for safe handling, storage, and disposal of hazardous materials (CT DPH, 2005).

Connecticut's source water protection program requires that local planning and zoning commissions consider protecting current and potential future source water areas in their land use plans and regulations. It also promotes model land use regulations for source water protection. The program encourages water suppliers and government officials to work together to reduce potential impacts of new development on source water. It also supports the implementation of compatible land uses in current and potential future source water areas, and identifies low to high-risk categories of land uses for source water areas (Table 2.2) (CT DPH, 2005).

The Connecticut Department of Public Health Drinking Water Division is responsible for overseeing the state SWAP. In 1997, the Drinking Water Division worked with DEEP to develop a work plan for the assessment of the state's drinking water sources. The assessment was completed in May of 2003 with approval from the U.S. Environmental Protection Agency (EPA) (CT DPH, 2005). At the time,

Table 2.2 Connecticut SWAP – Low to High Risk Land Use Categories

Low Risk	→			High Risk
Water company-owned land Federal, state, local & private nature preserves Open space with passive recreation Private land managed for forest products Public parks and recreation areas	Field crops – hay, pasture, orchards Low-density residential housing on 2 acre lots Churches, municipal offices	Agricultural production – vegetables, dairy, livestock, poultry, tobacco, nurseries Golf courses Medium-density housing on ½ to 1 acre lots	Institutional uses – schools, health care facilities, prisons High density housing on less than ½ acre lots Multifamily housing units Commercial facilities with only sewage discharges	High risk commercial sites – gas stations & auto services center, dry cleaners, photo processors, junk yards, machine shops Industrial manufacturing, chemical processing Waste disposal – lagoons, landfills, bulky waste sites

Source: Modified from CT DPH, 2005.

50 community surface water systems, including 166 sources, were assessed. The results showed that 69 percent of these systems were at low risk, 27 percent were at moderate risk, and 4 percent had high water quality risk susceptibility. Sources that had high susceptibility rankings typically had moderate to high density of potential contamination sources, high-intensity land development, no local source protection regulations, or high incidence of contaminants in source water before treatment (CT DPH, 2005).

2.6 CURRENT LAND MANAGEMENT EFFORTS TO ADDRESS WATER SUPPLY CHALLENGES IN CONNECTICUT

2.6.1 Current Management Approaches and Threats

Land management approaches are a critical tool for water quality protection and enhancement in both the SCCRWA and Aquarion drinking water supply systems. Despite progressive upstream and downstream management, however, these and other utilities face current and future threats to water quality, quantity, and delivery security. In this section we highlight key aspects of SCCRWA's and Aquarion's land management objectives as well as current water quality challenges specific to each utility.

2.6.1.1 SCCRWA: Land and Forest Management

Land and forest management are guided by the Connecticut DEEP's Best Management Practices as well as SCCRWA's land use and forest management plan (SCCRWA, 1989). The DEEP's *Best Management Practices for Water Quality While Harvesting Forest Products* includes a summary of laws affecting forestry operations in watersheds as well as recommendations for harvest management planning that reduces risks of water pollution from erosion, sedimentation, and other sources (CT DEP, 2007).

Land stewardship and management to protect and improve raw water quality are at the core of SCCRWA's operations. SCCRWA's Land Use Management Plan was first approved in 1983, and its Forest Management Plan in 1989 (SCCRWA, 1989). These plans articulate the following management goals driving SCCRWA's land stewardship:

1. *Watershed protection*: Protect and enhance the forest's role in providing high-quality water, flood control, and groundwater recharge
2. *Timber resource conservation*: Provide a continuing supply of forest products that return net revenue to the Regional Water Authority
3. *Wildlife resource protection*: Provide a refuge for the indigenous biota of the region
4. *Open-space preservation*: Provide an attractive environment for low-intensity forest recreation and maintain open space as scenic landscape for the region and as a buffer adjacent to developed areas

5. *Education and research*: Provide sites for environmental education and scientific research

Each goal is supported by a list of specific management objectives. The first and second goals are particularly consequential to SCCRWA's water quality maintenance efforts and are discussed in detail below.

Goal 1: Managing for watershed protection: The first goal of the Forest Management Plan, watershed protection, explicitly highlights the role of SCCRWA forestlands to filter and enhance the water supply and the importance of protecting SCCRWA's source water.

Objective 1: Managing to reduce algal blooms: This objective aims to limit reservoir algal blooms by controlling the flow of nutrients from the forest into water bodies. This is accomplished through a variety of strategies including the planting/management of buffer strips to protect watercourses, installing drainage improvements such as waterbars on roads and trails with moderate or high erosion potential, avoiding water crossings during timber harvests, removing trees that have fallen across or into streams, and enhancing the resilience of their forests by diversifying forest-stand ages and species composition (SCCRWA, 1989).

Objective 2: Reduce the organic material movement: SCCRWA aims to reduce the movement of organic materials by maintaining riparian buffers around their reservoirs to limit turbidity, color, nutrients, and suspended solids (SCCRWA, 1989). Because leaves dissolve in water to form dissolved organic carbon, which can react with chlorine to form potentially carcinogenic compounds known as trihalomethanes, evergreen trees are planted rather than broadleaf trees. In recent years, financial limitations have prevented SCCRWA from planting new evergreens, but the company is maintaining existing plantings (Hawley, 2010).

Objective 3: Limiting pesticide application: The third objective relates to the use of pesticide applications. Pesticide use on SCCRWA lands is extremely limited and is typically reserved for utility and highway rights-of-way and Christmas tree plantations. However, employee turnover and the potential use by independent contractors require ongoing training and supervision of pesticides on these lands. Hawley (2010) indicated that in recent years there has been limited pesticide use, and so few resources have been dedicated to these program elements.

Objective 4: Soil conservation and management: The fourth objective is to encourage groundwater recharge and minimize peak stream flows by protecting soil porosity and preventing erosion. The plan asserts that forest cover protects water quality by limiting soil erosion, promoting groundwater recharge, and moderating flooding and stream flow better than most other land uses. It also notes the forested areas have a low probability of containing hazardous or noxious materials (SCCRWA, 1989).

Objective 5: Catastrophic fire prevention: The last objective outlined in the 1989 plan is the coordinated protection of the forest from destructive fires. In years past, a coordinated effort to identify and suppress undesirable forest fires was a component of the SCCRWA forest management initiatives, a program element left over from a legacy of forest fires that have been declining in frequency and severity since the 1960s. Today, local fire departments have the lead in extinguishing all forest fires (SCCRWA, 1989; Hawley, 2010).

Goal 2: Timber resource conservation: The Forest Management Plan complements watershed management goals. SCCRWA utilizes active forest management to create resilient forests while simultaneously supplying the company with net revenue from the harvest. Annually, SCCRWA harvests approximately one million board feet per year (enough lumber to build more than 100 new three-bedroom homes).

Acquisition, outreach, and monitoring: In addition to forest management activities, SCCRWA's source water protection program includes land acquisition, education, water quality monitoring, and additional activities. In 1999, the authority created the Watershed Fund, a separate nonprofit entity that supports open-space acquisition and environmental education (SCCRWA, 2005–2006). Since its inception, the Watershed Fund has protected over 251 acres of watershed lands and has supported water quality education efforts in schools and the community at large (Watershed Fund, n.d.). Other resources and services SCCRWA offers to promote high water quality include a 24-hour emergency response program for hazardous materials spills in the watershed, reviews of plans for local development, and periodic inspections of homes, businesses, and industry to prevent pollution from occurring (SCCRWA, 2005–2006, 2007a). SCCRWA also runs a HazWaste Central program that helps people safely dispose of their hazardous household waste and offers environmental education programs at the Whitney Water Center to help people understand the effects of human land use activities on water quality (SCCRWA, 2011a).

2.6.1.2 SCCRWA: Current Threats

Maintaining water quality is an ongoing challenge for water utilities. The Source Water Assessment Report for SCCRWA includes several surface water systems and well fields operated by SCCRWA (CT DPH, 2003a). The report discusses the broad range of threats facing SCCRWA watershed lands, including development pressure, industrial and toxic contaminants, and pollution from residential and agricultural lands. Because of its overall high susceptibility rating due to environmental sensitivity, potential risk factors, and source protection needs, the Mill River Reservoir System, which includes the Whitney Reservoir, is a particularly noteworthy example of the challenges of maintaining high water quality in a watershed with growing use and development pressures.

The total watershed area of the Mill River is 23,000 acres, of which SCCRWA owns 5.4 percent and another 12.3 percent is publicly or privately protected. Land use in the watershed is divided between urban residential (31 percent), commercial or industrial development (6 percent), and agriculture (8 percent), and more than half of the watershed (55 percent) is classified as undeveloped land (CT DPH, 2003a). Nearly 30 percent of the undeveloped watershed may be susceptible to development. Additionally, the report noted that there are 115 "significant potential contaminant types" including waste storage, handling and disposal facilities, bulk chemical or petroleum storage sites, or industrial, commercial, or agricultural operations (CT DPH, 2003a).

The assessment indicates that the Mill River System has high susceptibility to environmental risk factors and would benefit from enhanced source protection. The identified strengths of the system are that there are no permitted point-source pollution discharges in the watershed and that the system has a comprehensive source water protection program in place. Identified weaknesses include potential contaminant sources in the watershed including impervious surfaces and intense development, the Lake Whitney Reservoir's eutrophic nutrient levels, and the utility's minimal landownership. To minimize future water quality challenges, the SWAP recommends that SCCRWA work directly with local officials and developers to ensure that only low-risk development occurs in the watershed. It also advocates increasing ownership or control of unprotected, undeveloped lands and expanding local plans and policies to increase protection of public drinking water sources (DPH, SCCRWA 2003). The report does not specifically detail SCCRWA's source water protection program, but given the recommendation to acquire additional land in the watershed, this program will remain critical for the utility's continuing water quality protection efforts in the Mill River System.

2.6.1.3 Aquarion Water Company: Land and Forest Management

Since the designation of the Centennial State Forest in 2002, the Conservation Lands Committee, made up of the Connecticut Department of Energy and Environmental Protection, the Nature Conservancy (TNC), and the Aquarion Water Company, has cooperatively managed over 15,000 acres of watershed land together (Forest to Faucet Partnership, 2008a).

The Centennial State Forest is managed with water quality in mind—particularly the 9,000 acres of Class I land, which are heavily regulated under Public Act No. 77-606 and subsequent laws governing watershed lands. The Conservation Lands Committee's (CLC's) Natural Resources Management Agreement (NRMA) clearly states that all management decisions are "subject to the statutory duties of BHC and DEEP to support and protect the public water supply and subject to BHC's Public Service Obligation" (NRMA, 2010). Protection and distribution of the water supply is "paramount," but the NRMA also lays out a number of additional management objectives aimed at increasing recreational access to the land and ensuring long-term ecological health. As such, the NRMA reflects the diverse values that the parties to the agreement—BHC, DEEP, and TNC—place on the Class I, II, and III lands surrounding Aquarion's main reservoirs.

Article I of the NRMA, which introduces the parties to the agreement, contains text regarding each party's primary interests in the subject land (emphasis added here to highlight each party's primary interests) (NRMA, 2010):

> Aquarion/BHC: *BHC's objective is to ensure that these lands and their associated resources continue to be managed in a manner that is consistent with and facilitates BHC's Public Service Obligation* while being permanently preserved as open space and that all uses are in concert with that primary purpose.

> DEEP: With respect to the lands encompassed by this Agreement, DEEP intends to manage the natural resources in accordance with generally accepted fish, wildlife, and forest management principles, conduct scientific investigations and assessments, and protect the land as open space by preserving in perpetuity its natural and open condition for the conservation of natural resources and public water supplies. This will include regulation, management, research, public education, and conservation law enforcement. *The goal is to conserve the property's natural resources and to provide the public with natural-resource-based recreational opportunities that are compatible with BHC's Public Service Obligation.*
>
> TNC: Portions of the conservation land occur within two functional conservation areas in Connecticut: the Saugatuck Forest Lands in the southwest and the Berkshire Taconic Landscape in the northwest. *TNC's goal is to manage these lands to promote and sustain their natural biological diversity in a manner that is compatible with BHC's Public Service Obligation.* (NRMA, 2010)

In addition to being guided by the NRMA, management decisions for the Centennial State Forest are informed by the U.S. Forest Service's Watershed Forest Management Information System (Forest to Faucet Partnership, 2008a), a spatial stewardship tool that supports the evaluation and planning of forest conservation, pollution mitigation, road maintenance, and silviculture (Forest to Faucet Partnership, 2008b). The Watershed Management Priority Indices (WMPI) uses commonly available GIS data and basic field measurements to quantify watershed sensitivity classes and construct a Conservation Priority Index, which can help water utilities and their stewardship partners prioritize particularly significant conservation areas.

2.6.1.4 Aquarion Water Company: Current Threats

There are several Source Water Assessment Reports for the Aquarion Water Company, including one for the Main System, which covers many company operated well fields and reservoirs (CT DPH, 2003b). As the SWAP details, the Easton, Hemlocks, and Trap Falls reservoirs are major assets of the Main System with low to moderate contamination susceptibility ratings. The Hemlocks Reservoir System, which includes the Hemlocks, Aspetuck, and Saugatuck Reservoirs, received a low overall susceptibility rating, with low environmental sensitivity, low potential risk factors, and moderate source protection needs. There are no point-source pollution discharges in the watershed, 36.8 percent of the 36,946-acre watershed is protected as open space, including the 25.7 percent owned by Aquarion, and the public water system has a comprehensive source water protection program. Weaknesses of the system include potential contaminant sources in the watershed and an absence of local regulations or zoning initiatives to protect public drinking water sources. Over 50 percent of the watershed is undeveloped and unprotected, which is both a risk and an opportunity. The SWAP recommends working with local officials and developers to ensure that only low-risk development occurs in the watershed. It also recommends working to establish local regulations and planning policies to protect public drinking water sources (CT DPH, 2003b).

Similarly, the SWAP encouraged additional watershed protection polices in the Easton and Trap Falls reservoirs. While the Easton reservoir was also assessed as having low overall susceptibility, the Trap Falls Reservoir received a moderate overall susceptibility rating, with low environmental sensitivity, moderate potential risk factors, and high source protection needs. Although Aquarion has source water protection programs in place and there are no permitted point source discharges within this watershed, the known contaminant release points within the watershed and minimal protection of the 9,883-acre watershed area present particular water quality management challenges to the Trap Falls Reservoir. These SWAP assessments emphasize the importance of watershed landownership for maintaining high-quality drinking water supplies.

2.6.2 Lessons Learned from SCCRWA and Aquarion

2.6.2.1 Public Opinion Impacts Land Management

The experiences of both SCCRWA and Aquarion elucidate the same basic conclusion: public opinion can significantly impact a utility's land management decisions. This lesson is relevant to varying degrees across the country, but Connecticut residents have routinely expressed a clear preference for maintaining publically accessible open spaces.

Connecticut residents' affection for open space may be due to the fact that they don't have much of it left. The average population density in the United States is about 87 people per square mile, but the 2010 Census reported an average density of about 737 people per square mile in Connecticut (USCB, 2010). Given the fact that Connecticut is one of the most densely populated states in the country, it makes sense that there is tremendous community support for preserving and expanding accessible open spaces. Such interests, be they expressed by individuals or grassroots coalitions of the public and nongovernmental organizations, can significantly impact utility land management decisions. Historically, effective political organization in opposition to land sales by the utilities has led to the establishment of the three classes of watershed lands and restrictions on its use and sale and the purchase of utility lands through perpetual conservation easements and outright ownership.

2.6.2.2 Compatibility of Public Interests with Watershed Land Management Objectives

Public interests have historically impacted and will continue to impact land management decisions in Connecticut. While watershed protection can yield decreased treatment costs for utilities, Connecticut citizens are particularly invested in preserving watersheds for the open space, recreational access, and nontimber forest products they offer. Other less tangible benefits include biodiversity conservation, climate mitigation and carbon sequestration. As the examples of SCCRWA and Aquarion demonstrate, water utilities, environmental groups, and state agencies can work together to manage watershed lands to support multiple values and uses.

SCCRWA's Forest Management Plan recognizes the diversity of interests in the utility's landholdings and acknowledges that allowing public access can promote good stewardship (SCCRWA, 1989). SCCRWA recognizes that both monetary and nonmonetary benefits of sustainable land management are valued by SCCRWA's constituents (Hawley, 2010). The existence of the CLC similarly highlights some of the benefits the public gains by owning and co-managing Aquarion's former and current landholdings. While water quality was not a primary goal for the DEEP or TNC, who financed the $90 million conservation project, the ecological and wildlife management goals of these groups are in line with Aquarion's management approach for reservoir water quality maintenance, and limited public access is permitted. The SCCRWA and Aquarion case studies illustrate that conservation and careful management of watershed lands supports multiple uses and can yield environmental, social, and economic benefits in addition to water quality protection.

2.6.2.3 *A Multibarrier Approach Supports High-Quality Drinking Water*

Connecticut is a salient example of how optimizing drinking water quality is not a choice between green infrastructure and gray infrastructure—it is a blend of these approaches. In a series of regulatory efforts, the state sought to require utilities to both filter their source water and simultaneously protect their most critical watershed lands, a multibarrier approach that encourages management interventions of upstream watersheds and downstream distribution. SCCRWA has continued to advance progressive land conservation goals within this legal framework, working to fulfill its legislative mandate to economically provide high-quality drinking water supplies. Aquarion has maintained watershed land management practices since the 1970s and has similarly broadened the scope of their stewardship activities following adoption of the CLC's Natural Resources Management Agreement. Protected, higher quality raw water upstream provides more quality security and consistency, as well as reduced maintenance costs (e.g., filter media replacement) downstream.

2.7 FUTURE CHALLENGES

2.7.1 Upcoming Regulatory Issues

Regulation can be a help or hindrance to water service providers. Unfortunately, much regulation is a reaction to existing deficient conditions. In these cases, regulation becomes the only option to ensure protection for the public and the environment. The continued degradation of the nation's water due to industrial and pharmaceutical pollutants along with the excessive demand for water for agriculture, industry, and growing populations is reason to believe new and emerging regulation will continue to require water companies do more with less. Foreseeable restrictions of future laws and policies may include limiting new contaminants of concern, establishing stream flow regulations, and requiring accounting and balancing of economic costs and benefits in an economy with greater water conservation and lower potential rate increases, as well as responses to threats to water security and infrastructure investments due to climate change. These challenges are discussed in more detail below.

2.7.1.1 Safe Drinking Water Act Amendments and Emerging Contaminants Regulations

The SDWA requires the EPA to review its National Primary Drinking Water Regulations at least once every six years and revise them based on current health effects assessments, changes in technology, or other factors (SDWA, 1974 et seq.). In 2011, Administrator Jackson reaffirmed EPA's goal to reassess and update laws and regulations in order to protect the health of the American people, including agency efforts to regulate classes of contaminants of concern (EPA, 2010). In a highly urbanized state such as Connecticut, with the potential for industrial, agricultural, and household pollutants in the watershed, water utilities should expect additional requirements to continue to be added through SDWA amendments and regulations.

New regulations for the microbial pathogen *Cryptosporidium* are already forcing both filtered and unfiltered water utilities to enhance monitoring for the pathogen and potentially upgrade their filtration technologies. The Long-Term 2 Enhanced Surface Water Treatment Rule (LT2) applies to all public water supply systems under the direct influence of surface water. Filtered systems in need of additional treatment will be required to reduce *Cryptosporidium* levels by 90 to 99.7 percent depending on existing levels (EPA, 2011a). The LT2 rule is expected to increase costs to public water systems, with nationwide cost estimates between $92 and $133 million (EPA, 2011a). In Connecticut, these costly upgrades will put additional pressures on utility operating budgets, which in turn may affect the funding available for source water protection.

Water utilities will also likely face additional regulation of other currently unregulated contaminants in the future. The EPA collects data on contaminants that are known or suspected to be present in drinking water, but do not yet have health-based standards under the SDWA through the Unregulated Contaminant Monitoring (UCM) program. This program includes pesticides, disinfection by-products, chemicals, waterborne pathogens, pharmaceuticals, and biological toxins (EPA, 2011b). One noteworthy group is PPCPs—pharmaceuticals and personal care products—which include human and veterinary prescription drugs, nutritional supplements, cosmetic products, bug repellant, sunscreen, and other chemical agents (NHDES, 2010). Although PPCPs are being detected in U.S. drinking water supplies at very low levels (on an average of 10 parts per trillion), they have been linked to developmental and behavioral deformities in fish and amphibians and this contaminant class is gaining public, and thus regulatory attention.

The UCM program applies to all large public water supply systems as well as a representative sample of public water systems that service less than 10,000 people. Every five years, the EPA reviews monitoring information, updates the Contaminant Candidate List (CCL), decides whether to regulate at least five contaminants, and prioritizes research and data collection efforts for the next monitoring cycle (EPA, 2011b). The most recent CCL was published in September 2009 (CCL 3) and includes 104 chemicals and 12 microbiological contaminants. These candidates were chosen from a total of 7,500 evaluated chemicals and microbes (EPA, 2009a).

2.7.1.2 *Minimum Stream Flow Regulations*

DEEP will be proposing Stream Flow Standards and Regulations in response to Public Act 05-142, enacted in 2005. The statute directed DEEP to develop minimum streamflow standards that balance the needs of humans to use water for drinking, washing, fire protection, irrigation, manufacturing, and recreation, with the needs of fish and wildlife (CT DEP, 2011b).

While the impacts of these new regulations are uncertain, water utilities such as SCCRWA and Aquarion along with the Connecticut Department of Public Utility Control have gone on record to voice their concerns. Maintaining streamflow minimums may require new unplanned capital intensive infrastructure costs (e.g., reservoir and well-field construction for providing additional storage and baseflow), severely limit or eliminate new connections, create supply deficits, challenge storage of reserves for crisis, and prolong duration of drought restriction conditions. The combined results of which may be, at a minimum, increased rates passed on to consumers. It could also force more significant economic decisions onto water utilities, such as changes to land ownership, management, and consideration of infrastructure and technology investments.

2.7.2 Upcoming Economic and Environmental Challenges

Connecticut drinking water utilities are facing a number of changing factors that will impact land and water treatment decisions. Aquarion and SCCRWA will need to apply the lessons discussed above to an operational landscape in which the rising cost of real estate, need for infrastructure improvements, and limitations on water rate increases will further restrict the range of "optimal" decisions with regard to gray versus green infrastructure.

2.7.2.1 *Cost of Acquiring Real Estate*

The cost of acquiring watershed real estate is a significant obstacle for Connecticut utilities. Prices vary around the state, but Aquarion attests that it would be nearly impossible to purchase open space or minimally developed land within its watersheds. A quick survey of real estate prices appears to support this conclusion: open space in and near the Hemlocks-Saugatuck system watershed ranges from about $79,000 to $119,999/acre (Table 2.3). Land values in the Stamford system are significantly higher: $323,000 to $675,000/acre for undeveloped land. Land prices based on historical transactions are much lower, but likely reflect bargain sales. Aquarion's 1999 sale of 730 acres went for about $16,400/acre, although the land had originally been sold to National Fairways for closer to $19,450/acre (Lomuscio, 2005). These

Table 2.3 Sample Land Prices across Selected Aquarion Watersheds

Transaction or Survey	Acreage	Total Sale Price	$/acre (unadjusted)
Aquarion, 1999	730 acres	$12 million	$16,438
SCCRWA Acquisitions (total)	5,000 acres	$17 million	$3,400
Saugatuck-Hemlock Weighted Avg.: Open Space	various	various	$92,692
Stamford Weighted Avg.	various	various	$676,435

Note: Prices adjusted using the 10-year composite of 4.18% from REIT Index from FTSE NAREIT U.S. Real Estate Index Series data.

adjusted figures would be much higher today—$33,100 and $28,000, respectively. SCCRWA's reported $17 million expenditure on 5,000 acres of watershed land works out to $3,400/acre, although that calculation may not include CT DEEP watershed land acquisition grants or other external funding.

High land prices make independent land acquisition an economically unfeasible strategy for water utilities. To help alleviate the financial burden of land acquisition, the Connecticut DEEP Open Space and Watershed Land Acquisition Grant Program provides funding for municipal and nonprofit land conservation organizations to jointly purchase watershed lands. The program provides partial funding for purchases of qualifying Class I and II open space that meets certain ecological or recreational criteria. Water companies can receive a grant for up to 65 percent of the fair market value for the land (CT DEP, 2012). This DEEP grant structure means that partnerships with land trusts and other co-beneficiaries of watershed land are ever more critical to utilities who wish to add permanent conservation protection to currently unprotected land.

The high price of both open space and low-density developed land also suggests that protection measures other than land acquisition should be an important part of a Connecticut utility's land and water protection toolkit. Examples of such measures include working closely with landowners to ensure septic system compliance and preventative livestock management on agriculture lands as well as purchasing development and management easements on land across zoning types.

2.7.2.2 Infrastructure Improvements

Across the country, water utilities are facing aging and deteriorating infrastructure that is in need of major maintenance or replacement. According to the EPA's 2007 Drinking Water Infrastructure Needs Survey and Assessment, the total 20-year capital improvement need is $334.8 billion (Figure 2.6) (EPA, 2009b). This includes investment costs for the installation of new infrastructure as well as the rehabilitation, expansion, or replacement of existing infrastructure.

Total national need is divided into five broad project categories: source, transmission and distribution, treatment, storage, and other, with the majority of infrastructure improvement need, around $200 billion, for transmission and distribution projects (EPA, 2009b). Most of this funding is required to replace or refurbish aging and deteriorating transmission and distribution mains. Treatment needs include

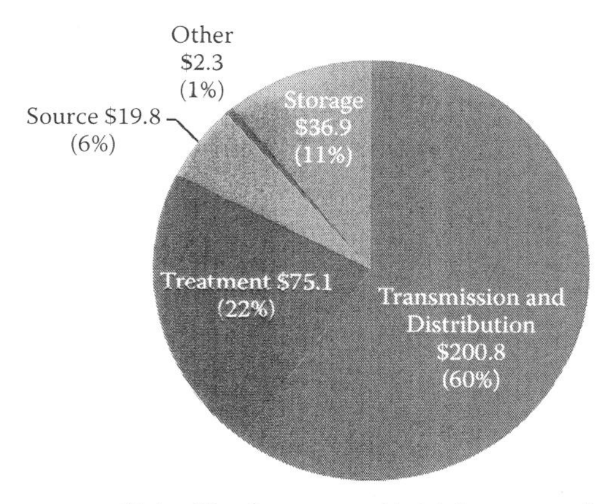

Figure 2.6 National projected 20-year costs of water system improvement needs, by project type (From EPA, 2009b).

construction, expansion, and rehabilitation of infrastructure, including filtration and disinfection facilities. Much of this funding need is to meet regulatory demands (EPA, 2009b). Source needs includes construction and rehabilitation of intakes, wells, and collectors. Storage needs include projects to construct, rehabilitate, or cover finished water storage tanks, but does not include dams and reservoirs. Other projects include data acquisition or system security (EPA, 2009b).

The EPA also assesses the findings of the 2007 survey by state. Connecticut has a total 20-year capital infrastructure need of $1.394 billion (Table 2.4) (EPA, 2009b). Although Connecticut falls well below the national average for each project category, the costs are still significant. This is reflected in projected infrastructure investment costs over the next five years for SCCRWA. The utility expects to spend over $100 million on distribution projects, as compared to $50 million for source projects (dams and land acquisition), and $30 million for treatment projects (Hudak, 2010).

In the coming years, it will be critical for water utilities to allocate funding and develop creative water quality protection solutions to address infrastructure

Table 2.4 Projected 20-Year Water Infrastructure Investment Needs for Community Water Systems in Connecticut and Nationally (in millions of $)

	CT	National Average
Transmission/Distribution	$807.1	$3,821
Source	$134.9	$375.8
Treatment	$280.6	$1,296.6
Storage	$151.6	$694.1
Other	$19.7	$43.2
Total	$1,394	$6,230.6

Source: EPA, 2009b.

improvement needs. These costs will burden the annual budgets of utilities, and will lead many to some difficult decisions regarding program priorities. Throughout the long-term water utility management planning process, the multiple, long-term benefits of green and gray infrastructure investments should be considered.

2.7.2.3 Rate Structures and Revenue Challenges

The United States has made huge strides in water conservation and efficiency. Despite a growing population, total water use across the country has held relatively constant over the past 20 years, meaning that per capita use has declined by roughly 30 percent since 1975 (Kenny et al., 2009). Water conservation and efficiency programs are incredibly important measures that can mitigate peak demand, increase available supply, and create a supply buffer for emergencies. However, if rate structures are not changed, these programs can also result in decreased revenues for water utilities. Given pending updates to the Safe Drinking Water Act, new minimum stream flow regulations, major infrastructure improvement needs, climate change, changing energy prices, and other challenges facing water utilities, the possibility of declining revenue is a major problem. To alleviate budget deficits, some utilities may choose to cut programs. In Connecticut, where filtration is required, watershed land acquisition may be one of the first programs to go. One alternative solution to this problem is for utilities to increase their rates, which would involve an application to the Department of Public Utility Control, as well as approval of individual utility boards. This decision is politically difficult, and customers are often upset and confused over the simultaneous request to reduce consumption, but pay more while doing so, effectively maintaining the cost of their water bills. A decision to increase rates would be especially difficult now and in the foreseeable future given the current economic downturn. To rectify this problem, creative rate structures can be implemented to return needed revenues, while at the same time rewarding conservation-minded consumers. Peter Glick, president of the Pacific Institute, remarks that "there's no reason why municipalities who implement conservation programs should have to raise their rates. If that happens it's a failure of rate design" (Walton, 2010). There are several alternatives to a uniform rate structure, including increasing block rate, which charges higher rates as use increases. Charges can also be applied to high-volume users during peak use periods, seasonal, or unexpected periods of droughts (Walton, 2010).

Another alternative that has been successfully applied by the Irvine Ranch Water District in California is to separately bill for capital and operational costs. As Walton describes, in this case, the district covers its capital costs through an allocation of the municipal property tax. Operational costs, which are more in line with water usage, are charged to the ratepayers (Walton, 2010). These creative rate structures may need to be applied in Connecticut to adequately address revenue shortfalls in the face of increasing costs and a suite of high priority needs for water utilities in the state.

2.7.2.4 Climate Change Impacts on Water Security and Investment Decisions

The state of Connecticut has already been experiencing changes in precipitation and temperature over the last several decades, and those trends are expected to continue as the climate changes across the globe (USGCRP, 2009). It is projected that Connecticut will experience increases in precipitation and extreme precipitation events, as well as more frequent and intense droughts, and changes in patterns of precipitation and snowmelt (Frumhoff et al., 2007). These changes are likely produce a very different water supply regime, creating water management and supply challenges for utilities.

In their recent report to the Governor's Steering Committee on Climate Change, the Connecticut Climate Change Adaptation Subcommittee warned that water availability may peak and drop at different times of the year and at more extreme levels. Dams, levees, sewers, pump stations, and coastal protections may be at risk of failing due to the stress of increased water levels. Extreme storm events may also damage treatment and distribution infrastructure, causing disruptions in water supply. Although annual precipitation is expected to increase, the state will experience water shortages, especially during the summer months when precipitation will be low, and high temperatures will increase both evapotranspiration as well as demand. Finally, snowpack is expected to decrease in the region, which will cause less spring runoff and affect the ability of managers to store water, as well as impact water quality and ecosystem health (Adaptation Subcommittee, 2010; Huntington, 2003).

Furthermore, increases in precipitation are expected to also increase runoff and alter wastewater patterns, which will likely cause higher pollutant loads in water systems throughout the state. Sea level rise will raise the water table, and along with increased precipitation, the usefulness of stormwater management best management practices such as swales and other infiltration techniques will be limited. Reservoirs will also face impacts, including potential saltwater intrusion, evaporation, and low quality water due to high temperatures and storm events (Adaptation Subcommittee, 2010).

These changes will require new solutions for storage, water delivery, flood control, environmental flow requirements, pollution control, water quality treatment, and watershed protection. The Adaptation Subcommittee notes that degradations in water quality due to increased temperature, low dissolved oxygen, concentrated contaminants, and pollution from increased runoff will complicate the water treatment process. However, the subcommittee also notes, "the quality of Connecticut's water supply will continue to remain high by maintaining ... watershed management practices that help diminish water supply treatment needs" (Adaptation Subcommittee, 2010).

The subcommittee recommends early planning and adaptation to climate change to reduce the severity of impact. Some potential solutions include using surface waters that are currently not allowed to be used for drinking water because of wastewater and industrial discharges, in order to augment supplies; reducing demand through conservation, water reuse, and efficiency measures; changing land management practices to reduce runoff; and increasing capacity and creating interconnections and redundancies to provide backup water supplies (Adaptation Subcommittee, 2010). These strategies will require long-term planning, research, adaptive management, and increased funding, all of which will be a challenge for water utilities to implement successfully.

2.7.2.5 *Cost and Reliability of Electricity*

After one of the worst economic declines in U.S. history, Connecticut residents and businesses have been encouraging efficient resource use to cut costs, resulting in lower average electric usage in 2009. Despite this trend, the Connecticut Siting Council (CSC), which regulates siting of power and waste facilities in the state, forecasts that "peak demand is expected to grow and is the value that must be used to weigh against resources in arriving at a forecast for long-term reliability" (CSC, 2009, p. 1). In their annual report to the Connecticut General Assembly, CSC indicated that normal weather peak load is expected to grow between 0.90 and 1.18 percent by 2018 (CSC, 2009). Although the peak load will grow, CSC predicts that energy conservation and efficiency efforts will keep Connecticut's energy supply stable, despite the planned retirement of several older oil-fired generating facilities. While some periods of deficit in generation may occur, in conclusion, the agency stated, "it is likely that electric resources will meet demand during the forecast period" (CSC, 2009).

Although it appears that electrical supply should be available for the next 10 years, the cost of this supply can have a crippling effect on business operations that have high electricity demands, such as the operation of a filtration treatment plant. As it stands, Connecticut ranks among the top three states in its cost of electricity (U.S. Energy Information Administration, 2011). While the explanation of the high price is complex, the effect of this on local water companies is that filtering polluted water can become an expensive endeavor. Uncertainty exists in predicting the future price of electricity. Although efforts to improve the grid and bring on additional sources of power may keep prices steady, or at best reduce rates slightly, Connecticut's electricity rates are not likely to drop significantly any time in the near future. To reduce current and future energy costs, utilities would be wise to invest in efficient, long-term water quality solutions.

2.8 RECOMMENDATIONS

Based on analysis of both Connecticut's general economic and biophysical condition of surface drinking watersheds and specifics of the South Central Connecticut Regional Water Authority and Aquarion utilities, it is apparent that future drinking water system management will require increased assessments of land use impacts on water quality and enhanced quantification of the costs and benefits of source water protection interventions. By embracing a triple bottom-line approach to long-term utility planning that includes environmental, social, and even potential future regulatory considerations, benefits of green and gray infrastructure investments can be assessed in more detail to ensure long-term viability of these companies. Whatever choices water utilities make to prepare for the myriad current and future water supply challenges, public education and engagement will be important in the maintenance and provision of high-quality drinking water.

2.8.1 Monitor Land Use Impacts on Water Quality

The most relevant report for Connecticut linking land use impacts to water quality was completed by the University of Connecticut (UCONN) Department of Natural Resources Management and Engineering in 2003 (Clausen et al., 2003). The study concludes that there was significant correlation between the water quality of 15 Connecticut streams and land use. The UCONN study examined 15 watersheds over a five-year period and analyzed water quality parameters including total nitrogen, total phosphorus, chloride, and fecal coliform bacteria. The researchers also took into account land use and land cover data from 1997 (Civco and Hurd, 1999). The satellite imagery provided 28 land use and land cover classifications, but the study only looked at the urban and agricultural land use types. A linear regression was used to relate either percent impervious or land use/land cover to water quality concentrations (Table 2.5).

Table 2.5 Summary of Studies Demonstrating Links between Land Use and Water Quality

Paper	Model	Data	Land Use(s)/Water Quality Assessed	Conclusion
Clausen et al., 2003	Linear regression	Land use data from satellite imagery Water quality data from USGS	% imperviousness,% urban,% agriculture Total residue, total nitrogen, total phosphorus, chloride, fecal coliform bacteria	Significant correlation between water quality of 15 Connecticut streams and urban and agriculture land use.
Herlihy et al., 1998	Multiple regression	Land use data from thematic mapper and USGS land use/land cover Water quality was sampled for 368 streams in mid-Atlantic US	Forest, agriculture, urban Nitrate, total phosphorus, chloride, base cations, and acid neutralizing capacity	Significant correlation between water quality and urban and agriculture land use.
Boyer et al., 2002	Linear regression	Land use data from the National Land Cover Database for 16 river basins in the northeast coast Water quality from USGS gauges	Forest, agriculture, urban Nitrogen	Nitrogen increased as percentage of urban and agriculture land increased and reduced as the percentage of forested land increased.

Source: Clausen et al., 2003.

Aquarion Case Study: The review of water quality data from Aquarion was limited to daily turbidity, total phosphorus, and total nitrate from 2002–2009 for their nine treatment plants (Galant, 2010). The water quality data was taken at the intake of each treatment plant. Land use data was taken from the Connecticut Department of Public Health's Source Water Assessment Reports for Community Public Water Systems (CT DPH, 2007) The land use types were categorized as urban (commercial, industrial, residential), agriculture, and undeveloped (forestlands, wetlands, and marshes). Existing literature relates higher percentages of agriculture and urban land cover to higher turbidity, total phosphorus, and total nitrate, but this can be very variable. Possible explanations are that land use percentages do not take into account effects of the location of specific land cover types. For example, even with a large portion of undeveloped land, the small portion of urban and agriculture lands may be located near the treatment plant intake. Furthermore, residence time, storage capacity, topographic relief, geology and soil type, and operational capacity of the source water can all add variability between land use and water quality (Barten, 2006; de la Cretaz and Barten, 2007).

2.8.2 Quantify Cost-Effectiveness of Source Water Protection

Poor water quality can lead to increases in both capital cost and variable costs (operation, maintenance, and treatment) for water suppliers. Varying levels of turbidity and Chlorophyll A impact the types of infrastructure required for water treatment (Janssens and Buekens, 1993) (Figure 2.7). A study of 430 U.S. water utilities found that most utilities with raw turbidity levels over 10 NTU had adopted conventional instead of direct filtration processes (Holmes, 1988).* This indicates there may be a sediment related water quality threshold for infrastructure investment (Table 2.6).

Aquarion Case Study: Aquarion's treatment plants are all conventional. The treatment process and water quality for each of Aquarion's treatment plants varies tremendously, but turbidity was low across all plants (i.e., 2002–2009 annual averages ranged from 0.3 to 2.3) (Table 2.7) (Galant, 2010). Unfortunately, the chemical costs for each treatment cost were not made available to determine water quality impacts to chemical costs for each treatment plant.

SCCRWA Case Study: Of SCCRWA's four water treatment plants (WTPs), those that require additional treatment processes have higher unit costs for every million gallons of water treated ($/MG) (Hudak, 2010). WTPs 4 and 2 have higher unit costs because both are conventional treatment plants that require the additional step of sedimentation (Table 2.8). On the other hand, WTPs 1 and 3 have lower unit costs because they are direct filtration plants that do not require the additional step of sedimentation (Hudak, 2010).

* Conventional filtration is a multi-stage operation that requires an additional settling phase to the direct filtration process (National Academy of Sciences, 2008).

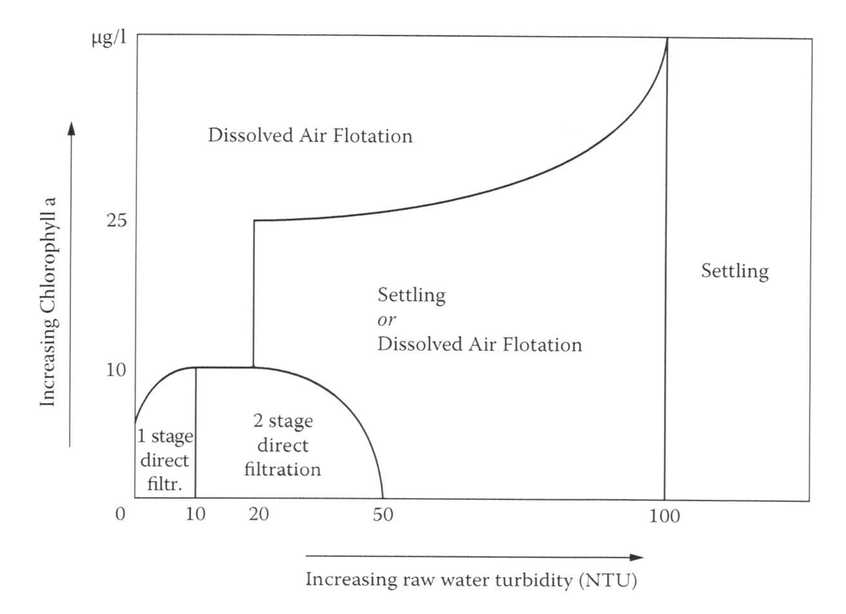

Figure 2.7 Turbidity and chlorophyll—impacts on water treatment (From Janssens and Buekens, 1993).

2.8.3 Moving beyond a Single Bottom Line

Public interest in SCCRWA and Aquarion's watersheds demonstrates that water company land is valued for more than simply its impact on water quality. Historically the coalitions that formed to appropriate these lands for the public were stakeholders with diverse interests. Though the representatives from open space and ecological conservation groups were the primary advocates and, ultimately, partially financed the land protection arrangements, they were certainly not the only interested parties. These lands are used by hikers, runners, and bikers, cross-country skiers, family

Table 2.6 Unit Costs for Various Treatment Plants

Treatment Process	Description	Upper Water Quality Limits
No Treatment (Filtration Waiver)	Disinfection	<0.075 oocysts/L for *Cryptosporidium* (EPA, 2008)
Direct Filtration	Filtration, Disinfection	Color < 40, Turbidity < 5 NTU, Algae < 400 asu/ml, iron <0.3 mg/L, manganese <0.05 mg/L (Faust and Osman, 1998) Turbidity < 10 NTU, chlorophyll a < 10 (Janssens and Buekens, 1993) Turbidity < 10 NTU (Holmes 1988)
Conventional Filtration	Sedimentation, Filtration, Disinfection	(Hansen et al., 1979)

Source: National Academy of Sciences, 2008.

Table 2.7 Treatment Process of Aquarion's Nine Treatment Plants

Treatment Plant Names	Treatment Capacity Mpd	Year in Service Year	Capital Cost $M	Treatment Process Technology	Average Turbidity (2002–2009, Jan–Dec) NTU	Average Phosphorus (2002–2009, Jan–Dec) mg/L	Average Nitrate (2002–2009, Jan–Dec) mg/L
TP 9	0.75	1996	4.1	DAF and filtration	0.26	0.009	0.01
TP 6	0.75	1996	4.1	DAF and filtration	0.49	0.009	0.03
TP 7	20	1993	27	High rate plate settler and filtration	0.53	0.009	0.31
TP 5	50	1997	47	DAF and filtration	0.69	0.048	0.11
TP 8	25	1980	18	High rate plate settlers and filtration	1.02	0.009	0.30
TP 4	30	1986, 2007	16	DAF and filtration	1.46	0.013	0.10
TP 2	4	1935, 2007	Unknown	Sedimentation and filtration	1.99	0.017	0.04
TP 3	20	1928, 1936, 1950, 1999	Unknown	Sedimentation and filtration	2.21	0.015	0.08
TP 1	8	1955	Unknown	Upflow clarifiers and filtration	2.28	0.025	0.47

Source: Galant, 2010.

Table 2.8 Unit Treatment Costs Compared to Percentage of Utility-Owned Watershed Land and Treatment Methods

Treatment Plant	Total Estimated cost (millions $/MG)	% Watershed Owned by SCCRWA	Solids Removal (prior to filtration)	Filtration	Advanced Oxidation	Disinfection
WTP #4	608.82	6%	X	X	X	X
WTP #2	655.50	23%	X	X		X
WTP #3	221.14	48%		X		X
WTP #1	305.19	55%		X		X

Source: Hudak, 2010.

picnickers, researchers, deer, turkey and mushroom hunters, birdwatchers, nature hobbyists, and others, as well as local residents who may not use the land directly, but appreciate its existence nonetheless. The values these groups place on the land should be factored into future management and acquisition decisions.

Aquarion and SCCRWA should incorporate these third-party or public values into the cost-benefit analysis for land use planning and utility management. When assessing potential acquisitions, why limit land value to the avoided costs of additional treatment steps borne by the water company, when history has shown that coalitions can form to purchase land for the benefit of multiple parties? Valuing non-water-related benefits of watershed lands is particularly important in Connecticut, where all utilities must filter their supplies. While maintaining forest cover and managing land for water quality may reduce the variable costs of treatment, land protection is, to some degree, redundant from a filtration perspective. This may increasingly be the case as treatment technologies improve and real estate prices increase. However, utilities should also consider the regulatory and social costs of a land sale proposal. If the many benefits of land conservation are quantified using a multiple bottom line approach that includes social, ecological, and economic values, utilities may find creative solutions to address current and future challenges.

Formally incorporating nonwater values into watershed management and land acquisition planning will likely be an intensive economic exercise. However, this process could also identify other parties whose management goals coincide with water quality conservation, such as wind or biomass energy developers. Some of the value of forestland can be quantified using past user transactions, such as the number and price of hunting and fishing permits issued in a given year. Other values may be established through willingness-to-pay surveys. The National Park Service has a thorough guide to identifying and quantifying values associated with open space, including the impact of conservation on real property values, recreational benefits, tourism, corporate relocation and retention, as well as public expenses in addition to those associated with drinking water provisions (National Park Service, 1995). The National Park Service recommends that such valuation include impacts of the open space on nearby property values, noting that "a positive correlation between proximity to open space and property values has been presented a number of times in peer-reviewed economic literature and by local planning boards" (National Park Service, 1995).

A multiple benefit approach will clarify who places value on the land, and what that value is, which aids in identifying who might be willing to finance purchases and ongoing management need. Previously, SCCRWA and Aquarion lands were purchased by well-funded public and private conservation and environmental groups. These buying consortiums could be expanded to include local property owners, sportsmen and recreational users. Even if such groups cannot pay for the bulk of a fee purchase, they may be willing to contribute capital for purchases or annual maintenance costs. For example, a local running club might finance brush removal or a mountain biking association might pay for or provide trail maintenance. Such creative solutions may can add economic value to landownership and optimize watershed management.

Finally, when thinking of which likely or unlikely "partners" may exist in collaborating for land protection, water companies would do well to not forget the energy sector. Small-scale wind and solar companies are actively looking for land to lease, and biomass startups are seeking partners in the forest and water resource sectors. The footprints of these systems can be small and their impact on watershed land negligible. Lease payments from such systems could provide guaranteed income for companies. With transmission infrastructure already running to dams and some hydroelectric facilities, the arrangement of additional alternative energy power sources in the vicinity could make water company lands appealing for this type of investment, not to mention these lands are not significantly subdivided, thus making access right-of-ways and operational agreements easier to obtain. Co-beneficial land use agreements should be fostered to encourage cooperative management.

2.8.4 Expanded Stakeholder Outreach and Education

When land acquisition fails, water companies may still have some options for directing protection measures. A utility—or a consortium of buyers that may include the utility—can purchase conservation easements on forested, residential and agricultural land within the watershed. New York City has had some success with this approach. Aquarion and SCCRWA also might do well to expand the resources they offer to watershed residents to manage individuals' impacts on water quality. Aquarion, for example, inspects neighborhood septic systems to ensure regulatory compliance, but might also consider assessing the cost effectiveness of actually replacing out-of-date and leaking systems. The utilities might also lend both technical support and capital to agricultural neighbors to implement best management practices to prevent livestock waste, pesticides, fertilizers and other contaminants from entering the streams that feed the reservoirs. Lobbying and education can also be effective in protecting watershed lands. Education may come from direct one-on-one interactions with adjacent landowners, users, schools, and even remotely through press releases to local media and through the websites that they maintain. Community partnerships will continue to play an important role in supporting water utility supply efforts.

2.9 CONCLUSIONS

Lake Whitney, one of SCCRWA's reservoirs, has been identified as "being adversely affected by nonpoint source pollution from urban land use" (Hudak and Ellum, personal communication, 2003). Watershed restoration efforts have been underway for years, with SCCRWA paying to acquire properties and construct stormwater treatment basins to address water quality concerns. SCCRWA's Environmental Planner John Hudak estimates that to date, approximately $1.6 million has been spent to treat about 600 acres of land in the roughly 23,000 acre Lake Whitney watershed. Given the relatively small treatment area, it is not surprising that "water quality monitoring has not detected any statistically significant changes in water quality" (Hudak, 2010). Even with water quality improvement expenditures by SCCRWA and low-impact design and retrofits in the community, some future watershed degradation is likely to occur. According to Hudak, a realistic goal is that these improvement efforts (constructed wetlands, site plan reviews, watershed inspections, improved land use regulations, open-space land acquisition, and education) will "hold things in check so that water quality will not worsen" (Hudak, 2010). The importance of the somewhat daunting task of managing land to prevent further degradation and work for water quality improvement is underscored by the fact that these efforts are being made despite the construction of a new filtration plant for Lake Whitney in 2005. Because of its significant operational costs, SCCRWA only runs the new plant as a last resort, and in 2008 and 2009 the new Whitney plant did not run at all (SCCRWA, 2011a).

Land use decisions at Lake Whitney appear to emphasize the need for the "precautionary principle" which states, "where there are threats of serious or irreversible damage, lack of full scientific certainty shall not be used as a reason for postponing cost-effective measures to prevent environmental degradation" (Rio Declaration Principle #15, 1992). Today, a substantial body of scientific evidence indicates that careful management of forested watershed can protect water quality in surface water reservoirs. While water treatment technologies are likely to continue to improve, operation costs will remain significant, and the evidence presented thus far indicates that treatment of polluted reservoir water is a difficult and expensive operation.

If water companies allow intensive watershed development and later learn that urbanization has untreatable adverse effects on water quality, then the reversal of such land management policies and its effects is costly and challenging, if not impossible. Conversely, the adherence to current scientific consensus, which favors source water protection, would have outcomes that may require increased land management expenditures but will not likely cause harm or peril. On the contrary, the protection of these lands and waters is likely to have benefits beyond high-quality drinking water. As illustrated by the example of Lake Whitney, efforts to improve degraded watersheds can yield additional benefits such as the education of students and decision makers, improved aquatic habitats, and aesthetics. As both private and public water companies ultimately should have the public's interest in maintaining high-quality water at heart, aggressive source water protection remains a wise precautionary approach to an uncertain future.

REFERENCES

Adaptation Subcommittee, (2010). The Impacts of Climate Change on Connecticut Agriculture, Infrastructure, Natural Resources and Public Health. Connecticut Climate Change Adaptation Subcommittee Report to the Governor's Steering Committee on Climate Change.

Aquarion Water Company (Aquarion), (2008). 2008 Water Quality Report For Customers in the Stamford System. http://www.aquarion.com/pdfs/Stamford08.pdf. Accessed October 11, 2011.

Aquarion Water Company (Aquarion), (2009a). 2009 Water Quality Report For Customers in the Greenwich System. http://www.aquarion.com/pdfs/Greenwich09.pdf Accessed October 11, 2011.

Aquarion Water Company (Aquarion), (2009b). 2009 Water Quality Report For Customers in the Mystic System. http://www.aquarion.com/pdfs/Mystic09.pdf. Accessed October 11, 2011.

Aquarion Water Company (Aquarion), (2010). Letter from John Herlihy, Director of Water Quality and Environmental Management, to the CT DEP Regarding Proposed DEP Stream Flow Standards and Regulations. http://www.ct.gov/dep/lib/dep/water/watershed_management/flowhearingcomments/flow_27.pdf. Accessed October 11, 2011.

Aquarion Water Company (Aquarion), (2011a). About Aquarion. http://www.aquarion.com/ct.cfm/section/About. Accessed October 10, 2011.

Aquarion Water Company (Aquarion), (2011b). Water Quality: Treatment. http://www.aquarion.com/CT.cfm/section/Quality/page/Treatment. Accessed October 10, 2011.

Aspetuck Land Trust, (n.d.). Trout Brook Valley Conservation Area Trail Map. http://mappery.com/maps/Trout-Brook-Valley-Map.pdf. Accessed October 11, 2011.

Barten, P.K., (2006). Why Forests Provide the Best Protection for Water Resources. http://www.forest-to-faucet.org/pdf/Whyforestsprovidethebestprotectionofwaterresources.pdf. Accessed October 10, 2011.

Belaga, J., W. Bliss, and D. Strait, (2001). Testimony of the Coalition for the Permanent Protection of Kelda Lands to the Connecticut General Assembly's Energy and Technology Committee. Testimony in support of S.B. 6172, H.B. 6172, and H.B. 6177. http://www.lwvweston.org/keldaet.htm. Accessed April 14, 2010.

Boyer, E.W., C.L. Goodale, N.A. Jaworski, and R.W. Howarth, (2002). Anthropogenic nitrogen sources and relationships to riverine nitrogen export in the northeastern U.S.A. *Biogeochemistry* 57/58, 137–169.

Budoff, C., (2001, February 7). Rowland Signs Deal to Buy Open Space. *Hartford Courant*. http://articles.courant.com/2001-02-07/news/0102071983_1_kelda-group-open-space-acres-of-watershed-land. Accessed November 1, 2011.

Business Wire, (1998, June 10). Land Trust Exercises Right to Acquire Trout Brook Valley.

Business Wire, (1999, September 2). BHC, Aspetuck Land Trust Close Trout Brook Valley Sale.

Civco, D.L., and J.D. Hurd, (1999). A hierarchical approach to land use and land cover mapping using multiple image types, *Proceedings of the 1999 ASPRS Annual Convention*, Portland, Oregon, 687–698.

Clausen, J.C., G. Warner, D. Civco, and M. Hood, (2003). Nonpoint Education for Municipal Officials Impervious Surface Research Final Report, University of Connecticut, Department of Natural Resources Management and Engineering.

Connecticut Attorney General's Office (CAGO), (1999, November 29). Press Release: Calls for Moratorium on Sales of Water Company Lands. http://www.ct.gov/ag/cwp/view.asp?a=1774&q=282910. Accessed April 15, 2010.

Connecticut Department of Environmental Protection (CT DEP), (2007). Best Management Practices for Water Quality While Harvesting Forest Products. Connecticut Field Guide. http://www.ct.gov/dep/lib/dep/forestry/best_management_practices/best_practices-manual.pdf

Connecticut Department of Environmental Protection (CT DEP), (2011a). Centennial Watershed State Forest. http://www.ct.gov/dep/cwp/view.asp?A=2716&Q=447970. Accessed October 11, 2011.

Connecticut Department of Environmental Protection (CT DEP), (2011b). Open Space and Watershed Land Acquisition Grant Program. http://www.ct.gov/dep/cwp/view.asp?A=2706&Q=323834. Accessed October 11, 2011.

Connecticut Department of Environmental Protection (CT DEP), (2012). Open Space and Watershed Land Acquisition Grant Program. http://www.ct.gov/dep/lib/dep/open_space/open_space_grant_round_application.pdf. Accessed November 23, 2012.

Connecticut Department of Public Health (CT DPH), (1999). Source Water Assessment Program Work Plan. Prepared by CT DPH & CT DEP, Approved by EPA November 1999. http://www.ct.gov/dph/lib/dph/drinking_water/pdf/SWAP_Workplan.pdf. Accessed October 20, 2011.

Connecticut Department of Public Health (CT DPH), (2003a). Regional Water Authority Source Water Assessment Report, North Cheshire Wellfield. http://www.dir.ct.gov/dph/Water/SWAP/Community/CT0930011.pdf. Accessed October 20, 2011.

Connecticut Department of Public Health (CT DPH), (2003b). Aquarion Water Company of Connecticut Main System Source Water Assessment Report. http://www.dir.ct.gov/dph/Water/SWAP/Community/CT0150011.pdf. Accessed October 20, 2011.

Connecticut Department of Public Health (CT DPH), (2005). Connecticut's Drinking Water Assessment and Source Protection Program. http://www.ct.gov/dph/lib/dph/drinking_water/pdf/SWAPWEB_05_12.pdf. Accessed October 20, 2011.

Connecticut Department of Public Health (CT DPH), (2007). Source Water Assessment Program. http://www.ct.gov/dph/cwp/view.asp?a=3139&Q=387342&dphNav_GID=1824. Accessed October 20, 2011.

Connecticut Executive Chambers (CT Executive Chambers), (2004, September 16). Governor Rell Announces the Dedication of Centennial Watershed State Forest. http://www.ct.gov/governorrell/cwp/view.asp?A=1720&Q=277614. Accessed October 11, 2011.

Connecticut General Assembly (CGA), (1974). Public Act 74-303. C.G.S. §§ 16-43; 25-32, Westlaw, 2011.

Connecticut General Assembly (CGA), (1975). Public Act 75-405. C.G.S. § 16-49 and C.G.S.A § 16-50, Westlaw, 2011.

Connecticut General Assembly (CGA), (1977a). Special Act 77-98. Charter of the South Central Connecticut Regional Water Authority. Effective July 25, 1977.

Connecticut General Assembly (CGA), (1977b). Public Act 77-414. C.G.S. §§ 4-54, 55, Westlaw, 2011.

Connecticut General Assembly (CGA), (1977c). Public Act 77-606. C.G.S. § 25-37, Westlaw, 2011.

Connecticut General Assembly (CGA), (1979). Public Act 79-294. C.G.S. §§ 25-32, 37, Westlaw, 2011.

Connecticut General Assembly (CGA), (1984). Public Act 84-502. C.G.S. § 25-32d, Westlaw, 2011.

Connecticut General Assembly (CGA), (1985). Public Health Code Regulation, C.G.S. § 19-13-B102. Standards for Quality of Drinking Water. http://www.ct.gov/dph/lib/dph/agency_regulations/sections/pdfs/title_19._health_and_safety/phc/chapter_ii/19-23._standards_for_quality....pdf. Accessed October 20, 2011.

Connecticut General Assembly (CGA), (1996). Public Act 96-180, C.G.S. § 25-37c. Regulations. Classification of Land Owned by or Acquired from a Water Company. http://www.cga.ct.gov/2011/pub/chap474.htm#Sec25-37c.htm. Accessed October 20, 2011.

Connecticut General Assembly (CGA), (1998). Public Act 98-157. C.G.S. §§ 7-131d-g; 12-498; 16-43;16-50c; 23-74; 23-78; 25-32d; 25-33k. Amendments. Westlaw, 2011.

Connecticut General Assembly (CGA), (2002). Public Act No. 02-85, House Bill 5210. C.G.S. §§ 22a-358, 518. An Act Concerning the South Central Connecticut Regional Water Authority Concerning the Sale of Water to Community Water Systems. http://www.cga.ct.gov/2002/act/Pa/2002PA-00085-R00HB-05210-PA.htm. Accessed October 20, 2011.

Connecticut General Assembly (CGA), (2006). Public Act 06-53, Substitute Senate Bill No. 313. C.G.S. § 25-32. An Act Concerning Protection of Public Water Supply Sources. http://www.cga.ct.gov/2006/ACT/PA/2006PA-00053-R00SB-00313-PA.htm. Accessed October 20, 2011.

Connecticut General Assembly (CGA), (2007). Public Act 07-252,.Substitute House Bill No. 7163. C.G.S. §§ 16-43 et seq. An Act Concerning Revisions to Statutes Related to the Departments of Public Health and Social Services and Town Clerks. http://www.cga.ct.gov/2007/ACT/PA/2007PA-00252-R00HB-07163-PA.htm. Accessed October 20, 2011.

Connecticut Siting Council (CSC), (2009). 2009 Forecast of Loads and Resources. Docket No. F-2009. http://www.ct.gov/csc/lib/csc/f2009/f-2009final[2].pdf

Connecticut Water Works Association (CWWA), (2010). Connecticut Water Works Association Position Paper: Proposed Stream Flow Regulations. From E. Gara, CWWA Executive Director to CT DEP Bureau of Water Protection and Land Reuse. http://www.ct.gov/dep/lib/dep/water/watershed_management/flowhearingcomments/flow_297.pdf. Accessed October 11, 2011.

de la Cretaz, A.L., and P.K. Barten, (2007). *Land Use Effects on Streamflow and Water Quality in the Northeastern United States*. CRC Press, Boca Raton, FL.

Environmental Protection Agency (EPA), (2004). Understanding the Safe Drinking Water Act. http://www.epa.gov/safewater/sdwa/pdfs/fs_30ann_sdwa_web.pdf. Accessed October 20, 2011.

Environmental Protection Agency (EPA), (2008). Long-Term 2 Enhanced Surface Water Treatment Rule: A Quick Reference for Schedule 4 Systems. http://www.epa.gov/safewater/disinfection/lt2/pdfs/qrg_lt2_qrg_sch4_final.pdf. Accessed October 20, 2011.

Environmental Protection Agency (EPA), (2009a). Fact Sheet: Final Third Drinking Water Contaminant Candidate List (CCL 3). http://www.epa.gov/ogwdw/ccl/pdfs/ccl3_docs/fs_cc3_final.pdf. Accessed October 20, 2011.

Environmental Protection Agency (EPA), (2009b). Drinking Water Infrastructure Needs Survey and Assessment: Fourth Report to Congress. Office of Ground Water and Drinking Water, Drinking Water Protection Division.

Environmental Protection Agency (EPA), (2010, March 22). Administrator Lisa P. Jackson, Remarks to the Association of Metropolitan Water Agencies, as Prepared. http://yosemite.epa.gov/opa/admpress.nsf/8d49f7ad4bbcf4ef852573590040b7f6/6bfef816f3107ea9852576ee004df76e!OpenDocument. Accessed October 20, 2011.

Environmental Protection Agency (EPA), (2011a). Safe Drinking Water Act—Basic Information http://www.epa.gov/safewater/sdwa/basicinformation.html. Accessed October 9, 2011.

Environmental Protection Agency (EPA), (2011b). Basic Information: Long-Term 2 Enhanced Surface Water Treatment Rule. http://water.epa.gov/lawsregs/rulesregs/sdwa/lt2/basicinformation.cfm. Accessed October 9, 2011.

Environmental Protection Agency (EPA), (2011c). Unregulated Contaminant Monitoring Program. http://water.epa.gov/lawsregs/rulesregs/sdwa/ucmr/. Accessed October 9, 2011.

Ernst, C., R. Gulick, and K. Nixon, (2004). Protecting the Source: Conserving Forests to Protect Water. *Opflow*. American Water Works Association. 30(5), 1-5.

Faust, S.D., and A.M. Osman, (1998). *Chemistry of Water Treatment*. Boca Raton, FL: CRC Press.

Ferrucci, M., and T. Walicki, LLC (2009). Forest Management Plan for the Aspetuck-Hemlock Watershed Forest, 2009–2018. http://www.fwforesters.com

Forest-to-Faucet Partnership, (2008a). University of Massachusetts Amherst, http://www.forest-to-faucet.org/projects_tools6.html. Accessed October 11, 2011.

Forest-to-Faucet Partnership, (2008b). Watershed Forest Management Information System. University of Massachusetts Amherst, http://www.forest-to-faucet.org/projects_tools2.html. Accessed October 11, 2011.

Fotos, M.G., (2010, February 23). Professor of Environmental Policy, Trinity College, Hartford, CT. Personal communication.

Freeman, J., R. Madsen, and K. Hart, (2008). Statistical Analysis of Drinking Water Treatment Plant Costs, Source Water Quality, and Land Cover Characteristics. U.S. Environmental Protection Agency, U.S.D.A. Forest Service, The Trust for Public Land White Paper. http://wren.palwv.org/library/documents/landnwater_9_2008_whitepaper.pdf. Accessed October 11, 2011.

Frumhoff, P.C., J.L. McCarthy, J.M. Melillo, S.C. Moser, and D.J. Wuebbles, (2007, July). Confronting Climate Change in the U.S., Northeast: Science, Impacts, and Solutions. Synthesis Reports of Northeast Climate Impacts Assessment. Cambridge, MA: Union of Concerned Scientists. http://www.climatechoices.org/assets/documents/climatechoices/confronting-climate-change-in-the-u-s-northeast.pdf. Accessed October 20, 2011.

Galant, P., (2010, March 15). Vice president, Tighe and Bond, Management for Aquarion Water Company, Bridgeport, CT. Personal communication.

Greenland, P. (2007). Aquarion Company, *International Directory of Company Histories*, Encyclopedia.com. (December 13, 2012). http://www.encyclopedia.com/doc/1G2-3480000011.html.

Hansen, S.P., R.C. Gumerman, and R.L. Culp, (1979). Estimating Water Treatment Costs, Volume 3: Cost Curves Applicable to 2,500 gdp to 1 mgd Treatment Plants, EPA-600/2-79-162c.

Haines, G., (2010, February 24). Personal communication.

Hawley, T., (2010, March 22). Natural Resources Manager, SCCRWA, New Haven, CT. Personal communication.

Hanover, B., (2010, March 15). Water Quality Manager of Aquarion's Connecticut Division. Personal communication.

Herlihy, A.T., J.L. Soddard, and C.B. Johnson, (1998). The relationship between stream chemistry and watershed land cover data in the mid Atlantic region, US. *Water Air and Soil Pollution* 105(1–2), 377–386.

Holmes, T.P., (1988). The offsite impact of soil-erosion on the water-treatment industry. *Land Economics* 64, 356–366.

Holstead, J.R., (2002). Class I and II Water Company Lands. Connecticut General Assembly. Office of Legislative Research Report. 2002-R-0460. http://www.cga.ct.gov/2002/rpt/2002-R-0460.htm. Accessed October 10, 2011.

Hopper, K., and E. Pepper. (2003). A Toolkit for Communities: Protecting Land to Safeguard Connecicut's Drinking Water. The Trust for Public Land, New Haven, CT.

Hudak, J., (2010). Environmental planning manager, South Central Regional Water Authority. Presentation to Yale School of Forestry and Environmental Studies, March 22.

Hudak, J.P., and M.E. Ellum, (2003). Effectiveness of stormwater treatment systems within a highly urbanized watershed. *American Water Resources Association* 32, 511–519.

Huntington, T., (2003). Climate warming could reduce runoff significantly in New England, USA. *Agricultural and Forest Meteorology* 54, 193–201.

Janssens, J.G., and A. Buekens, (1993). Assessment of process selection for particle removal in surface water treatment. *Aqua: Journal of Water Supply Research and Technology* 42(5), 279–288.

Kelda Group, (2000). Annual Report and Accounts. Bradford, UK.

Kelda Group, (2006). Annual Report and Accounts. Bradford, UK.

Kenny, J.F., et al., (2009). Estimated Use of Water in the United States in 2005. U.S. Geological Survey Circular 1344, 52 p.

Kinsman, S.E., (2000, January 8). British Kelda Group Takes Over Aquarion. *Hartford Courant.* http://articles.courant.com/2000-01-08/business/0001080792_1_aquarion-s-bridgeport-kelda-group-plc-largest-investor-owned-water-utilities.

Liberante, M., (2010, April 1). Chief operator of Aquarion's Connecticut division, Bridgeport, CT. Personal communication.

Lomuscio, J., (2005). *Village of the Damned: The Fight for Open Space and the Flooding of a Connecticut Town*. Lebanon, NH: University Press of New England.

McCarthy, K.E., (2001, April 27). Regulation of Water Company Land. Connecticut General Assembly. Office of Legislative Research Report. 2001-R-0431. http://www.cga.ct.gov/2001/rpt/2001-R-0431.htm. Accessed October 25, 2011.

McCluskey, D.S, and C.C. Bennitt, (1996). *Who Wants to Buy a Water Company?* Bethel, CT: Rutledge Books.

McCluskey, D.S., and C.C. Bennitt, (1997). Partnerships protect watersheds: The case of the New Haven Water Company. *Land Lines* 9, 1.

Miner, M., (2009). Trout Brook Valley Forever Yours or Forever Gone. Aspetuck Land Trust. http://www.aspetucklandtrust.org/html/tbv-story-2.html. Accessed April 10, 2010.

Moreau, C., (2000, April 21). CEO's Firing Stirs Fears of Huge Land Sale. *Hartford Courant.*

National Academy of Sciences, (2008). Safe Drinking Water Is Essential: Filtration Systems. http://www.drinking-water.org/html/en/Treatment/Filtration-Systems.html. Accessed October 25, 2011.

National Park Service, (1995). *Economic Impacts of Protecting Rivers, Trails, and Greenway Corridors,* Fourth Edition Revised. Rivers, Trails, and Conservation Assistance, U.S. Department of the Interior National Park Service.

Natural Resources Management Agreement (NRMA), (2010). Conservation Lands Committee.

Nature Conservancy, (2006, March). An Introduction to the Saugatuck River Watershed and the Saugatuck River Watershed Partnership. Unpublished report.

New Haven Water Company, (NHWC), (1941). Water: Lifestream of a Community. New Haven, CT: New Haven Water Company.
Office of Policy and Management (OPM), (2008, December 31). Managing Water in Connecticut: A Report on the Study of Water Resources Planning in the State. Prepared by the Office of Policy and Management for the General Assembly's Energy and Technology and Appropriations Committees. http://www.ct.gov/opm/lib/opm/igp/pubreps/water_resources_report-1-09.pdf. Accessed October 25, 2011.
Patton, G., (2010, March 23). Forester, Aquarion Ltd, Bridgeport, CT. Personal communication.
Roach, B., (2010, March 23). Forester, Aquarion Ltd, Bridgeport, CT. Personal communication.
Safe Drinking Water Act (SDWA), (1974) December 16. 42 USC §§ 300(f), 88 Stat. 1660.
Safe Drinking Water Act Amendments (SDWA), (1986, June 19). Pub. L. No. 99-359.
Safe Drinking Water Act Amendments (SDWA), (1996, August 6). Pub. L. No. 104-182.
Safe Drinking Water Act Regulations (SDWAR), (1989). Surface Water Treatment Rule of 1989, 40 CFR §§ 141.70 et seq. United States Environmental Protection Agency.
SCCRWA, (1989). Forest Management Plan for the South Central Regional Water Authority.
SCCRWA, (2005–2006). Connecting with Customers: Annual Report 2005–2006. http://www.rwater.com/ar2005-06/AR_2005-06_Med.pdf. Accessed October 11, 2011.
SCCRWA, (2007a). Protecting the Sources of Your Drinking Water. *Waterlines*, Winter–Spring 2007. http://www.rwater.com/consumer-info/waterlines/2007_1/index.html. Accessed October 30, 2011.
SCCRWA, (2007b). The Land We Need for the Water We Use. http://www.rwater.com/our-land/pdfs/Land_We_Need_Large.pdf. Accessed October 30, 2011.
SCCRWA, (2010a, January 21). SCCRWA Testimony to the CT DEP Regarding Proposed DEP Stream Flow Standards and Regulations. http://www.ct.gov/dep/lib/dep/water/watershed_management/flowhearingcomments/flow_38.pdf. Accessed October 30, 2011.
SCCRWA, (2010b). About the Regional Water Authority: Regional Water Authority Statistics as of 05/31/2010. http://www.rwater.com/who-we-are/pdfs/Regional-Water-Authority-Statistics.pdf. Accessed October 30, 2011.
SCCRWA, (2010c). South Central Connecticut Regional Water Authority Service Area. http://www.rwater.com/who-we-are/service-area-map.html. Accessed October 30, 2011.
SCCRWA, (2010d, March). Land Protection Criteria. Internal document provided by Dianne Tompkins, SCCRWA Senior Land Use Manager.
SCCRWA, (2011a). New Plant Technology. http://www.whitneydigs.com/Tech/index.html. Accessed October 20, 2011.
SCCRWA, (2011b). Regional Water Authority: Recreation Rules, Regulations and Tips. http://www.rwater.com/recreation/rules-regs-tips.html. Accessed October 11, 2011.
Staff Reports, (1999, November 30). Lawmakers May Consider Land Sale Moratorium. *Hartford Courant.* http://articles.courant.com/1999-11-30/business/9911300934_1_aquarion-kelda-group-public-utility-control. Accessed October 20, 2011.
Tompkins, D., (2010, April 20). SCCRWA Senior Land Use Manager. Personal communication.
United States Census Bureau (USCB), (2010). Connecticut Quick Facts from US Census Bureau, Geography. Persons per square mile. http://quickfacts.census.gov/qfd/states/09000.html. Accessed October 13, 2011.
United States Energy Information Administration, (2011). Table 5A. Residential Average Monthly Bill by Census Division, and State 2011. http://www.eia.gov/electricity/sales_revenue_price/pdf/table5_a.pdf

United States Geological Survey (USGS), (2004). Water Use, Groundwater-Recharge and Availability, and Quality of Water in the Greenwich Area, Fairfield County, Connecticut and Westchester County, New York, 2000–2002.

United States Global Change Research Program (USGCRP), (2009). Global Climate Change Impacts in the U.S.: Northeast. http://www.globalchange.gov/publications/reports/scientific-assessments/us-impacts/regional-climate-change-impacts/northeast. Accessed October 20, 2011.

Walton, B., (2010, April 19). U.S. Urban Residents Cut Water Usage; Utilities Are Forced to Raise Prices. Circle of Blue Water News. http://www.circleofblue.org/waternews/2010/world/u-s-urban-residents-cut-water-usage-utilities-are-forced-to-raise-prices/. Accessed October 30, 2011.

Watershed Fund, (n.d.). Previous Watershed Fund Grants. http://www.thewatershedfund.org/The-Watershed-Fund-Previous-Grants/. Accessed October 20, 2011.

CHAPTER 3

Source Water Protection in Massachusetts

Lessons from and Opportunities for Worcester and Boston

Emily Alcott, Peter Caligiuri, Jennifer Hoyle, and Nathan Karres

CONTENTS

3.0 EXECUTIVE SUMMARY

This chapter will compare and contrast the history, evolution, and current design of the Massachusetts Water Resources Authority (MWRA) in Boston and the Worcester water supplies. In so doing, we highlight common strategies and tools employed to meet federal drinking water regulations. We also underscore the key differences between the systems that led one system, MWRA, to pursue filtration avoidance and the other, Worcester, to install filtration to ensure high-quality drinking water.

Both MWRA and Worcester rely on a multiple barrier approach including source water protection and an engineered treatment component to ensure high-quality drinking water. Federal regulation has repeatedly emphasized the importance of source protection to generate high raw water quality from source water watersheds. This rationale is based on the principle of forest hydrology that forested watersheds function as a biological filter of natural and anthropogenic contaminants.

The comparison of watershed management and protection efforts of the Worcester and Boston water supply systems demonstrate that all stakeholders invested in drinking water supply believe, that a multibarrier approach, with emphasis on source water protection, is essential to optimize the delivery of safe drinking water. Furthermore, this leveraging of forests as a natural water filter creates a unique opportunity for the land conservation community and drinking water suppliers to collaborate on source water protection. It is clear that the approach taken by both Boston and Worcester will become increasingly important as our ability to detect contaminants improves, as regulations for safe drinking water become more stringent, and as public health concerns continue to require the safest water at the lowest possible cost. However, as these trends continue, more pressure will be placed on water suppliers to either quantitatively prove that land conservation and management can protect against emerging contaminants or choose instead to invest in engineered technologies for contaminant removal (or more likely, both). Ultimately, to maintain any sustainable program in land management for source water protection, water suppliers need an all-encompassing optimization tool, integrating complete asset management for natural and built assets to account for all possible benefits of watershed management.

3.1 INTRODUCTION

The ability of forests to serve as a natural water filter generating higher raw water quality creates a unique opportunity for the land conservation community and drinking water suppliers to collaborate. Through the protection and careful management of drinking water source watersheds, these two groups can simultaneously act in the interest of public health, economics, and natural resources. This chapter presents a comparative analysis of two Massachusetts public water suppliers, Boston's Massachusetts Water Resources Authority (MWRA) and Worcester's Department of Public Works and outlines strategies for the continued, concurrent preservation of the Massachusetts water supply as well as financial and natural resources.

3.1.1 Source Water Watersheds in Massachusetts

We present a brief overview of the location of surface water source areas as they relate to the major population centers in the state (Figure 3.1). Population density is considerably higher in the eastern half of Massachusetts, leaving the central and western regions of the state more sparsely populated and less developed. Consequently, more than one-third of the state's total population is served by surface water from watersheds in the rural central part of the state. As a result, the greatest opportunities to mitigate the impacts of urban growth via source water protection measures exist in central and western Massachusetts.

Despite this overarching theme, the development of two of the state's largest drinking water supply systems, MWRA and Worcester, are noticeably different. These two case studies provide insight into alternative approaches to achieve the dual objectives of satisfying demand for water quantity while providing safe,

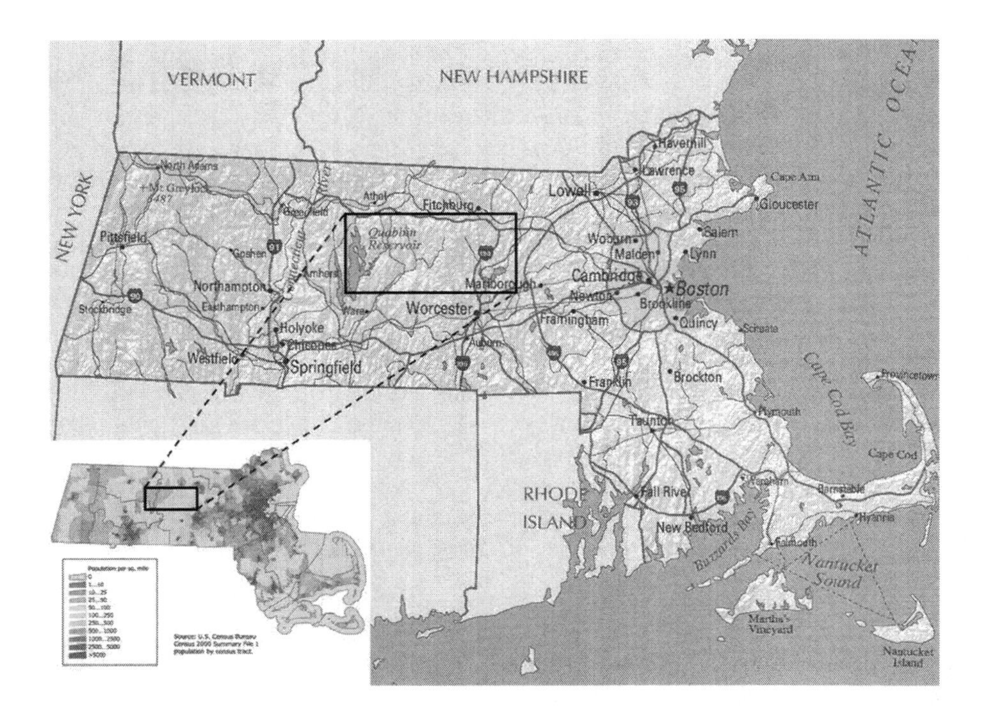

Figure 3.1 An overview of population density in the state of Massachusetts.

high-quality drinking water. Both MWRA and Worcester rely on a multibarrier approach including source water protection and an engineered treatment component to ensure high-quality drinking water. Federal regulation has repeatedly emphasized the importance of source protection to generate high raw water quality from source watersheds. This rationale is based on the principle of forest hydrology that forested watersheds function as a biological filter of natural and anthropogenic contaminants (e.g., Satterlund and Adams, 1992; Likens and Bormann, 1995).

Protection of drinking supply watersheds focuses on safeguarding and improving the resilience and resistance of watershed forests to enhance their natural filtration function (e.g., resistance to pests, disease). Effective management of watershed forests requires knowledge of factors—both past and present—that have contributed to the current composition, structure, and function of forested watersheds. This underscores the need to understand the land use patterns and land cover change in central New England over time (Barten et al., 1998).

3.1.2 Land-Use History and Watershed Management

The present-day New England forested landscape is a result of influences ranging from ancient glaciations and climatic shifts to prescribed burnings by Native Americans and, more recently, European settlement, intensive agricultural production, and their subsequent abandonment. Like much of the New England forested landscape, the native forests of Massachusetts were nearly eliminated by the early 1800s. By the mid-nineteenth century, as much as 80 percent of presettlement native forestland was cleared for agricultural production, pastureland, construction materials, fuel, and charcoal (Raup, 1966; Foster, 1992; Howard and Lee, 2002; DCR, 2007).

Over time, intensive land use depleted New England's fertile, mesic forest soils and forced farmers to head westward in search of more productive land. This abandonment led to colonization of these "old fields" by conifer species, primarily eastern red cedar (*Juniperus virginiana* L.) and eastern white pine (*Pinus strobus* L.) (Raup, 1966; Foster, 1992; Howard and Lee, 2002). By providing shade, increased soil organic matter, and nutrients, these conifer species helped ameliorate the harsh open-field microclimate and provided suitable microsites for less-tolerant native hardwoods (e.g., oak, hickory, maple) (Howard and Lee, 2002).

By the mid-1900s, the white pine dominant forests had given way to the second and third growth even-aged, mixed-species hardwood/conifer forest that dominate the New England landscape today. The watersheds of the MWRA and Worcester water supply systems are a product of this land use history. For example, the Quabbin Reservoir watershed is estimated to have been as much as 75 percent cleared during the height of agricultural settlement (around 1850) and nearly 100 percent forested by 1950 (DCR, 2007). Today, the Quabbin, Ware, and Wachusett Reservoir watersheds, which comprise the active segment of the MWRA system, are more than 81 percent forested.

Watershed forests are the first barriers in a multiple-barrier approach to water supply protection. However, not all forests are equally effective as a biological filter of potential water supply contaminants. The physical and biological characteristics of a forest (e.g., composition, structure, phase of stand development, soil integrity), as well as the overall susceptibility of the forest to natural and anthropogenic disturbance, dictate how effective the forest is as a filter.

Paired watershed studies from long-term ecological research sites like Coweeta Hydrologic Laboratory (Swank and Crossley, 1988) and Hubbard Brook Experimental Forest (Likens and Bormann, 1995) have demonstrated that forest alteration (e.g., the removal of vegetation) directly impacts the landscape's ability to serve as a biological filter. This research suggests that proactive, contextual management aimed at increasing the resiliency, resistance, and redundancy of watershed forests can maintain and even enhance water quality. Effective watershed management must be place specific and consider forest characteristics, land use history, geology, climate, hydrology, and natural disturbance regime.

Due to the active management necessary to maintain watersheds for source water protection, water supply managers must recognize, anticipate, and mitigate risks to the watershed and water supply. The degree to which the water supply system is designed, built, and managed to respond to natural and human threats varies from system to system. Strategies range from forest thinning to prevent catastrophic fires to increased redundancy in engineered systems, or a combination of both. The following sections describe the historical evolution and current profile of the MWRA and Worcester water supply systems, including both their natural and built assets of their systems. This is followed by a discussion of the alternative approaches to dealing with existing and imminent threats to water quality in the MWRA and Worcester systems.

3.2 HISTORY AND CURRENT STATUS OF DRINKING WATER SUPPLY IN MASSACHUSETTS

3.2.1 Boston

3.2.1.1 History of Boston's Drinking Water Supply

The history of Boston's water supply system (Figure 3.2) began in 1795 when private water suppliers began delivering water through wooden pipes from Jamaica Pond to the city of Boston. In the mid-nineteenth century, a lack of water to fight several catastrophic fires and multiple disease epidemics escalated public water quantity and quality concerns. This led to the formation of the Cochituate Water Board in 1845 and the construction of Lake Cochituate to augment supply and improve water quality (MWRA, 2006; DCR, 2007, 2008).

As Boston's population continued to grow, the Boston Water Board was forced to divert the Sudbury River in 1878 to create a series of seven reservoirs—Sudbury, Whitehall, Hopkinton, Ashland, Stearns, Brackett, and Foss. In 1895, the Metropolitan Water Board replaced the Boston Water Board and began considering new water supply sources to keep pace with a rapidly growing population. In 1897, the Wachusett Dam was constructed to impound the Nashua River and fill the Wachusett Reservoir. With a storage capacity of 65 billion gallons and a yield of 118 million gallons per day, the Wachusett Reservoir became the largest public water supply reservoir in the world (MWRA, 2006; DCR, 2007, 2008).

In 1919, the Metropolitan Water Board, Sewer Board, and Parks Commission merged to form the Metropolitan District Commission (MDC). To satisfy growing water demand, MDC began construction on an aqueduct to divert additional Ware River water into the Wachusett Reservoir. In the 1930s, the aqueduct was extended from the Ware River to the Swift River. In 1939, the Winsor Dam was constructed to impound the Swift River, which created Quabbin Reservoir. The reservoir was filled in 1946 and provided an additional 412 billion gallons of storage capacity and a yield of 155 million gallons per day to the MDC system (MWRA, 2006; DCR, 2007, 2008).

In the 1960s a major drought renewed water quantity concerns within the MDC system. This prompted the MDC to test several strategies to increase water yield from the system, including the harvesting of red pine stands to establish grasslands, reduce evapotranspiration, and increase yield. System demand peaked in the early 1980s at 340 million gallons per day (MGD). This led MDC to initiate an aggressive water conservation program that included leak detection and repair, universal metering, and public education (MWRA, 2006; DCR, 2007, 2008). In less than 10

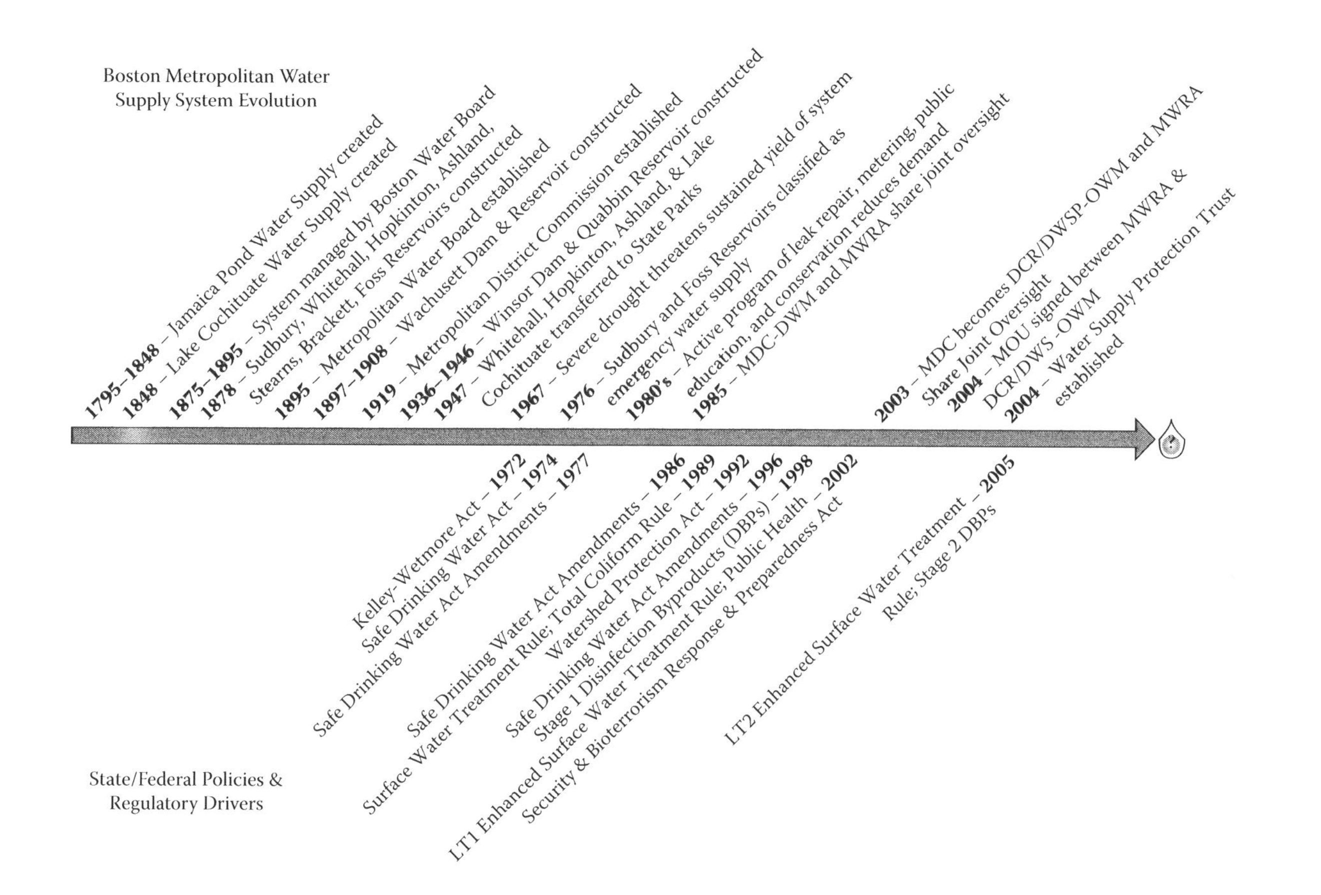

Figure 3.2 Historical timeline of major events and policies in the formation of the current Boston water system.

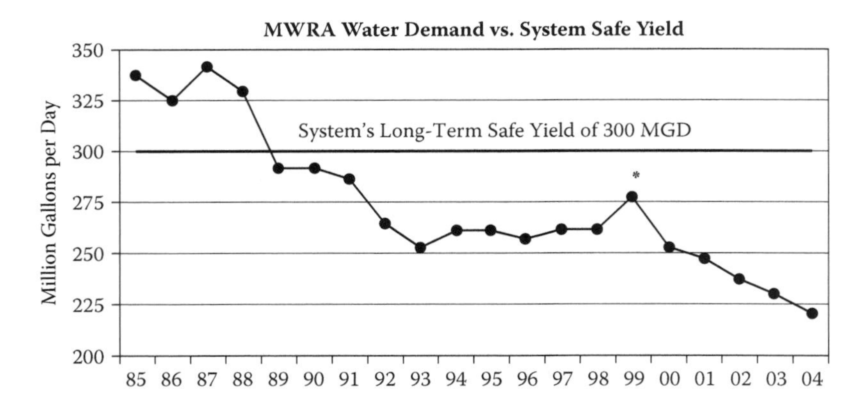

Figure 3.3 Demonstration of declining per capita demand for the Boston system over the past two decades.

years, these efforts reduced water demand by nearly 100 million gallons per day (Figure 3.3).

In 1985, management of the Boston water supply system was divided between the MDC-Division of Watershed Management (MDC-DWM) and the MWRA. MDC-DWM was delegated control of reservoir watershed management and operation, while MWRA took responsibility for treatment, transmission, and distribution. This arrangement continued until 2003 when the Office of Watershed Management (OWM), of the State Department of Conservation and Recreation's Division of Water Supply Protection subsumed MDC-DWM. A 2004 Memorandum of Understanding (MOU) between these agencies reaffirmed the shared management roles and responsibilities of each, as well as established the terms by which MWRA ratepayer revenues are used to fund the operations of OWM (MWRA, 2006; DCR, 2007, 2008).

3.2.1.2 Boston's Current Drinking Water Supply

Today the MWRA water supply system delivers water to more than 2.2 million people and 5,500 industrial users in 48 communities in the Boston metro area and three communities in central Massachusetts (Chicopee, South Hadley Fire District #1, and Wilbraham). It also serves as a backup water supply in three additional communities, including the city of Worcester (DCR, 2008). The current system comprises three linearly connected watersheds—Quabbin, Ware, and Wachusett (Figure 3.4).

From Quabbin, water is transferred to the Wachusett Reservoir via the Quabbin Aqueduct. When necessary (e.g., drought conditions) the Ware River provides seasonal water transfers to Quabbin Reservoir. These transfers are rare, however, occurring in only 3 of the past 10 years. Contribution from the each of the three

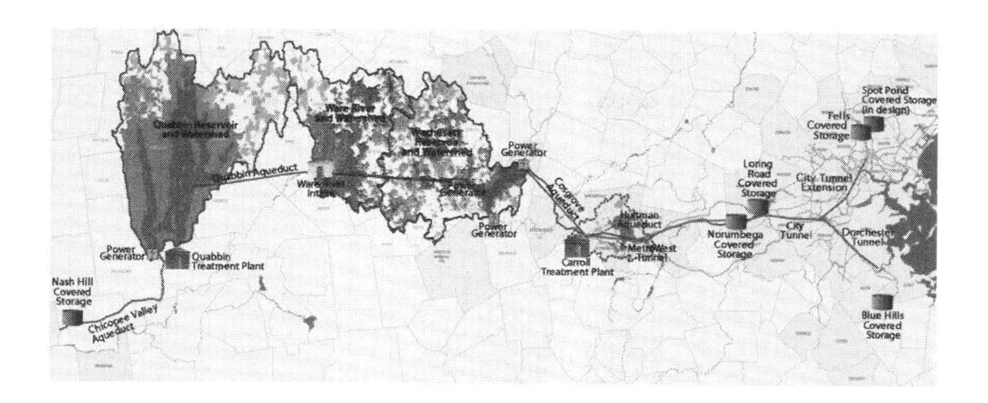

Figure 3.4 A schematic showing the general layout of the watersheds and infrastructure for the Boston water supply system. (See color insert.)

Table 3.1 Watershed Areas and Annual Outflows Showing the Relative Contribution of Each Reservoir System (MWRA, 2008)

Source	Watershed Area[a]		Average Annual Outflow[b] (mgd)	Average Annual Withdrawal (mgd)
	Square Miles	Acres		
Ware River (MWRA Intake)	96	61,740	110	8.08[c]
Quabbin Reservoir	187	119,940	195.2	137.9
Wachusett Reservoir	117	74,890	127.4	123.1
Total DCR/MWRA Water Supply System	401	256,570	432.6	261

Source: Watershed Statistics, GIS, Office of Watershed Management of the Massachusetts Department of Conservation and Recreation's Division of Water Supply Protection; Water Withdrawal Statistics: MWRA, 2003, MWRA, 2008.

[a] Including area of reservoir surface for Quabbin Reservoir and Wachusett Reservoir.
[b] Outflow includes withdrawals and downstream releases.
[c] This is not a supply but a transfer to Quabbin Reservoir.

watersheds varies intra-annually (Table 3.1). From the Wachusett Reservoir, water enters the Cosgrove intake and is transferred through the Cosgrove Aqueduct into the municipal distribution system (DCR, 2007, 2008). There are two water treatment facilities in the MWRA system: the John J. Carroll Water Treatment Plant, serving metropolitan Boston, and the Ware Water (Quabbin) Treatment Facility, serving the communities on the Chicopee Valley Aqueduct. The Carroll Treatment Facility provides primary and residual disinfection through ozonation and chloramination as well as corrosion control and fluoridation treatments. The Ware facility utilizes only chloramination for primary disinfection and residual protection (DCR, 2008; MWRA, 2006).

3.2.1.3 Land Use Patterns in the MWRA System

The MWRA water supply system is relatively unique for its size in the United States because it relies entirely on unfiltered surface water. The Safe Drinking Water Act (SDWA) Amendments of 1986 created a provision for public systems that utilize surface water supplies to pursue a Filtration Avoidance Determination (FAD) (hereafter, filtration waiver). To acquire and maintain a filtration waiver, surface water supply systems must demonstrate compliance with 11 regulatory requirements (e.g., fecal coliform and turbidity). The ability of MWRA to maintain its filtration waiver depends on its ability to manage and mitigate threats to water quality across its source watersheds (DCR, 2003).

As of 2008, the MWRA source watersheds together were 82 percent forested and 57 percent protected (DCR 2008). However, considered separately, these three watersheds have important ownership and land use differences, which have considerable implications for water quality (Figures 3.5 and 3.6). As these charts illustrate, intensive land uses (e.g., commercial) represent a much larger percentage of land area in the Ware and Wachusett watersheds. Moreover, the percentage of watershed land that is protected through ownership, watershed protection restriction, or other programs decreases significantly from the Quabbin to Wachusett watersheds, while both percent private ownership and population density increase dramatically (DCR, 2008). In total, the Wachusett watershed contains 12 towns, roughly 34,000 people, approximately 2,250 domestic farm animals, 118 miles of highway, and 17.5 miles of railway, all of which represent real threats to water quality (Stearns, 2000).

MWRA began formulating the necessary documentation to support their FAD application for both the Quabbin and Wachusett reservoirs in 1986. Because Quabbin Reservoir serves a separate population along the Chicopee Valley Aqueduct, the Massachusetts Department of Environmental Protection (DEP) required MWRA to develop two separate watershed protection plans. The plan for Quabbin Reservoir was approved and filtration avoidance granted in 1989. However, because all water

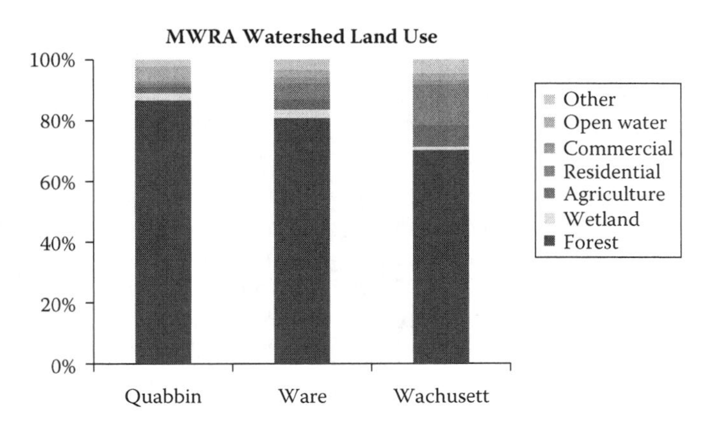

Figure 3.5 Land use type percent areas for each of the major watersheds.

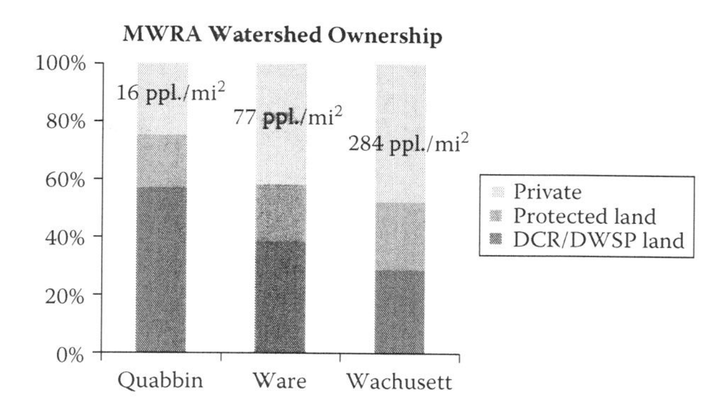

Figure 3.6 Population densities for various land use types for each of the major watersheds.

for the Boston metropolitan area must pass through the Wachusett Reservoir before entering the transmission and distribution network, the entire system is at risk from this single, highly susceptible watershed. Moreover, fecal coliform levels had been problematic in the Wachusett Reservoir due to higher levels of human development in the watershed (Figure 3.7). Consequently, the MWRA board of directors decided to vote against seeking a filtration waiver for Wachusett Reservoir (Stearns, 2000).

In 1998 OWM and MWRA elected to pursue a FAD for Wachusett Reservoir, claiming that watershed protection efforts had been widely successful. These efforts included an aggressive land acquisition made possible by a $135 million bond created through the 1992 Watershed Protection Act, as well as wildlife management activities and enhanced disinfection. These programs enabled DCR and MWRA to meet the regulatory requirements for filtration avoidance. As a result, DEP granted the filtration waiver for the Wachusett Reservoir in November of 1998 (Stearns, 2000).

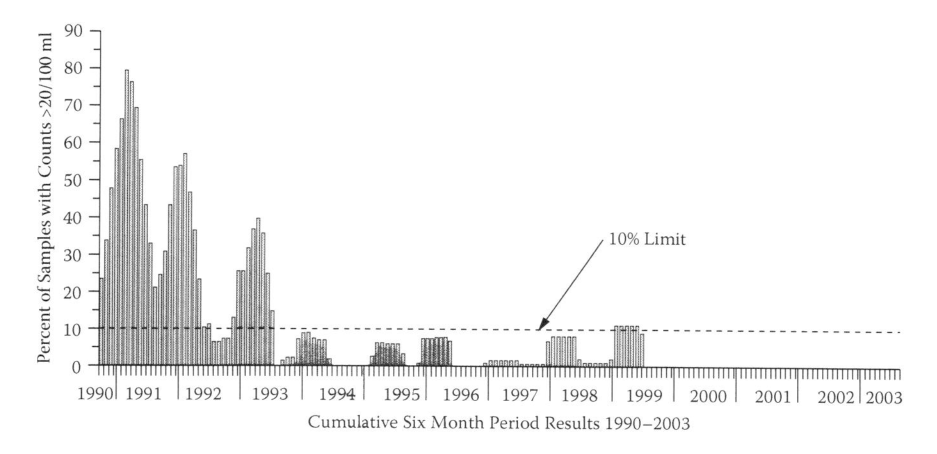

Figure 3.7 Water quality data for the Wachusett Reservoir showing fecal coliform exceedances that precipitated EPA action.

The EPA, however, disagreed with DEP's decision. The EPA argued that because Wachusett Reservoir had previously failed water quality testing, engineered filtration was the only option for the MWRA system. In 2000, the case was decided in favor of MWRA, upholding the filtration waiver. According to the Judge Stearns, who wrote the ruling, "the 'ozone-only' option favored by the MWRA is a sound alternative to [filtration] when competing demands for limited resources and the level of risk from all potential threats to the safety of MWRA water are considered." Judge Stearns concluded that the MWRA watershed protection efforts had been highly successful to date and shifting expenses toward active filtration would negatively impact such efforts (Stearns, 2000).

3.2.1.4 MWRA Watershed Protection Efforts

MWRA's Division of Water Supply Protection (DWSP) has identified 13 broad categories of land use protection programs. Some of the details of the program efforts are discussed below (Table 3.2) (DCR, 2008).

Land procurement. Land procurement, whether "in fee" or through acquisition of conservation easements, has been a central mechanism for source water protection, especially within the Wachusett watershed. The program is funded primarily through ratepayer billing and state bonds; $128 million has been invested in land procurement efforts between 1985 and 2008. Approximately $109 of the total $128 million was expended on lands within the Wachusett watershed. However, since funds are limited, land procurement decisions are made by a dedicated DWSP group known as the Land Acquisition Panel, which incorporates biophysical, land use, and population data to prioritize which decisions are the most influential in terms of water quality impact.

Land preservation/watershed protection restriction. DCR employs the use of conservation easements to expand the area of land protected from development. Known as Watershed Protection Restrictions in Massachusetts, these agreements involve the transfer of future development rights to DCR while the property owner still maintains possession of the land. The primary advantage of this arrangement is that the parcel in question is removed from any potential development, thus protecting against future potential risk to surface water quality. In terms of absolute water quality, in-fee transactions would be preferred since DCR can control all activities within the property boundaries. However, low-density residential activity is believed to have a low impact of water quality.

An additional advantage of Watershed Protection Restrictions is that the Department of Conservation and Recreation (and in turn MWRA) can avoid payments in lieu of taxes (PILOT). This requires MWRA to pay PILOT funds directly to local communities and alleviates concerns of stakeholders that DCR ownership of land would decrease the property tax revenue base. Massachusetts implemented the PILOT program, which mandates state entities including MWRA to pay property taxes based on the highest and best use of a parcel. Essentially, Watershed Protection Restrictions allow DCR to circumvent this restriction to maximize land procurement funds.

Table 3.2 Matrix Showing the Major Sources and Control Programs Utilized in the Boston System

	Control Program or Measure												
Source	**Land Procurement**	**Land Preservation**	**Land Management**	**Wildlife Management**	**Public Access Management**	**Watershed Security**	**Infrastructure**	**Watershed Protection Act**	**Technical Assistance & Community Outreach**	**Interpretive Services**	**Water Quality Monitoring**	**Environmental Quality Assessment**	**Emergency Response**
Wildlife			•	•	•						•	•	
Public Access/ Recreation			•	•	•	•	•			•	•	•	
Timber Harvesting			•				•				•	•	•
Wastewater	•	•						•		•	•	•	
Roadways/Railways/ ROWs (right of ways)											•	•	•
Agriculture	•	•						•	•		•	•	
Construction	•	•						•	•	•	•	•	
Commercial, Industrial, and Governmental Sites	•	•						•	•	•	•	•	•
Residential Sites	•	•						•	•	•	•	•	
Solid Waste Facilities	•	•						•		•	•	•	
Future Growth	•	•	•	•	•			•	•	•	•	•	
Climate Change	•	•									•	•	•

Source: DCR 2008.

Land management. A guiding principle in watershed management throughout the MWRA system is creating a resilient and redundant natural forest biofilter. DWSP manages the forest to build ecological resilience (e.g., managing for many tree species) and resistance to threats (e.g., thinning to prevent pest outbreak) in the system. This concept is similar to redundancy incorporated in built treatment systems through multiple levels of filtrations and disinfection. These efforts help to mitigate the effects of disturbance (human and natural) and maximize the potential of the forest to sequester contaminants before entering the water storage and supply system. Additionally, all activities taking place on DCR lands face stringent environmental requirements for best management practices.

Wildlife management. Wildlife management comprises two program elements: (1) the protection of valuable species and (2) the protection of source water quality. Species protection involves a number of best practice measures to avoid land use/management actions that have the potential to negatively impact rare or endangered species. The second wildlife program type, protection of water from wildlife, involves a number of programs to prevent the introduction of fecal coliform into source waters. Two prominent examples of this type of program are the gull harassment program, which employs a number of tactics to keep birds from introducing new sources of pathogens near reservoir intake areas, and the beaver management program, which controls beaver problems on a case-by-case basis with a range of interventions from elimination to relocation.

Public access management and watershed security. Especially in highly urbanized settings, such as the Wachusett watershed, illicit or unauthorized access to watershed property constitutes a significant risk. Outright bans on public access are highly unpopular, so DCR must balance the interests of safe drinking water with public access. The greatest flexibility exists for lands that do not have a direct linkage to source waters.

Watershed Protection Act. The Watershed Protection Act created a system for regulating activities that can impact source watershed health. Examples of restrictions include buffer zones along watercourses, limits on the amount of impervious surface, and restrictions on the types of chemicals that can be used or stored. The act also included a significant funding mechanism to increase the capacity of DCR to acquire lands, whether through purchase or through development restrictions. The Watershed Protection Act created a $135 million bond for land acquisition, allowing $8 million per year to be spent to purchase critical watershed lands (DCR, 2007).

Technical assistance and community outreach. Community outreach includes various *ad hoc* programs to either inform the public of DCR/MWRA work or elicit behavioral changes to achieve water quality protection goals (e.g., invasive species education). DCR provides technical assistance to communities and forestland owners through land use and master plans, protective zoning regulations, and site design, among other activities. Private forestland owners can also receive technical assistance in the form of best management silvicultural practices and education about tax and other incentives for land preservation.

Water quality monitoring. Separate from normal system monitoring for regulatory compliance, DCR conducts additional monitoring to track progress toward source protection improvement and to better understand pollutant-loading dynamics. Sampling takes place across all three watersheds with particular attention to special operations such as the gull harassment program. According to DCR, this water quality monitoring comprises an important piece of the management system by informing adaptive decision-making and management activities. For DCR, monitoring is integral to the successful implementation of other program elements.

3.2.2 Worcester

3.2.2.1 History of Worcester's Drinking Water Supply

Worcester's first settlers drew their water supply directly from the area's numerous small streams and rivers. When Worcester was incorporated as a city in 1848, a large river running through the center of town was dammed and became Worcester's first public water supply (Bell Pond). As the city expanded and drinking water demand increased, the City of Worcester used eminent domain to clear the way for construction of a series of reservoirs. The city's first reservoir, Lynde Brook Reservoir, was built in 1860. As water demand continued to increase, the city continued to periodically add reservoirs until the construction of its 10th and final reservoir in 1950, Quinapoxet (Figure 3.8). Until 1945, Worcester's water supply went completely untreated. In 1945, the city began disinfecting their water supply by adding a chloramine/ammonia mix. This was later replaced by chlorine, which continued to be Worcester's only form of treatment for the next 40 years.

The passage of the Safe Drinking Water Act in 1974 brought increased public and regulatory scrutiny to Worcester's water supply. This escalated national attention surrounding safe drinking water, and was concurrent with multiple turbidity and fecal coliform violations and visibly poor quality drinking water, it brought about a public outcry over the unfit condition of Worcester's water supply. This public crisis led to the emergence of a number of citizen and Department of Public Works (DPW) committees in an effort to clean up Worcester's drinking water supply. These water committees recognized and highlighted imminent threats posed by land development, aging infrastructure of water distribution systems, and poor quality raw water. This public water crisis and subsequent action fostered a progressive attitude surrounding drinking water supply management that still exists in Worcester today.

In anticipation of the Safe Drinking Water Act (SDWA) Amendments of 1986, in 1984 Worcester DPW and the public water crisis committees entered the planning stages for Worcester's first water treatment plant. Based on Worcester's historical turbidity and coliform violations, DPW officials and the city council recognized that the city was unlikely to receive a filtration waiver, so the city never pursued one. Worcester residents and DPW officials also recognized the threat posed by the city's aging water distribution infrastructure and decided to budget $3 million annually for infrastructure repairs and improvements. The Worcester Water Treatment plant went online in 1997 and remains operational. In 2002, with state grants available for land acquisition for source water protection, Worcester began its land acquisition program. The city has budgeted $300,000 annually for the purchase of priority lands within Worcester's water supply reservoir watersheds.

3.2.2.2 Current Profile of Worcester Drinking Water Delivery

Worcester's public water supply is derived from a system of 10 reservoirs located in the nearby towns of Holden, Leicester, Paxton, Princeton, and Rutland. These 10 reservoirs have a cumulative storage capacity of 7.38 billion gallons and are designed to operate as two separate systems of five reservoirs, one *low-service* series and one *high-service* series (Tables 3.3 and 3.4). The low-service series produces a safe yield of 18 million gallons per day and is composed of Lynde Brook Reservoir and Kettle Brook Reservoirs 1–4. Water in this series flows from Kettle Brook Reservoir 4 southward through the remaining Kettlebrook Reservoirs, then to Lynde Brook Reservoir via a pipeline. The high-service series produces a safe yield of 3.6 million gallons per day and is composed of Holden 1 and 2, Kendall, Pine Hill, and Quinapoxet, respectively. Water collected in the Pine Hill Reservoir flows downhill (fed by gravity) to the Kendall Reservoir, where it is met by water pumped from the lower-elevation

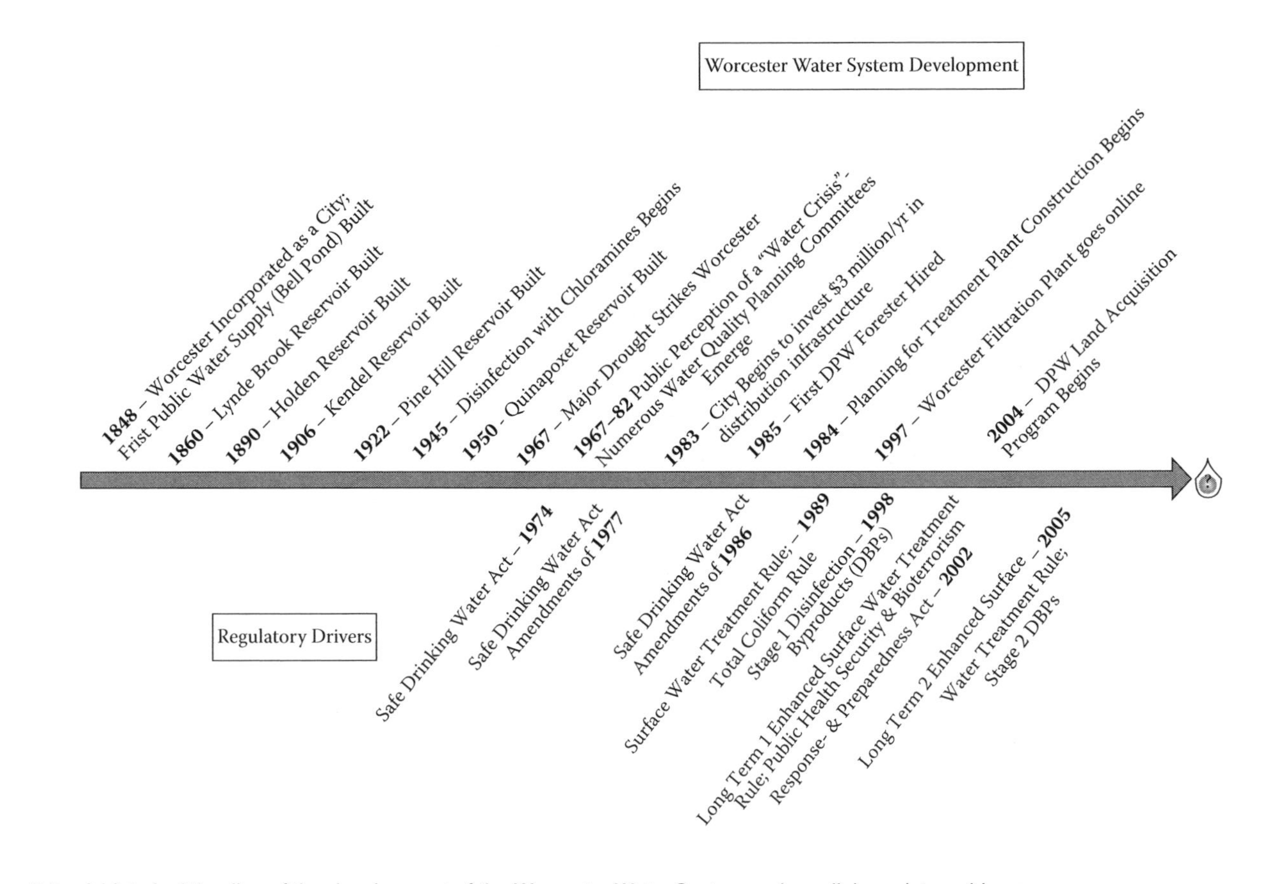

Figure 3.8 A historical timeline of the development of the Worcester Water System and parallel regulatory drivers.

Table 3.3 Storage Capacity of Each of the City of Worcester's 10 Reservoirs

Reservoir	Location (Town)	Service Series	Capacity (Millions of Gallons)
Lynde Brook	Leicester	High	717.4
Kettle Brook #1	Leicester	High	19.3
Kettle Brook #2	Leicester	High	127.3
Kettle Brook #3	Leicester, Paxton	High	152.3
Kettle Brook #4	Paxton	High	513.7
Holden #1	Holden	Low	729.3
Holden #2	Holden	Low	257.4
Kendal	Holden	Low	792.2
Pine Hill	Paxton, Holden, Rutland	Low	2,971.0
Quinapoxet	Holden, Princeton	Low	1,100.0

Source: Worcester Water Quality Report 2008.

Quinapoxet Reservoir. Water collected in the Kendall Reservoir then flows south through Holden Reservoirs 1 and/or 2. Water collected in Holden Reservoir 1 is pumped into the Worcester water filtration plant, while water collected in #2 is pumped back into Holden 1 via a recirculating pump at Olean Street Pump Station. These 10 reservoirs are located in a combined watershed area of approximately 26,000 acres. Through purchase, eminent domain, and conservation easements, the City of Worcester owns roughly 8,000 acres, or 30 percent of this area.

For emergency supply, the City of Worcester also has two inactive groundwater wells, Coal Mine Brook Well and the Shewsbury Well, and two additional reservoirs that can be used as water supply in case of emergency. In addition to these emergency supplies, the City of Worcester has tied into the MWRA's Wachucett Reservoir. Worcester entered a cooperative agreement with MWRA in 1911 and built a pump station on the shores of the Wachusett reservoir. These reserve supplies were last activated during a major drought that struck Worcester in the 1960s.

3.2.2.3 Threats to Worcester's Water Supply

With private lands comprising approximately 70 percent of Worcester's water supply watersheds, the Environmental Protection Agency ranks Worcester's water supply as highly susceptible to contamination (MDEP, 2002). Worcester's surface water reservoirs face threats typical of any New England water supply. These threats include residential land use (primarily improper disposal of hazardous materials), transportation corridors (primarily salting of roadways), agriculture (primarily excessive fertilizer application and livestock contamination), and riparian area activity. In addition to these common threats, the Worcester system has experienced two additional major threats that have had formative, lasting influence on the Worcester water supply system. These are described in detail below.

The first major threat to Worcester's water supply came in the form of a severe drought in the late 1960s. The city was forced to rely on its emergency supply agreement (signed in 1911) with MWRA and pump from Wachusett Reservoir. This

Table 3.4 The Current Land Use and Land Ownership for Each of the 10 Drinking Water Reservoir Watersheds

Reservoirs	Location in Worcester Surface Water Collection System	Watershed Size (acres)	Reservoir Watershed Land Ownership Profile	General Land Use Description
Kendall Holden #1 Holden #2	Downstream end of "low service" series; contributes 1/5th of Worcester's water supply.	Kendall: 1,498 Holden #1: 2,895 Holden #2: 474	95% of Kendall watershed lands, 90% of Holden #2, and >50% of Holden #1 are controlled and managed by Worcester DPW. Several private landowners have large parcels of land enrolled in Chapter 61 programs.**	Hilly land, stony, well-drained soils common in central Mass. Most soils have groundwater more than 6 feet below the surface. Few small wetlands scattered throughout watershed. Mostly in forest cover except for a few small hayfields. Low population density, limited land available for development of new homes.
Pine Hill	First collection reservoir in "low service" series, 2nd largest reservoir in Worcester water supply system; contributes 17% of Worcester's water supply.	4,292	More than half of this watershed is permanently protected open space. Primary landowners are Worcester Water Department and Metropolitan District Commission, City of Paxton. There are some private agricultural restrictions on farmlands in the northern half of the watershed. Anna Maria College owns a large area abutting water department property. Several private owners have large parcels of land enrolled in the Chapter 61A program.**	Hilly land, stony, well-drained soils common in central Mass. Nearly all soils have seasonally high water tables 1.5–3.5 feet below the surface during the spring. There are several large wetlands scattered throughout the watershed.

(continued)

Table 3.4 The Current Land Use and Land Ownership for Each of the 10 Drinking Water Reservoir Watersheds (continued)

Reservoirs	Location in Worcester Surface Water Collection System	Watershed Size (acres)	Reservoir Watershed Land Ownership Profile	General Land Use Description
Kettle Brook Lynde Brook	"High service" reservoir series; contribute 20% of Worcester's water supply.	Kettle Brook #4: 1206 Kettle Brook #3: 484 Kettle Brook #2: 321 Kettle Brook #1: 690 Lynde Brook: 1791 Total: 4492	Half of watershed lands are permanently protected open space. Worcester water department owns extensive lands around each reservoir. Greater Worcester Land Trust owns several permanently protected parcels. Several landowners have large parcels in Chapter 61 programs.** Large undeveloped area beside Lynde Brook is owned by Worcester Municipal Airport, at risk for development. Only a couple of small farms remaining. Large subdivisions on east side of reservoirs. Development of new homes and roads is the primary concern for this watershed.	Stony, well-drained soils. Small wetlands scattered throughout watershed. Seasonally high water tables 1.5–3.5 feet below surface in spring, generally groundwater is more than 6 feet below ground surface.
Quinapoxet	Low service series. It is at a lower elevation than the other reservoirs in this series and is pumped to Kendall Reservoir and contributes approximately 25% of Worcester's Water supply.	12,607	Largely rural. Many public and private conservation areas. Low-density development throughout but watershed does not face high development threats.	Rolling hills with stony well-drained soils. Large wetlands scattered throughout. Nearly all soils have seasonally high water tables 1.5–3.5 feet below the surface during the spring.

(continued)

Table 3.4 The Current Land Use and Land Ownership for Each of the 10 Drinking Water Reservoir Watersheds (continued)

Reservoirs	Location in Worcester Surface Water Collection System	Watershed Size (acres)	Reservoir Watershed Land Ownership Profile	General Land Use Description

Source: Worcester SWPP's 2003.
Notes: *Watershed area = reservoir area + Zones A, B, and C. **Massachusetts Current Use Tax Program (e.g. Ch. 61, Ch. 61A, and Ch. 61B). Taxes for those properties enrolled in a Current Use Tax Program are determined based on current use of property (i.e. productive potential of growing trees) (Massachusetts Department of Environmental Protection, 2002, Source Water Protection and Assessment for Worcester DPW, Water Supply Division. http://www.mass.gov/dep/water/drinking/swapreps.htm).

drought also forced the city to withdraw 30 percent of the total water holding capacity from its emergency groundwater wells. The 1960s drought led the city to develop a Drought Contingency Plan in June of 1990 (Worcester Department of Public Works, 1990). This plan outlines four stages of water supply drought planning ranging from voluntary water demand reductions (Stage I) to severe drought emergency demand reductions of 30–40 percent (Stage IV).

The construction of the Worcester Regional Airport presented a second major threat to Worcester's water supply. Built in the Lynde Brook Reservoir watershed between 1961 and 1966, airport construction included the building of a massive runway raised 60 feet above ground level. The runway was constructed out of sediment and cut off a major tributary to the reservoir. This resulted in a sharp increase in sediment transport into Worcester's water supply and led to DPW taking the reservoir offline during and immediately following construction. This sediment erosion still accounts for nearly all of the turbidity present in Worcester's water supply. Moreover, much of the airport's drainage system is outdated and deteriorating, resulting in extensive erosion. The airport owns large tracts of undeveloped land adjacent within the Lynde Brook reservoir and plans to expand in the future. This expansion includes the addition of a parallel taxiway, which could potentially introduce more sediment into the Lynde Brook Reservoir, both during construction and operation. Worcester DPW is working with the airport to remedy these problems.

3.2.2.4 Worcester Water Filtration Plant

In anticipation of the SDWA Amendments of 1986, specifically the Surface Water Treatment Rule, the City of Worcester began construction of the Worcester Water Filtration Plant in 1984 (Worcester DPW, 2008a). Built to a filtration capacity of 50 million gallons per day, the filtration plant went online in 1997 and filters water for all of Worcester's 182,000 residents. The plant utilizes ozone for primary disinfection. Primary disinfection is followed by rapid mixing and coagulation (using aluminum sulfate and cationic polymer), flocculation, filtration by anthracite coal and sand, pH adjustment, and the addition of chlorine as a secondary disinfectant (Figure 3.9.). The plant has a treated-water storage capacity of 5.5 million gallons

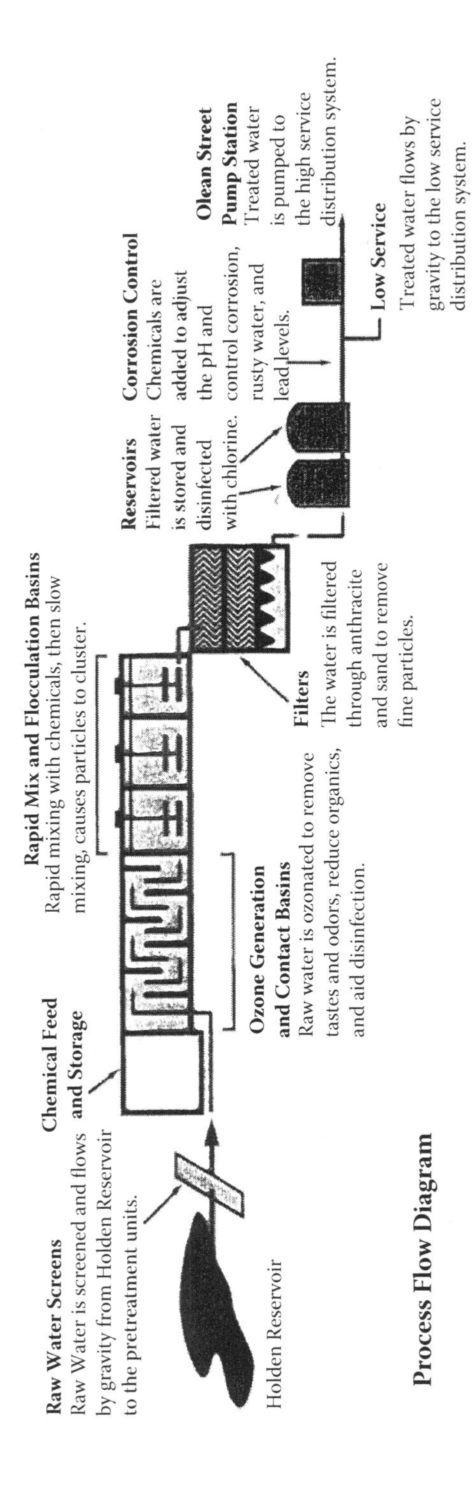

Figure 3.9 Worcester treatment train schematic.

(in two separate 2.75-million-gallon storage tanks). To ensure security and redundancy, the plant was designed and built as two identical water treatment processes in a mirror image of itself. Typically, only one side of the plant is in use.

3.2.2.5 Worcester Watershed Protection Efforts

Forestry. Worcester began managing its watershed with the hiring of the first DPW forester in the early 1980s. During this time watershed management and protection was minimal, but active forestry of the 10 reservoir watersheds did occur. Worcester DPW is in the process at developing their first comprehensive forestry management plan for their 10 reservoirs.

Land acquisition and partnership with the Trust for Public Land (TPL). The City of Worcester's 2008 Annual Water Quality report notes that the first barrier of water supply protection is to have clean source water. The report goes on to say that to preserve this high-quality source water, the water supplier must control the land within the source water watershed. The inclusion of this progressive approach to source water protection speaks to the culture in Worcester that recognizes that their treatment plant is only a second barrier of a multibarrier approach. Worcester DPW and the citizens of Worcester recognize that their treatment plant was built to treat relatively clean water (no sedimentation basins), and that they can extend the life of their treatment plant if they start with cleaner water. Since 2002, Worcester has strategically been acquiring land within their 10 reservoirs' watersheds to gain control over water supply lands. The city has budgeted approximately $300,000 a year using DPW money and State Watershed Protection Act funds.

In 2003, the City of Worcester produced Source Water Protection Plans for each of their 10 reservoirs. The city worked with the Massachusetts Watershed Coalition to identify the areas of most risk to water supply and are therefore top conservation priorities. These priority areas were identified primarily by their proximity to water supply and reservoir tributaries, but also by land use type and threats to water supply (e.g., roads, inappropriate access) (Worcester DPW, 2008b). Worcester partnered with the Trust for Public Land (TPL) in 2004 to increase the effectiveness of their land acquisition program. TPL can provide Worcester with the expertise of land acquisition and staff for outreach. Worcester DPW, with the help of TPL, purchases land based on a combination of opportunity (affordability, availability) and priority (those properties identified as top conservation priorities). Between 2002 and 2008 the City of Worcester acquired 516 acres across six different parcels, each of which contained Zone A areas (Stone, 2009). The main challenges to Worcester's land acquisition program are the looming elimination of the state's watershed protection program and the extensive size of Worcester's reservoir water supply system.

3.2.2.6 Profile of Water Quality

Worcester has high-quality raw water. Worcester's raw water has very low turbidity, and as a result, the Worcester construction of filtration plant did not require the installation of costly sedimentation basins. Worcester DPW and the citizens of the city of Worcester recognize that high-quality raw water will extend the life of their treatment plant and also allow the city to use less chlorine (see below).

Half-log disinfection credit. In 2008, Worcester DPW applied to the state's Department of Environmental Protection for a one-half log credit toward the disinfection and removal requirements of the Surface Water Treatment Rule (Guerin, 2008). In their application, the City of Worcester cited their Source Water Protection Plans, Land Acquisition Program, and the fact that this would enable them to reduce their chlorination levels, which are higher than desirable during winter months. However, this would decrease *giardia* inactivation that Worcester DPW has to remove from their water supply from 3.0 log to 2.5 log (Guerin, 2008). Citing Worcester's 2003 DEP-approved Source Water Protection Plans and direct filtration plant, the DEP granted Worcester this one-half log credit (Stone, 2009) because of progress made in source watershed protection.

3.3 STATEWIDE CHALLENGES: TRENDS FACING WATER UTILITIES

3.3.1 Land Use Change and Population Growth

As documented earlier, the New England region faces significant pressures with respect to population growth. Trends of increasing ex-urbanization appear to be continuing with little abatement (Hall et al., 2002). With a fragmented and decentralized land management structure and in the absence of regulatory incentives, Massachusetts is poorly positioned to prevent further sprawl. This is in direct conflict with the primary mechanisms of watershed protection: land conservation and management. Growing populations not only increase the quantity and sources of contaminants but also decrease the capacity for water supply systems to protect their watersheds through direct ownership. As demonstrated in the MWRA and Worcester systems, human activities within watershed boundaries inevitably increase the risks of water contamination. Exclusion of harmful land use activities, through ownership, regulation, or disincentives, is essential for effective watershed protection.

Traditional land acquisition strategies will be met with a decreased financial capacity to control increasingly expensive real estate. The Wachusett watershed demonstrates how fee-simple transactions are being phased out in favor of watershed protection restrictions to reduce acquisition program costs. Similarly, while Worcester residents are currently supportive of land conservation programs for the protection of water quality, such funding will be both less efficient and more controversial as land use prices rise and revenues fall. These increasing land prices will require utilities to adopt creative land protection land conservation strategies (e.g., partnerships with conservation organizations) and/or require utilities to find ways to generate additional revenue from their conserved lands. Revenue sources could include existing markets for timber harvest sales, low-impact recreational passes (e.g., hiking), and nontimber forest products (e.g., maple syrup). Comparably, as mitigation banks and ecosystem service markets continue to take shape, utilities could generate revenue from the sale of mitigation credits or ecosystem service credits generated on watershed lands (e.g., carbon sequestration, clean water, biodiversity conservation).

3.3.2 Forest Management

As previously mentioned, the first barrier in a multiple barrier approach to water supply protection is the management and conservation of source watersheds. For both MWRA and Worcester, the majority of reservoir watersheds are covered by public and private forestland. The ability of watershed forests to produce higher quality water than more intensive and/or developed land uses is undisputed (e.g., de la Cretaz and Barten, 2007).

Watershed foresters for MWRA contend that active management can reduce the threat of water quality impacts (DCR, 2007). Under the advice of the Watershed Management Science Committee,* these foresters have developed zones for active and passive management of the Quabbin and Ware River watershed forests to increase resiliency against catastrophic natural disturbances (e.g., fire, pest outbreak) (e.g., Kyker-Snowman, 2000). The goals are to protect riparian and ecologically sensitive areas while moving an even-aged 100-year-old forest that originated postagriculture to one that is multi-aged, with a vigorous regeneration and groundstory stratum. Such a multi-aged multiple strata forest is thought to be capable of immediately responding to catastrophic overstory disturbances such as tornadoes and hurricanes by stabilizing soils, preventing erosion and release of sequestered pollutants and nutrients (DCR, 2007). In addition, having a younger and more diverse species composition represented within the forest is thought to provide greater resiliency to chronic disturbance impacts such as air pollutants (i.e. nitrogen, mercury) by binding pollutants to organic matter and/or sequestering within plant tissues (DCR, 2007). Diversifying composition through forestry also makes such forests less susceptible to species-specific insects and diseases that can impact monospecific stands (e.g., hemlock woolly adelgid, gypsy moth, beech-bark disease, Asian long-horned beetle) (DCR, 2007). To help meet the management goals of the MWRA watershed system, the Office of Watershed Management has generated some revenues from the timber harvests on 1–2 percent of the watershed annually. These funds are used to support land acquisition and other watershed protection activities.

Conversely, some researchers argue that the elimination of harvesting and its associated impacts will better preserve and improve water quality in the face of continued forest disturbances (e.g., Foster and Orwig, 2006). In this case their main argument was based on the ability of forests to recover from insects and disease without human help. They point to the recruitment of new stands of birch beneath adelgid-infested hemlock stands. Though these hypotheses warrant additional research, this debate has sparked a larger conflict between multiple stakeholder groups involved in the Massachusetts Forest Futures Visioning Process. This conflict centered on

* The Science Committee comprises researchers and experts from universities, the U.S. Forest Service, and nongovernmental organizations knowledgeable about forest dynamics, hydrology, wildlife ecology, invasives, and socioeconomic policy of forestlands.

regulations for forest management and the ability to actively manage public forestland in the state. The outcome has been to increase the forest reserve areas (i.e. no management zones) with new restrictions on harvesting and silvicultural treatments to limit the size, type, and location of management activities. This may have important implications on the ability to manage publicly owned watershed forests for water quality. Currently, however, the MWRA system is exempt from these rules because of the oversight of the nonpolitical Watershed Management Science Committee.

3.3.3 Decreasing Demand and Revenues

Despite their public ownership, the water supply utilities like MWRA and Worcester DPW are subject to the same cost and efficiency constraints as private entities. While public systems undoubtedly have more mechanisms at hand for capital investment (e.g., bonds), these systems must still recoup operating costs in addition to expenses for infrastructure upgrades. Both the MWRA and Worcester systems are generally unsubsidized and consequently must finance their budgets through revenue collection. In the case of MWRA, residents pay some of the highest per-capita rates in the country (approximately $1200 per household annually) (Figure 3.10) (MWRA, 2009a).

With costs already nearing the top of currently acceptable rate structures for metropolitan water supply districts, the MWRA system would likely face greater constraints for increasing revenues through rate hikes. Water and sewer rates for the Worcester system are considerably lower ($360 per household annually) and, without conservation-inducing pricing, may have more latitude for cost recovery.

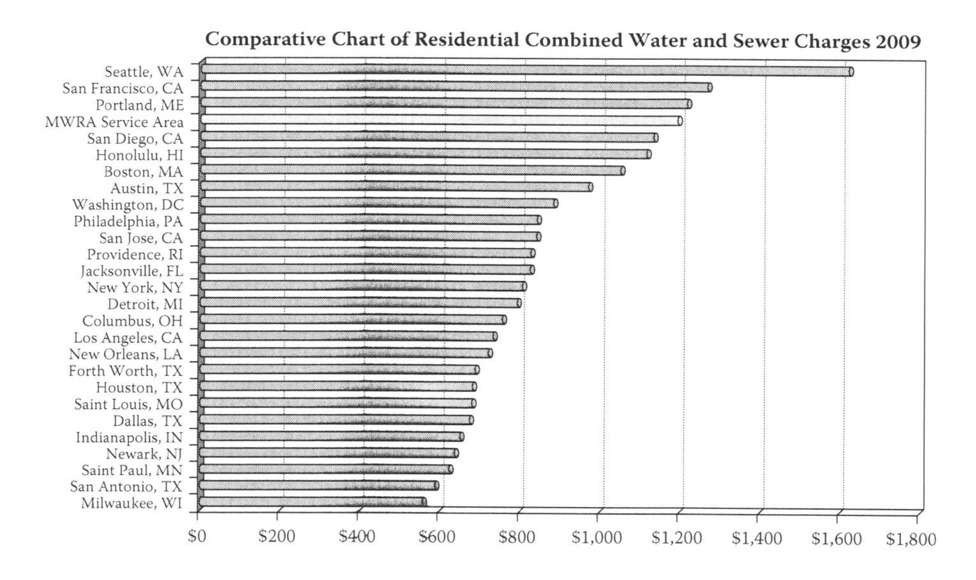

Figure 3.10 Rate comparison of combined water and sewer charges for major U.S. cities. Boston/MWRA (fourth line from the top) has an average yearly household charge of $1,200 (From MWRA 2009).

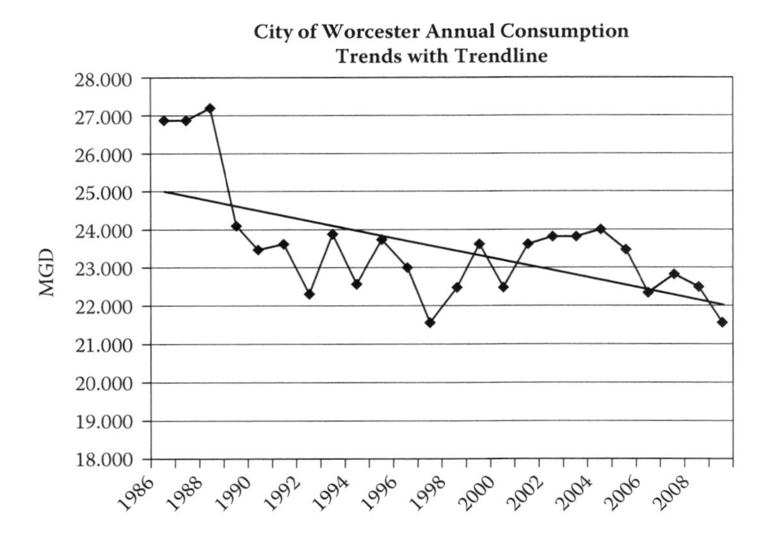

Figure 3.11 Decreased demand for Worcester water supply over time (From Guerin 2010).

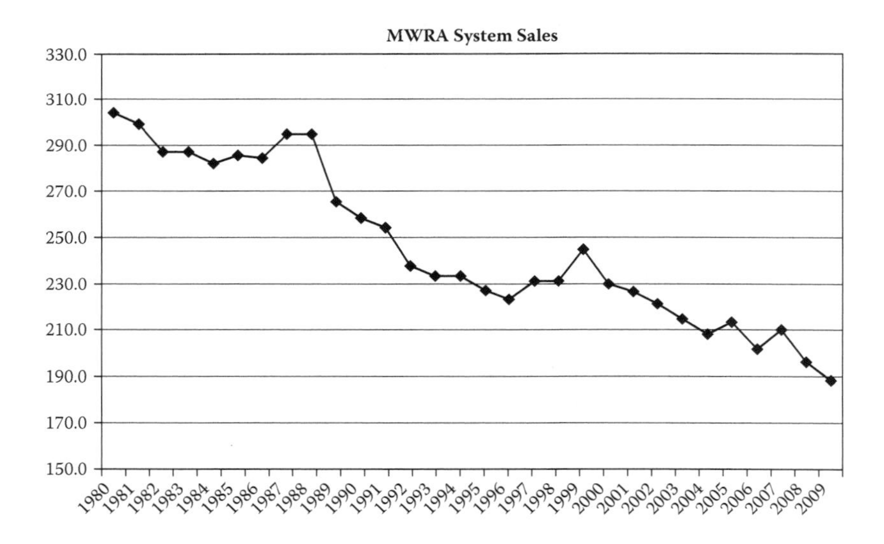

Figure 3.12 Boston and most other water systems in Massachusetts have seen total water demand decrease significantly over the past decades (From MWRA 2009a).

But whatever the size of the system, water suppliers across the country are facing decreasing revenues in response to long-term trends of decreasing per capita water consumption (Figures 3.11 and 3.12).

The result of this trend is the need for water suppliers to increase cost rates simply to maintain the same level of service to customers. This challenge means that it is in the economic interests of water districts to promote measures that increase demand (Walton, 2010). This conflict of supply and demand greatly increases the complexity of water management in Massachusetts and also increases the opportunities for failure.

3.3.4 Changing Climate

Changes in water quantity and quality concerns are perhaps some of the most pressing issues associated with climate change. In the case of Massachusetts, climate change is expected to have variable impacts relative to the size of the water systems (Kirshen et al., 2005). Regardless of size, all water systems will face the dual challenges of reduced consumption (as discussed above) and likely increased precipitation that will occur in larger, more episodic events (Keim et al., 2003). Changes in precipitation patterns may also affect storage supplies where smaller local supply systems could face reductions as large as 10 percent while the larger MWRA system is expected to remain well within its safe yield capacity through 2100 (Kirshen et al., 2005). The impact of climate change in the New England area will likely be attenuated by decreasing per capita household water usage. However, this projection accounts only for population growth in line with current trends. Given our collective inability to reliably downscale models of future water supply or water demand, the supply-impacts of climate change will exacerbate problems of uncertainty.

Outside of the direct impacts of climate change on precipitation quantities and water supply, secondary effects may prove at least as problematic, if not more so. Future storms in New England are likely to deliver larger total quantities of water, but they are also expected to be of greater intensity as a result of climate change (Kirshen et al., 2005; Nearing, 2005; Nearing et al., 2004). This change in storm profiles has the potential to significantly increase erosion rates and thus impact drinking water quality. These storms are also expected to present risk to New England forests through blowdowns. More intense storms, combined with a greater percentage of precipitation falling as rain rather than snow, have the potential to impact sediment transport and erosion to roughly the same degree as land use change (Keim et al., 2003, Nearing, 2005).

3.3.5 Increase in Emerging Contaminants

The debate surrounding the potential to regulate "emerging" contaminants is most accurately a question of *when* rather than *if.* Scientific reports dating back to the mid-1980s document the emergence of new classes of contaminants cycling from human waste streams into source drinking waters (Daughton, 2004). The proliferation of these compounds is tremendous; as of March 2004, more than 23 million organic and inorganic substances were indexed by the American Chemical Society. The prospect of regulating such a diverse and ubiquitous number of chemical compounds is a daunting task. Daughton points out that what is actually emerging is our *concern* for, or recognition of, these compounds and the daily creation of novel chemicals. Thus some of these so-called emerging contaminants, such as hexavalent chromium, have actually been inventoried for years or decades (Daughton, 2004).

EPA concern with emerging contaminants dates back to at least the 1990s when, in partnership with the U.S. Geological Survey (USGS), it began surveying water sources across the United States for unregulated contaminants of concern (Daughton, 2004). The most recent USGS national reconnaissance of pharmaceuticals and other organic contaminants, collectively referred to as PPCPs (pharmaceuticals and personal care products), demonstrated the ubiquity of these contaminants in both ground and surface sources of drinking water (Focazio et al., 2008). An explanation for the apparent ubiquity of PPCPs is likely due to the large number of pathways for introduction. Anthropogenic sources include "manufacturing releases, waste disposal, accidental releases, purposeful introduction (e.g., pesticides, sewage sludge application), and consumer activity (which includes both the excretion and purposeful disposal of a wide range of naturally occurring and anthropogenic chemicals such as PPCPs)" (Daughton, 2004, p. 713; EPA, 2006).

For drinking water suppliers, the increasing prevalence of anthropogenic environmental contaminants poses a serious threat to current methods of water delivery. Especially for unfiltered systems or filtered systems reliant upon unprotected source water, the new regulatory rules over PPCPs could dramatically alter current management strategies. Already, new rules are being implemented by the EPA to mitigate the potential effects of disinfection by-products (Richardson, 2007). The EPA is expanding its monitoring requirements for unregulated compounds and also incorporating an increasing number of chemical classes into its National Drinking Water Contaminant Occurrence Database (NCOD) (Richardson, 2007). Without incentives for changes for Massachusetts water suppliers, the solutions for source protection are likely to remain limited to the more traditional strategies (e.g., land conservation, disinfection). Whether such strategies can cope with the challenging diversity of chemicals and sources of PPCPs is a question of serious concern. Given the current trajectory of attention to PPCPs and other emerging contaminants, it is imperative that water suppliers give serious consideration to such planning efforts and seek to identify all available alternatives for protecting drinking water supplies and subsequently the health of water consumers.

3.4 LAND USE POLICY, MONITORING, AND ASSET MANAGEMENT

As evidenced by the challenges facing water suppliers, described previously, there is increasing pressure on water managers to prove the value of further investment in watershed management for surface water protection. This means that water managers will need to analyze, understand, and compare the incremental costs of source water protection versus engineering water treatment solutions for delivery of safe drinking water.

Attempts to demonstrate incremental costs linking amount of watershed forested to engineered treatment costs have been unable to provide statistically convincing results (Ernst, 2004). A combination of variability between watersheds and inadequate monitoring data make this relationship challenging to generalize across watersheds. Though an established incremental cost comparison between forest cover versus treatment cost would provide a valuable policy tool, it is not required. Water quality engineers agree better raw quality water is easier to treat, and better raw water quality water comes from protected and well-managed forests. Efforts should be redirected from attempts to demonstrate a generalizable incremental cost relationship to attempts at a localized level to deliver high-quality drinking water more effectively and efficiently. These efforts must combine top-down and bottom-up approaches for optimizing the delivery of high-quality drinking water. There are three strategies to improve integration between bottom-up and top-down approaches: (1) better coordination between state-level land use policy and the management of drinking water resources (Trust for Public Land, 2009); (2) better monitoring to enable more effective and efficient resource management for drinking water suppliers; and (3) a comprehensive approach to drinking water supply through the use of asset management and the triple bottom line.

In this section we describe components of these three strategies—policy, monitoring, and asset management efforts—that are already available and utilized by water utilities in Massachusetts.

3.4.1 Policies Currently in Place

At the state level, a number of policies are in place to assist water suppliers with the task of source water protection. Massachusetts has long recognized the importance of watershed protection for maintaining source water quality, and some of these measures stem directly from this stance. Other measures are simply extensions of or the reframing of policies or incentives that were originally created for alternative purposes. A number of special measures exist for the Boston system also, given the proportionately larger size of the system and number of its customers.

3.4.1.1 Watershed Protection Act

The Watershed Protection Act of 1992 established a protective zoning system surrounding surface waters of Quabbin, Ware, and Wachusett watersheds (DCR, 2003). The purpose of this act was to establish minimally protective measures for ensuring drinking water safety while maintaining the property rights of private citizens. This act established three zones of protection: (1) lands adjacent to reservoirs were prohibited from making any alterations within 400 feet of regulated waterways; (2) lands adjacent to tributaries and lakes were prohibited from making any alterations within 200 feet of regulated waterways; and (3) lands adjacent to tributaries and lakes were subject to several land use restrictions within 200–400 feet of regulated waterways (Figure 3.13). By legislative decree, Massachusetts mandated a minimum threshold of protection for the state's water supply system that serves the most customers (MWRA). However, these protective regulations fail to restrict land use activities for the remainder of the state's water supply systems. Seemingly the state legislature has determined that separate standards are warranted for the MWRA system, presumably due to its significance and

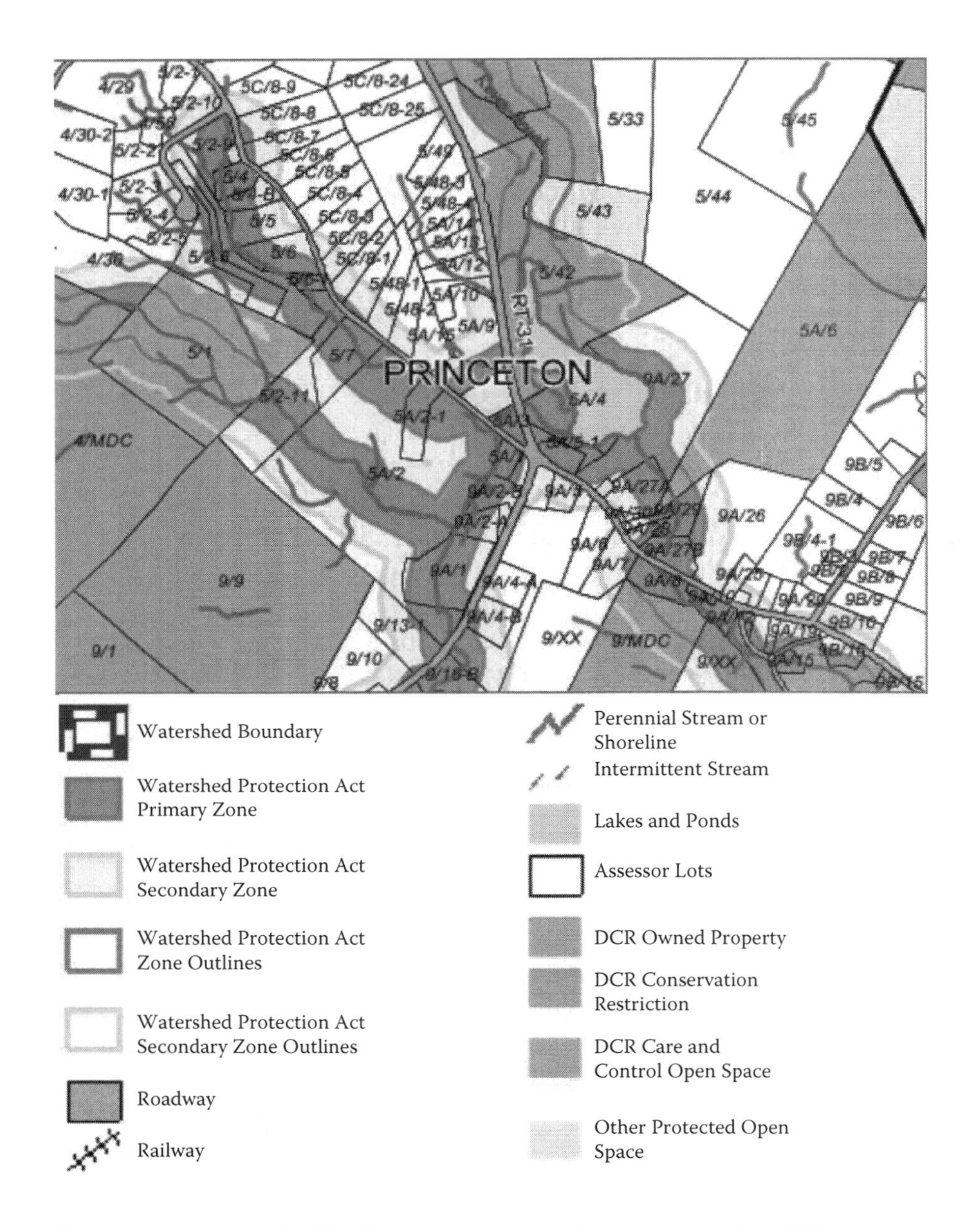

Figure 3.13 An example of the Watershed Protection Act zones taken from DEP interactive web resource for public access (From MassGIS, 2010). (See color insert.)

unfiltered status. Applying these measures to protect other watersheds, including filtered systems such as Worcester, is an issue not currently being addressed by Massachusetts.

3.4.1.2 Tax Incentives

The Massachusetts General Laws (MGL) contain several tax provisions that pertain to watershed protection. Chapter 59 of the MGL defines the PILOT (payment in lieu of taxes) program that stabilizes local property revenues by requiring state-owned land to be taxed at a fair market value. This program ensures local support for land acquisition programs since municipalities can be assured that no perceptible revenue will be lost through the sale of private land to state owners (DCR, 2008; MGL, 2009).

Chapter 61 provides direct tax reductions to landowners who keep their land in an undeveloped state such as forest, farmland, or open space. In exchange, the landowners receive reduced property tax rates. While Chapter 61 does provide an effective mechanism for conserving land, it is by no means a secure method. Landowners can rescind their commitment by paying a penalty equivalent to three years of property tax reductions and are then free to sell their land on the market. Accordingly, such a protection mechanism is effective due to the bureaucratic and financial disincentive for landowners to divest from the program. The program is a useful tool for watershed managers and is used extensively throughout the DCR/MWRA watersheds (DCR, 2008).

3.4.1.3 Drinking Water Supply Protection Grants

While the regulatory arm of the Watershed Protection Act does not offer a parallel institution for non-MWRA watersheds, the bonds for land acquisition do have a small system complement. The Watershed Protection Grant Program, financed through the Massachusetts capital budget, provides funds for public water systems seeking to protect watershed resources through land acquisition. The maximum grant amount is $500,000 and cannot exceed 50 percent of the total project budget. Land trusts cannot participate in this program except when in partnership with a municipality or public utility. The program does allow grantees to choose between fee-simple or easement purchases. Because the program is a capital budget line item, it is subject to changes in discretionary funding. Accordingly, the program is subject to significant fluctuations due to changes in the economic climate. Since the program's inception in 2005, annual funding has decreased from more than $3 million to just over $500,000 (MDEP, 2010). Such volatility in funding decreases the willingness of municipalities to conduct longer-term planning of land conservation.

3.4.1.4 Coordinating State and Local Policy and Strategy to Protect Drinking Water Sources

Funded by a grant from the EPA, the Trust for Public Land (TPL), the Smart Growth Leadership Institute, the Association of State Drinking Water Administrators, and the River Network joined forces in a program called Protecting Drinking Water Sources. This program is aimed at better integrating state land use policy, incentives, and drinking water programs (Protecting Drinking Water Sources, 2010). The program's mission is to "better align planning, economic development, regulation and conservation to protect drinking water sources at the local and watershed levels" (Protecting Drinking Water Sources, 2010, website homepage). Protecting Drinking Water Sources selects pilot states with aims to improve collaboration among state agencies, local managers, and concerned stakeholders. This collaboration can maximize incentives and efforts aimed at protecting drinking water quality. Alumni states include Maine, Ohio, New Hampshire, Oregon, Utah, and North Carolina.

In an attempt to maximize and prioritize efforts, program staff works to coordinate state-level land use policies with land management efforts by drinking water suppliers. The program develops a list of short-term and long-term action items for each state that range from streamlining state geographic information systems (GIS) layers, to targeting priority watersheds for creating watershed-based land commissions, to creating a dedicated funding program for source water protection land conservation, to developing guidelines to integrate recreation with ways that are compatible to source water protection. As a whole, this program works to reduce duplicated efforts and build coalitions when and where they make sense.

3.4.2 Upstream Monitoring and Mapping Strategies for Integrating Natural Resources

To optimize the use of built and natural assets to deliver high-quality drinking water, drinking water suppliers need an understanding of when and why water quality is changing and what is driving these changes. Watershed managers must be able to prioritize their management efforts on both natural and built assets to optimize their limited time and resources. Monitoring of water quality within engineered treatment plants is both commonplace and required. This real-time monitoring and analysis of water quality data enables drinking water suppliers to respond effectively and efficiently to water quality problems. Conversely, monitoring of raw water quality (in reservoirs and tributaries) is currently driven by minimal requirements and does not enable cost- and resource-efficient land management. More thorough and strategic monitoring of raw water quality can aid real-time response and planning. Additionally, analysis of these trends over time can help drinking water suppliers understand how different disturbances (e.g., land use change, fire, wind storm) impact raw water quality. An upstream understanding of what is influencing water quality will allow water suppliers to focus both their management and outreach efforts on the highest priority areas and management actions.

3.4.2.1 Where, When, and What to Monitor

Based on attempts to quantify the incremental cost between acres of forested watershed and engineered treatment costs, strategies have been developed for effective and efficient monitoring to link raw water quality with engineered treatment costs (Freeman et al., 2008). This *water quality index* is a combination of strategic sampling locations with the primary water quality metrics that drive treatment costs and decisions (Freeman et al., 2008). The water quality index recommends that sampling should take place (1) at all major tributaries; (2) at sites strategically targeted for specific land use (e.g., near agricultural sites, near developments); and (3) during dry and wet weather. Monitoring parameters should include (1) total organic carbon, (2) dissolved organic carbon, (3) turbidity, (4) alkalinity, (5) conductivity, (6) temperature, and (7) pH. The combination of these strategic monitoring locations, times, and parameters may help watershed managers develop a more holistic picture of their watershed and how different events (e.g., land uses, weather changes, management activities) impact raw water quality (Freeman et al., 2008).

3.4.2.2 Real-Time Monitoring and Mapping

Within treatment plants, water quality engineers actively monitor real-time water quality. The same must be true for raw water quality for surface water supplies. Real-time monitoring of water quality can help watershed managers respond quickly to an adverse event within the watershed (e.g., a chemical spill) and address and alleviate a problem before it escalates. This data can then be archived and analyzed to help water suppliers identify activities that have a noticeable impact on water quality. Real-time monitoring will allow water quality engineers to better anticipate what types of treatment changes they may need to make (e.g., chlorine levels).

Case Study: The Philadelphia Water Department and USGS

An example of real-time monitoring is a joint effort between the Philadelphia Water Department (PWD) and the U.S. Geological Survey (USGS). They partnered to make real-time water quality monitoring a priority (Figure 3.14). PWD and USGS are using a combination of old USGS stream gauges and strategically placed new stream gauges to understand water quality changes in real time. This enables PWD staff to understand how real-time events (e.g., rain storms) are influencing water quality, which in turn allows them to focus public outreach efforts. Data for seven different water quality parameters are available online from each stream gauge, which is indicated on a map based on its location within the city. The software uses a color-coding system for each water quality parameter to indicate *good* (green), *undesirable changes* (yellow), or *bad* (red) conditions. This data is available to the public through the USGS or PWD website.

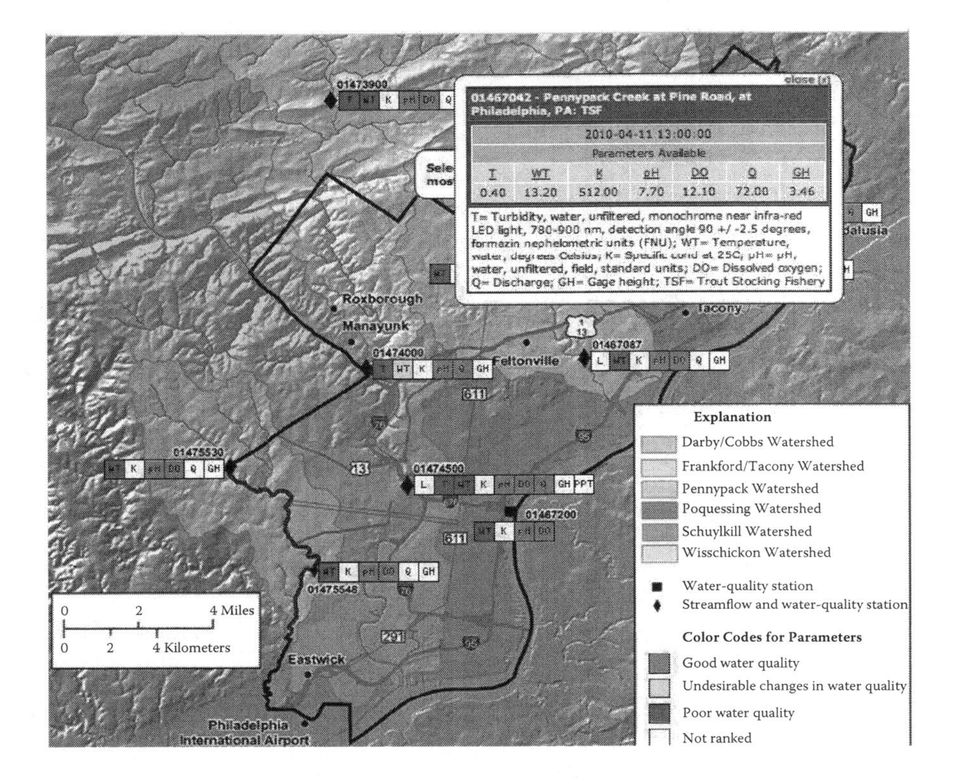

Figure 3.14 The Philadelphia Water Department (PWD) and U.S. Geological Survey (USGS) real-time monitoring network. The public can access data in real time on the PWD and USGS websites.

3.4.2.3 Monitoring and mapping for Prioritizing Management and Conservation

In addition to real-time response, monitoring can help identify where and how drinking water suppliers should prioritize their management and conservation efforts. Incorporating water quality monitoring data (e.g., dissolved organic carbon [DOC], turbidity) into existing GIS layer data (e.g., soil type, slope, land use) allows watershed managers to better understand where to focus management efforts (e.g., thinning, replanting) and outreach efforts, or prioritize land for purchase or conservation easements (Trust for Public Land, 2005).

Case Study: Forest-to-Faucet Partnership

An example of a mapping–monitoring project is the Forest-to-Faucet Partnership. This is a collaborative initiative between UMASS Amherst and the U.S. Forest Service and aims to foster a better understanding of the relationship between forests

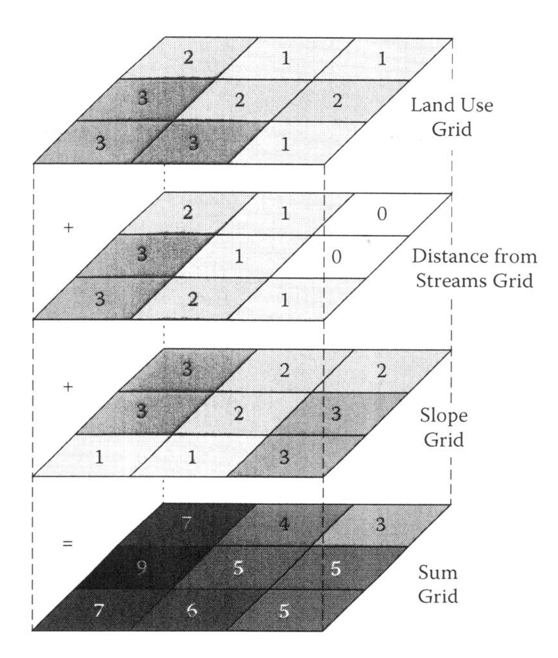

Figure 3.15 An example of summed prioritization grid overlaying data layers critical for water quality (From Trust for Public Land 2005).

and water quality/quantity. Through the use of GIS data layers, the Forest-to-Faucet Partnership created the Watershed Forest Management Information System (Barten et al., 2010). This system allows users to overlay GIS data layers critical to water quality (e.g., slope, soil type, forest cover type, proximity to water bodies) and water quality management (e.g., political boundaries, relevant stakeholders). Overlaying this data creates a spatially based sum grid that prioritizes areas critical for water quality management and/or conservation, which could be used by water utilities to inform watershed land protection priorities (Figure 3.15).

3.4.3 Asset Management Programs That Address Built Assets

The EPA defines *asset management* as "a process for maintaining a desired level of customer service at the best appropriate cost" (EPA, 2008, PowerPoint slide 8). Application of asset management to the water utility industry traditionally involves careful tracking of assets, or physical resources used in the operation of a water utility. This includes tracking events (e.g., maintenance) throughout asset life cycles. These assets include people, tools, equipment, machinery, and buildings used to deliver drinking water to utility customers. In a presentation about the basic concepts behind asset management, the EPA outlines the goal of an asset management program: "The more a utility understands their assets—the demand for the assets, their condition and remaining useful life, their risk and consequence of failure, their feasible renewal options (repair, refurbish, replace), and the cost of those options—the higher the confidence everyone can have that the utility's investment decisions are indeed the lowest life cycle cost strategies for sustained performance at a level of risk the community is willing to accept" (EPA, 2008, PowerPoint slide 6). The advantages of a well-supported and maintained asset management program include ongoing support for operational planning and budget projections, increased system efficiency, better use of staff time, and improved customer service.

Currently the Worcester DPW does not operate a stand-alone asset management program. The MWRA initiated a Facilities Asset Management Plan in 2000, which is outlined in detail by Fortin (2004). According to the case study, benefits to the MWRA from the program include heightened communication among departments, new and standardized procedures for equipment management and replacement, development and adoption of performance improvement metrics, optimization of project management program and documented cost savings, introduction and adoption of predictive maintenance programs and documented cost savings, and enhanced communication and teamwork between management and workforce (Fortin, 2004). The application of asset management to MWRA's built assets has been very successful. In addition, the land or real estate owned by MWRA/DCR is currently rolled in the statewide asset management program Division of Capital Asset Management (DCAM), which maintains accountancy of state-owned real estate properties and buildings (Quabbin LMP., 2007). However, there is no evidence of the application of this tool to the natural assets of the drinking water delivery service, and no reason why this type of careful management cannot be applied to green infrastructure to better understand these assets.

3.5 COMPREHENSIVE APPROACH TO DRINKING WATER SUPPLY MANAGEMENT

Drinking water protection and delivery can become much more efficient if systemwide analyses are conducted. Furthermore, the actual costs and benefits of watershed management for surface water quality could be better realized if systemwide assets were managed more comprehensively. Here we present the concept of using the triple bottom line strategy to contextualize asset management that will, in turn, inform decision making and long-term planning at water utilities. This integrative approach to drinking water supply management will lead to more cost-effective and efficient allocation of scarce financial resources to address a growing list of threats to drinking water systems.

To enable water utility managers to support the case for further watershed protection given the increasing pressures described above, managers should adopt a tool that captures the incremental costs and benefits of investments in green infrastructure. Monitoring and data (as described in Section 3.4) will show that not all land assets are created equal and deserve different treatment in terms of mitigating risks to that asset, similar to traditional challenges faced by built asset management plans. An integrative tool has the potential to indicate how to channel and direct funding more effectively to protect water quality throughout the system. Ultimately, to maintain any sustainable program in land management for source water protection, water suppliers need an all-encompassing tool to assess incremental cost and benefits of source water protection in conjunction with engineered solutions. This section will provide some background on one such integrative framework, the triple bottom line approach, and outline a scenario illustrating how this dynamic could play out in the case of the MWRA and Worcester water supplies.

3.5.1 The Triple Bottom Line Approach

In a recent report on the application of the triple bottom line (TBL) tool to controlling combined sewer overflow by the Philadelphia Water District (PWD), Stratus Consulting discusses the advantages of the triple bottom line approach over traditional cost benefit analyses in the context of public water supply:

> "The TBL approach reflects the fact that society and its enterprises—including the institutions that work specifically in the public interest (for example, water and wastewater utilities)—typically are engaged in activities intended to provide the greatest total value to the communities they serve. These values extend well beyond the traditional financial bottom line that portrays only cash flows (i.e., revenues and expenditures) of a standard financial analysis. PWD and similar utilities that serve the public interest also need to consider their stewardship and other responsibilities, and to thus account for how they may generate values that contribute towards the "social" and "environmental" bottom lines. Hence, a more complete and meaningful accounting of PWD activities needs to provide a TBL perspective that reflects all three bottom lines: financial, social, and environmental." (Stratus Consulting, 2009, p. 3–1)

One utility using the TBL approach is the Milwaukee Metropolitan Sewerage District (MMSD). MMSD implemented the TBL approach to access the full array of benefits from use of green infrastructure (e.g., rain gardens) in storm water management. MMSD used a TBL framework to understand which sections of the MMSD service area green infrastructure can have a particularly positive effect, provide the groundwork for a life-cycle costs assessment of the green infrastructure projects (Table 3.5), and access true cost and benefit analysis grounded in Milwaukee (MMSD, 2009).

Milwaukee's summary report shows the TBL benefit categories evaluated for all of the green infrastructure options and the gray infrastructure options available. The report outlines the need to look holistically at the benefits of using green infrastructure to address combined sewer overflow and storm water runoff issues. For example, TBL considers ancillary benefits such as recreational benefits, potential employment, and energy reduction beyond the economic bottom line. Because these ancillary benefits are weighted differently by different people, MMSD used public meetings to determine what factors should be considered in TBL analysis (Shafer, 2010). It is important to note that this TBL analysis is qualitative.

Table 3.5 An Example of a Qualitative Application of the Triple Bottom Line Approach to Combined-Sewer Overflow Project Options in Milwaukee, Wisconsin

	Gray Infrastructure Storage	Greenways	Rain Gardens	Wetlands
ENVIRONMENTAL				
Reduces Volumes of CSOs and SSOs	●	●	●	●
Reduces Amount of Polluted Stormwater Runoff	●	●	●	●
Reduces Energy Use	○ tunnel pumping assumed	◎	◎	◎
Reduces Greenhouse Gas Emissions and/or Stores Carbon	◎	●	●	◎
Reduces Flood Management Facility Size or Improves Drainage Issues	◎	● needs massive implementation	● needs massive implementation	● needs massive implementation
Enhances Groundwater Recharge and/or Evapotranspiration	○	●	●	●
Improves Air Qualify	◎	●	●	●
Reduces Urban Heat Island Effects	◎	●	●	●
Effective Substantial Runoff Reduction (Water Quality)	◎	●	●	●
ECONOMIC				
Creates Green Jobs	●	●	●	●
Reduces Infrastructure and Site Costs (relative to gray infrastructure)	N.A.	● assumes on-site space available	● assumes on-site space available	● assumes on-site space available
Economical (relative to tunnel, capital costs only)	◎	●	○	●
Increase Property Values	◎	●	●	●
SOCIAL				
Improves Community Quality of Life	●	●	●	◎
Reduces Days Beaches Close	●	●	●	●
Improves Aesthetics	◎	●	●	●

(continued)

Table 3.5 An Example of a Qualitative Application of the Triple Bottom Line Approach to Combined-Sewer Overflow Project Options in Milwaukee, Wisconsin (continued)

	Gray Infrastructure Storage	Greenways	Rain Gardens	Wetlands
Provides Recreational Amenity	○	●	○	○

Source: Milwaukee Metropolitan Sewerage District, 2009.
Note: CSO: combined sewer overflow; SSO: sanitary sewer overflow
Key: ● yes; ◎ partial yes; ○ no

There are examples that are more analytical than the MMSD case, such as the present value benefits for the Philadelphia watershed. This case considered two options to address Philadelphia's combined server overflow needs (Table 3.6): first, a 50 percent green infrastructure approach and second, a built infrastructure approach (construction of a 30-foot tunnel to separate sewer flows). It is important to note that the TBL analysis of Philadelphia watersheds (Table 3.6), though different from the MMSD analysis (Table 3.5), comes to very similar conclusions. Both case studies demonstrate that the use of green infrastructure to address combined sewer overflow retrofits has a higher net value when the social and environmental benefits are included in the analysis.

Table 3.6 An Example of a Quantitative Application of the Triple Bottom Line Approach to Combined-Sewer Overflow Project Options in Philadelphia

Citywide Present Value Benefits of Key CSO Options: Cumulative through 2049 (2009 million US$)		
Benefit Categories	**50% LID Option**	**30' Tunnel Option[a]**
Increased recreational opportunities	$524.5	
Improved aesthetics/property value (50%)	$574.7	
Reduction in heat stress mortality	$1,057.6	
Water quality/aquatic habitat enhancement	$336.4	$189.0
Wetland services	$1.6	
Social costs avoided by green collar jobs	$124.9	
Air quality improvements from trees	$131.0	
Energy savings/usage	$33.7	$(2.5)
Reduced (increased) damage from SO_2 and NO_x emissions	$46.3	$(45.2)
Reduced (increased) damage from CO_2 emissions	$21.2	$(5.9)
Disruption costs from construction and maintenance	$(5.6)	$(13.4)
Total	**$2,846.4**	**$122.0**

Source: Stratus Consulting, 2009.
[a] 28-foot tunnel option in Delaware River Watershed.

While these examples of the TBL approach to combine sewer overflow (CSO) cases are relevant to the effort to access incremental costs of source water protection cases, there are significant differences. Primarily, in CSO applications the water user is the downstream ecosystem, not human drinking water consumers. This difference can serve as an advantage to the use of TBL in source watersheds, because the end use of the water impacts public health. And traditionally, though the two are not mutually exclusive, human health is valued over the health of the environment. Drinking water holds direct impacts on public health, and as a result, public health benefits could be an additional fourth bottom line.

Stratus Consulting is currently developing two case studies involving application of the TBL for drinking water reservoirs. One involves the reservoir of a drinking water supplier with a waiver to filtration, analyzing the benefits of moving a boat ramp away from a sensitive zone in the reservoir. The second case investigates the benefits of closing a drinking water reservoir watershed to recreation for security purposes (Henderson, 2010). These case studies differ from the Milwaukee and Philadelphia projects in that they involve changes to systems that are already in place, not projected costs and benefits of planned projects. Further, these cases involve direct links to public health concerns related to drinking water quality protection. Stratus is making every effort to adjust their analysis to address these differences. For example, they are presenting results as a net benefit rather than straight benefits, and they are considering (1) quantifying people's impressions of the importance of the benefit for inclusion in the analysis, (2) recategorizing public health benefits as a forth bottom line, and (3) weighting the bottom lines differently to reflect project participant values (Henderson, 2010).

It is important to note that in one of the two cases, a public health agency is the lead contact for the project. This indicates the agency's recognition of the advantage of application of the TBL approach to drinking water supply protection scenarios to incorporate all possible benefits of source protection. While it is clear that the TBL approach can be used to analyze and externally promote the multiple benefits of green infrastructure, it remains to be seen how these green infrastructure options can be managed in parallel to their built counterparts. The next section presents an option for improving the internal management of natural assets.

3.5.2 Natural and Built Asset Management

To fully realize the value of the TBL framework, water utilities in Massachusetts need a tool to encourage and shape more monitoring and evaluation of the incremental benefits of watershed protection. One potential tool is an asset management program that integrates costs and benefits of natural and built infrastructure (Figure 3.16). Ideally, an asset management program would include all watershed lands and the maintenance and water quality implications of events on that landscape. For example, this program would quantify and record sampling results from storm water runoff events or forest management actions for each watershed property. This would enable the water utility to collect and track information about ownership and management of land for water quality.

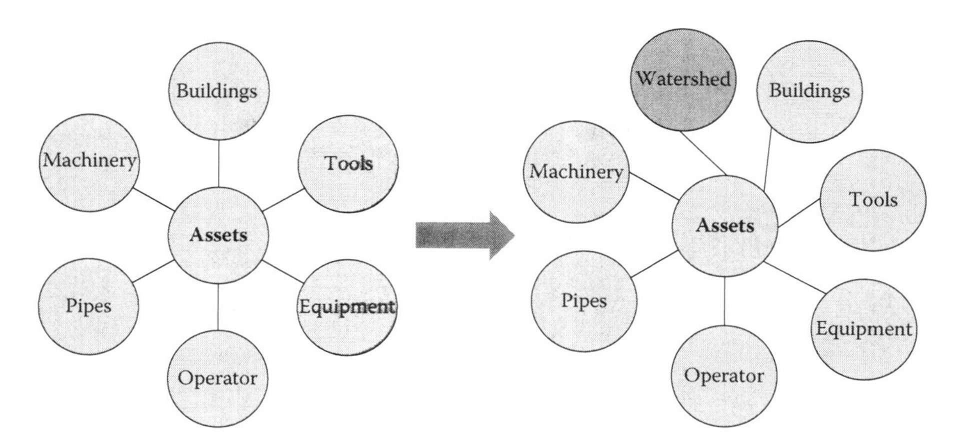

Figure 3.16 A conceptual model of current assets considered in water utility asset management plans and inclusion of watershed properties in a more holistic water utility asset management plan (right) (From adapted from EPA 2010).

Linking assets with events and accurately calculating risks to the system by internalizing true social, environmental, and financial costs and benefits of alternative management scenarios via the TBL approach should allow water suppliers to better address the incremental costs of watershed protection versus built water treatment options.

3.5.3 Fully Integrated Management of a Hypothetical Watershed Property in Massachusetts

To illustrate the TBL approach for source watershed protection, a hypothetical case of a small farm in Massachusetts is presented. Natural assets and events/parameters that impact raw water quality can be monitored by a water utility using monitoring systems described previously. An example of an asset management database entry for a hypothetical watershed property in Massachusetts (Ashton Farm) demonstrates that a strategically located gauge on the property allows the water utility to track change in runoff water quality during various storm events, and before and after certain land use changes (Table 3.7). This would allow the utility to better understand the connection between how they manage their specific watershed and the resulting surface water quality.

This asset management approach, combined with the traditional asset management of built assets, can create a broader application of the TBL tool (Table 3.8). This hypothetical TBL analysis shows, qualitatively, that management of this property as a forested landscape would have greater overall net benefits to water users than development of the property. Inherent in this assessment is the use of the data collected in the asset management program to inform the analysis—for example, without a clear understanding of turbidity trends in surface water runoff from Ashton Farm, managers could not accurately assess the effect of development on the property value for improved downstream water quality.

Table 3.7 Sample Asset Management Database Entry for a Hypothetical Privately Owned Small Farm in a Source Water Watershed in Massachusetts

Sample Watershed Property Asset Management Entry Ashton Farm Asset and Events[a]	
Property Profile	
Address:	120 Carnel Hump Road, Petersham, MA[b]
Size:	25 acres
Ownership:	Private, In Fee
Contact and Communication Details	TBD
Special Deed Encumbrances	None
Jurisdictional Boundaries	Town of Petersham, MA; Worcester County
Stakeholder Groups	Trustees of Reservations, Massachusetts Forest Watch, Audubon Society, Harvard Forest
Built Characteristics	
Existing Structures	House, garage
Sewerage Condition	Septic
Past Projects/Renovations	Septic installed 1973, garage apartment added 1985
Land Use Characteristics	
Average Riparian Buffer Width	150 feet
Landcover	Second growth hardwood forest
Forest	60%
Inactive Cropland	10%
Active Cropland	0%
Impervious Cover	1%
Riparian	28%
Wetland	1%
Historical Use	Pastureland
Current Use	Residential
Surface Water Characteristics	
Watershed	Quabbin
Sub-Watershed	East Branch of Fever Brook
Watershed Location	Eastern bank of East Branch of Fever Brook
Soil Type	Montauk-Canton, extremely stony
Slope	2%
Underlying Geology	Northbridge granite
Water Bodies	Pond, brook, bog
Miles of Water Body Frontage	1/2 mile
Ground Water Characteristics	No wells, no wellhead protection areas
Stream Classification	Riffle pool, free flowing[c]

(continued)

Table 3.7 Sample Asset Management Database Entry for a Hypothetical Privately Owned Small Farm in a Source Water Watershed in Massachusetts (continued)

Surface Water Quality Characteristics	
Real-time Monitoring	Ashton Gauge Link
Monitoring Trends (high and low flow)	
Turbidity	Link to Trend Chart
Nitrogen	Link to Trend Chart
Phosphorus	Link to Trend Chart
Total Organic Carbon	Link to Trend Chart
Dissolved Organic Carbon	Link to Trend Chart
Turbidity	Link to Trend Chart
Alkalinity	Link to Trend Chart
Conductivity	Link to Trend Chart
Temperature	Link to Trend Chart
pH	Link to Trend Chart

[a] This is a hypothetical property and list of assets designed to show incorporation of natural water utility assets in an asset management database. This list is not exhaustive, and all the data in this table are fictional.

[b] Link to Real Estate databases for property value and zoning details.

[c] Link to theoretical statewide standardized classification system designed to assess river health.

3.6 SUMMARY TRENDS, RECOMMENDATIONS, AND CONCLUSIONS

The comparison of watershed management and protection efforts of the Worcester and Boston water supply systems demonstrates that stakeholders invested in drinking water supply believe, qualitatively, that the multibarrier approach, with emphasis on source water protection, is the best method toward achieving the production of safe drinking water. These two case studies provide insight into alternative approaches to satisfying demand for water quantity and quality. Both MWRA and Worcester rely on integrating source water protection and engineered treatment to ensure high-quality drinking water.

Watershed forests are the first barriers in this multibarrier approach. Leveraging forests to serve as a natural water filter creates a unique opportunity for land conservation organizations, public governments providing recreation, and drinking water suppliers to collaborate. We believe this approach will be increasingly important as our ability to detect contaminants improves, as regulations for safe drinking water become more stringent, and as public health concerns continue to require the safest water at the lowest cost. As these trends continue and public sector dollars remain limited, pressures on water suppliers will be made to quantitatively prove that land use restrictions will protect against emerging contaminants. If not, an alternative will be to invest in emerging technologies for contaminant removal, and maybe both. Ultimately, to maintain a sustainable program in land management for source water protection, water suppliers need an optimization tool that integrates asset management with TBL assessment to account for all possible benefits of watershed management.

Table 3.8 Sample Triple Bottom Line Analysis for a Hypothetical Privately Owned Small Farm in a Source Water Watershed in Massachusetts; Hypothetical Results Show That the Management of the Property for Source Water Protection Would Have Greater Overall Benefits to the Water Utility's Public Constituency

Sample Watershed Property Triple Bottom Line Analysis Ashton Farm Land Use Analysis[a]	Developed[b]	Forested[c]
Social Benefits		
Cultural Preservation	–	+
Recreation	–	+
Open Space	–	+
Community Aestetics	–	+
Property Value	+	–
Environmental Benefits		
Fish and Wildlife Habitat	–	+
Riparian and Wetland Preservation/Restoration	–	+
Carbon Sequestration	–	+
Increased Resiliency/Resistance to Disturbance	–	+
Public Health Benefits		
Improved Water Quality	–	+
Improved Air Quality	–	+
Improved Mental Health	–	+
Economic Benefits		
Increased Housing	+	–
Avoided Costs of Gray Infrastructure	–	+
Decreased Operation and Maintenance (O&M) Costs	–	+
Energy	–	+
Treatment Chemicals	–	+
Labor	+	–
Service Interruptions	–	+

[a] This is a hypothetical property designed to illustrate application of a Triple Bottom Line (TBL) analysis to a drinking water quality protection case. This analysis is qualitative, but for further information on a qualitative TBL analysis see the referenced Stratus Consulting report.

[b] Hypothetical scenario in which Ashton Farm is developed. + indicates net positive effect to society in given benefit category, – indicates net negative effect to society in given benefit category.

[c] Hypothetical scenario in which Ashton Farm is converted to managed forest. + indicates net positive effect to society in given benefit category, – indicates net negative effect to society in given benefit category.

3.6.1 Main Points regarding the MWRA

- The watersheds of the MWRA are a product of abandoned agriculture. The height of agricultural settlement was around 1850 and was nearly 100 percent forested in 1950. Today, the Quabbin, Ware, and Wachusett Reservoir watersheds are more than 81 percent forested.
- Today the MWRA water supply system delivers water to more than 2.2 million people and 5,500 industrial users in 48 communities in the Boston metro area and three communities in central Massachusetts.
- The plan for Quabbin Reservoir was approved and filtration avoidance granted in 1989. However, because all water for the Boston metropolitan area must pass through Wachusett Reservoir, the entire system is at risk from this single, highly susceptible watershed.
- The Watershed Protection Act created a system for regulating activities (e.g., buffer zones along watercourses, limits on impervious surface, and restrictions on hazardous chemicals) that can impact source watershed health. The Watershed Protection Act also included a significant funding mechanism to acquire lands, whether in-fee or through development restrictions. Land procurement has been a central mechanism for source water protection within the Wachusett watershed. The program is funded primarily through ratepayer billing and state bonds.
- A guiding principle in land and forest management throughout the MWRA system is a focus on building redundancy into the existing natural forest biofilter. DWSP manages the forest to build resistance and resilience in the system, similar to redundancy incorporated in built treatment systems through multiple levels of filtrations and disinfection.

3.6.2 Main Points regarding Worcester DPW

- The city's first reservoir was built in 1860. As water demand continued to increase, the city continued to periodically add reservoirs until the construction of its 10th and final reservoir in 1950. These 10 reservoirs have a cumulative storage capacity of 7.4 billion gallons.
- The passage of the Safe Drinking Water Act (SDWA) in 1974 brought increased public and regulatory scrutiny to Worcester's water supply. This escalated attention combined with multiple turbidity and fecal coliform violations led to a public outcry.
- With private lands comprising approximately 70 percent of Worcester's water supply watersheds, the EPA ranks Worcester's water supply as highly susceptible to contamination. The construction of the Worcester Regional Airport presented a second major threat to Worcester's water supply.
- Worcester first began managing its watershed in the early 1980s. Worcester DPW is in the process at developing their first comprehensive forestry management plan for their 10 reservoirs.
- Since 2002, Worcester has strategically been acquiring land within their 10 reservoirs' watersheds to gain control over water supply lands. The city, through the state's Watershed Protection Act, has acquired 516 acres between 2002 and 2008.

- The main challenges to Worcester's land acquisition program are the looming elimination of the state's watershed protection program and the extensive size of Worcester's reservoir water supply system.
- Worcester's raw water has very low turbidity, and as a result, the construction of filtration plant did not require the installation of costly sedimentation basins.

3.6.3 Water Supply Trends into the Future

- Traditional land acquisition strategies will be met with a decreased financial capacity to control increasingly expensive real estate.
- These increasing land prices will require utilities to adopt creative land protection land conservation strategies (e.g., partnerships with conservation organizations) and/or require utilities to find ways to generate additional revenue from their conserved lands.
- Despite their public ownership, MWRA and Worcester DPW are subject to the same cost and efficiency constraints as private entities. Public systems do some mechanisms at hand for capital investment (e.g., bonds). However, they must still recoup operating costs in addition to expenses for infrastructure upgrades. Both the MWRA and Worcester systems are generally unsubsidized. In the case of MWRA, residents pay some of the highest per capita rates in the country.
- Water suppliers across the country are facing decreasing revenues in response to long-term trends of decreasing per capita water consumption.
- Changes in water quantity and quality concerns are perhaps some of the most pressing issues associated with climate change. In the case of Massachusetts, climate change is expected to have variable impacts relative to the size of the water systems changes within increases in gross annual precipitation but with greater degrees of variability—stronger dry periods and wet periods.
- There is a growing concern with emerging unregulated contaminants (pharmaceuticals and other organics), with demonstrated ubiquity of these chemicals in both ground and surface sources of drinking water.

3.6.4 Some Current and Suggested Improvements in Watershed Protection and Management

- Trust for Public Land, the Smart Growth Leadership Institute, the Association of State Drinking Water Administrators, and the River Network joined forces in a program called Protecting Drinking Water Sources, aimed at better integrating state land use policy, incentives, and drinking water programs.
- Monitoring raw water quality is currently driven by minimal requirements and does not enable cost- and resource-efficient land management. More thorough and strategic monitoring of raw water quality within drinking water supplies can aid in developing strategies to link raw water quality with engineered treatment costs.
- Application of asset management (this includes people, tools, equipment, machinery and buildings) to water suppliers involves careful tracking of assets and events throughout asset life cycles. Worcester Water Department does not operate a stand-alone asset management program. The MWRA initiated a Facilities Asset Management Plan in 2000.

- The TBL approach reflects the fact that society and its enterprises typically are engaged in activities intended to provide the greatest total value to the communities they serve. These values extend well beyond the traditional financial bottom line that portrays only cash flows. MWRA and Worcester DPW also need to account for how they may generate values that contribute toward the social and environmental bottom lines.

REFERENCES

Barten, P.K., Y. Zhangm, P. Gregory, et al., (2010). Watershed Forest Management Information System: Forest-to-Faucet Partnership. http://www.forest-to-faucet.org/software_downloads1.html. Accessed June 11, 2012.

Barten, P.K., T. Kyker-Snowman, P.J. Lyons, T. Mahlstedt, R. O'Connor, and B.A. Spencer, (1998). Massachusetts: Managing a Watershed Protection Forest. *Journal of Forestry*. http://foreststofaucets.info/wp-content/uploads/2010/09/Barten_etal_JF98.pdf. Accessed June 11, 2012.

Daughton, C.G., (2004). Nonregulated Water Contaminants: Emerging Research. *Environmental Impact Assessment Review*, 24: 711–732.

DCR (Department of Conservation and Recreation, (2003). Watershed Protection Plan Update for the Wachusett Reservoir Watershed. Commonwealth of Massachusetts, Executive Office of Environmental Affairs, Bureau of Watershed Management.

DCR, (2007). Quabbin Reservoir Watershed System: Land Management Plan 2007–2017. Massachusetts Department of Conservation and Recreation, Office of Watershed Management.

DCR, (2008). Watershed Protection Plan Update. Massachusetts Department of Conservation and Recreation, Office of Watershed Management.

de la Cretaz, A. L., and P.K. Barten, (2007). *Land Use Effects on Streamflow and Water Quality in the Northeastern United States*. CRC Press, Boca Raton, FL.

EPA (U.S. Environmental Protection Agency), (2006, March). Origins and Fate of PPCPs in the Environment. Office of Research and Development. http://www.epa.gov/ppcp/pdf/drawing.pdf. Accessed June 11, 2012.

EPA, (2008). Presentation: Asset Management 101. http://water.epa.gov/infrastructure/drinkingwater/pws/cupss/upload/presentation_cupss_am101.pdf. Accessed June 11, 2012.

Ernst, C., (2004). Protecting the Source. Trust for Public Land. http://cloud.tpl.org/pubs/water-protecting-the-source-04.pdf. Accessed June 11, 2012.

Focazio, M.J., D.W. Kolpinb, K.K. Barnesb, E.T. Furlongc, M.T. Meyerd, S.D. Zauggc, L.B. Barbere, and M.E. Thurman, (2008). A National Reconnaissance for Pharmaceuticals and Other Organic Wastewater Contaminants in the United States—(II) Untreated Drinking Water Sources. *Science of the Total Environment*, 402: 201–216.

Fortin, J., (2004). *MWRA'S Facility Asset Management Program: A Case Study*. MWRA White Paper. http://www.bcwaternews.com/assetmgr/MWRA_paper.pdf

Foster, D.R., (1992). Land-Use History (1730–1990) and Vegetation Dynamics in Central New England, USA. *Journal of Ecology*, 80: 753–772.

Foster, D.R., and D.A. Orwig, (2006). Preemptive and Salvage Harvesting of New England Forests: When Doing Nothing Is a Viable Alternative. *Conservation Biology*, 20:959–970.

Freeman, J., R. Madsen, and K. Hart, (2008). Statistical Analysis of Drinking Water Treatment Costs, Water Quality, and Land Cover Characteristics. White Paper, Trust for Public Land. http://wren.palwv.org/library/documents/landnwater_9_2008_whitepaper.pdf. Accessed June 11, 2012.

Guerin, P.D., (2008, November 18). Letter to Massachusetts Department of Environmental Protection for Half-Log Disinfection Credit Request. Worcester Department of Public Works and Parks.

Guerin, P.D., (2010, April 6). Personal communication. City of Worcester Annual Consumption Trends with Trendline.

Hall, B., G. Motzkin, D.R. Foster, M. Syfert, and J. Burk, (2002). Three Hundred Years of Forest and Land-Use Change in Massachusetts, USA. *Journal of Biogeography*, 29:1319–1335.

Henderson, J., (2010, April 9). Personal communication. Stratus Consulting.

Howard, L.F., and T.D, Lee, (2002). Upland Old-Field Succession in Southeastern New Hampshire. *Journal of the Torrey Botanical Society*, 129: 60–76.

Keim, B.D., R.W. Dudley, G.A. Hodgkins, and T.G. Huntington, (2003). Changes in the Proportion of Precipitation Occurring as Snow in New England (1949–2000). *Journal of Climate*, 17 (13): 2626–2636.

Kirshen, P., M. Ruth, and W. Anderson, (2005). Climate Change in Metropolitan Boston. *New England Journal of Public Policy*, 20(2): 89–103.

Kyker-Snowman, T., (2000). Managing the Shift from Water Yield to Water Quality of Boston's Water Supply Watersheds. In: Dissemeyer, G.E. (Ed.), *Drinking Water from Forests and Grasslands*. USDA Forest Service General Technical Report SRS-39. Asheville, NC.

Likens, G.E., and F.H. Bormann, (1995). *Biogeochemistry of a Forested Ecosystem*. Springer-Verlag, New York.

MassDEP (Massachusetts Department of Environmental Protection), (2002). Source Water Protection and Assessment for Worcester DPW, Water Supply Division. http://www.mass.gov/dep/water/drinking/swapreps.htm. Accessed June 11, 2012.

MGL (Massachusetts General Laws), (2009). General Laws of Massachusetts. http://www.mass.gov/legis/laws/mgl/. Accessed June 11, 2012.

MMSD (Milwaukee Metropolitan Sewerage District), (2009). Fresh Coast Green Solutions: Weaving Milwaukee's Green and Gray Infrastructure into a Sustainable Future.

MWC (Massachusetts Watershed Coalition), (2003a). Kendall, Holden #1, and Holden #2 Source Water Protection Plan. Prepared for the Worcester Department of Public Works.

MWC, (2003b). Kettle and Lynde Brook Source Water Protection Plan. Prepared for the Worcester Department of Public Works.

MWC, (2003c). Pine Hill Source Water Protection Plan. Prepared for the Worcester Department of Public Works.

MWC, (2003d). Quinapoxet Source Water Protection Plan. Prepared for the Worcester Department of Public Works.

MWRA (Massachusetts Water Resources Authority), (2006). Metropolitan Boston's Water System History. http://www.mwra.state.ma.us/04water/html/hist1.htm. Accessed June 11, 2012.

MWRA, (2008, January 30). MWRA Receives the New England Water Environment Association's Asset Management Award. http://www.mwra.com/01news/2008/013008amaward.htm. Accessed June 11, 2012.

MWRA, (2009a). Annual Water and Sewer Retail Rate Survey. Community Advisory Board to the Massachusetts Water Resources Authority, 2009. http://archives.lib.state.ma.us/handle/2452/46899. Accessed June 11, 2012.

MWRA, (2009b). Facts about MWRA's FY2010 Rates and Budget. http://www.mwra.state.ma.us/finance/rates/fy2010/ratefacts/ratesfacts.htm. Accessed June 11, 2012.

MWRA, (2010). Board of Directors Report on Key Indicators of MWRA Performance for Second Quarter FY2010. http://www.mwra.state.ma.us/quarterly/orangenotebook/fy2010/q2.pdf. Accessed June 11, 2012.

Nearing, M.A., (2005). Modeling Response of Soil Erosion and Runoff to Changes in Precipitation and Cover. *Catena,* 61(2–3): 131–154.

Nearing, M.A., F.F. Pruski, and M.R. O'Neal, (2004). Expected Climate Change Impacts on Soil Erosion Rates: A review. *Journal of Soil and Water Conservation*, 59(1): 43–50.

Protecting Drinking Water Sources, (2010). Enabling Source Water Protection. http://www.landuseandwater.org/. Accessed April 20, 2010.

Quabbin, L.M.P. 2007. Quabbin Reservoir Watershed System: Land Management Plan 2007–2017. Section 5: Management Plan Objectives and Methods.

Raup, H.M., (1966, April). The View from John Sanderson's Farm: A Perspective for the Use of Land. *Forest History*, 10(1): 2–11.

Richardson, S.D., (2007). Water Analysis: Emerging Contaminants and Current Issues. *Analytical Chemistry*, 79: 4295–4324.

Satterlund, D.R., and P.W. Adams, (1992). *Wildland Watershed Management*. Wiley, New York

Shafer, K., (2010, February 11). Personal communication. Milwaukee Metropolitan Sewerage District.

Stearns, D.J., (2000). United States District Court of Massachusetts, *United States of America v. Massachusetts Water Resources Authority, and Metropolitan District Commission*.

Stone, M., (2009, April 15). Re: Letter to Massachusetts Department of Environmental Protection for Half-Log Disinfection Credit Request. Massachusetts Department of Environmental Protection.

Stratus Consulting, (2009). A Triple Bottom Line Assessment of Traditional and Green Infrastructure Options for Controlling CSO Events in Philadelphia's Watersheds, Final Report.

Swank, W.T., and D.A. Crossley Jr. (Eds.), (1988). Forest Hydrology and Ecology at Coweeta. *Ecological Studies,* 66: 17–31.

Trust for Public Land, (2005). Source Protection Handbook: Using Land Conservation to Protect Drinking Water Supplies. http://cloud.tpl.org/pubs/water_source_protect_hbook.pdf. Accessed June 11, 2012.

Trust for Public Land, (2009). An Action Plan to Protect Maine's Drinking Water Sources: Aligning Land Use and Source Water Protection. http://www.smartgrowthamerica.org/documents/maines-drinking-water-sources.pdf. Accessed June 11, 2012.

Walton, B., (2010, April 19). U.S. Urban Residents Cut Water Usage; Utilities Are Forced to Raise Price. Circle of Blue Waternews. http://www.circleofblue.org/waternews/2010/world/u-s-urban-residents-cut-water-usage-utilities-are-forced-to-raise-prices/. Accessed June 11, 2012.

Worcester Department of Public Works, (1990). Water Operations. Annual Report ending June 30, 1990.

Worcester Department of Public Works, (2008a). Water Treatment Plant. http://www.ci.worcester.ma.us/dpw/water-sewer-operations/water-treatment-plant. Accessed June 11, 2012.

Worcester Department of Public Works, (2008b). Water Quality Report, Water Operations.

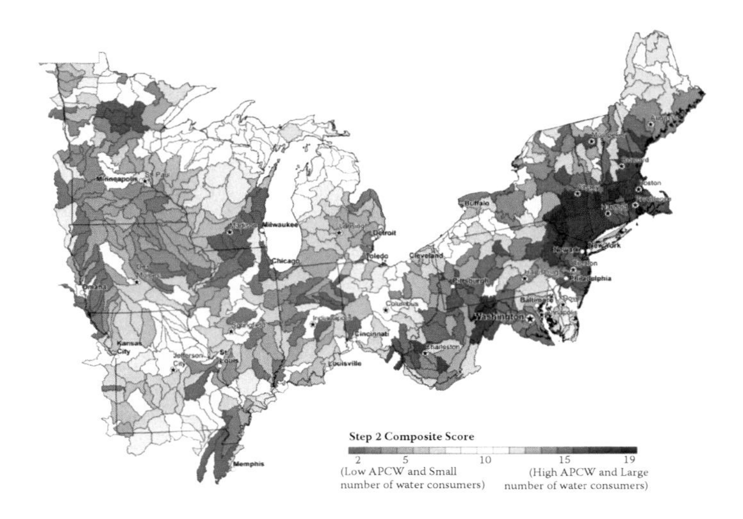

Figure 1.1 A map illustrating the importance of watersheds for drinking water supplies for each of the 540 watersheds in the Northeast and Midwest. It highlights those areas that provide surface drinking water to the greatest number of consumers. The higher a watershed's ability to provide drinking water, the darker brown it appears on the map and the higher its Ability to Produce Clean Water (APCW) score (From USFS, 2009).

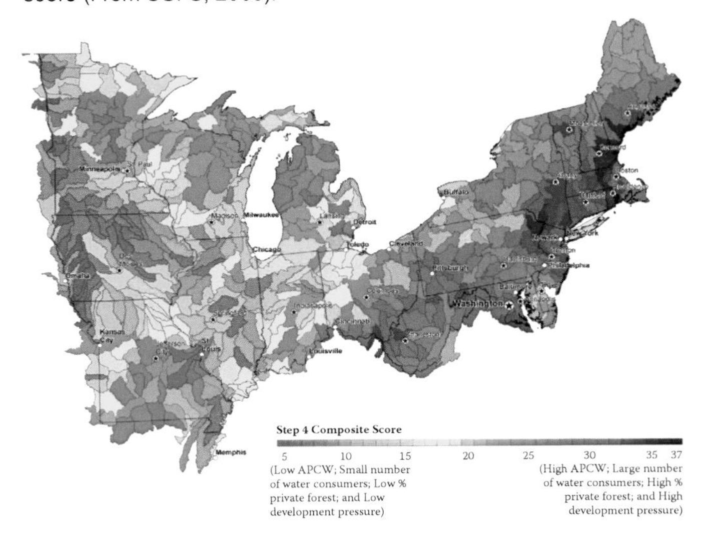

Figure 1.2 A map showing the development pressure on forests and drinking water supplies. The map combines data on the ability to produce clean water, surface drinking water consumers served, percent private forest land, and housing conversion pressure, to highlight important water supply protection areas that are at the highest risk for future development. The greater a watershed's development pressure, the more blue it appears on the map, and the higher its Ability to Produce Clean Water (APCW) score (From USFS, 2009).

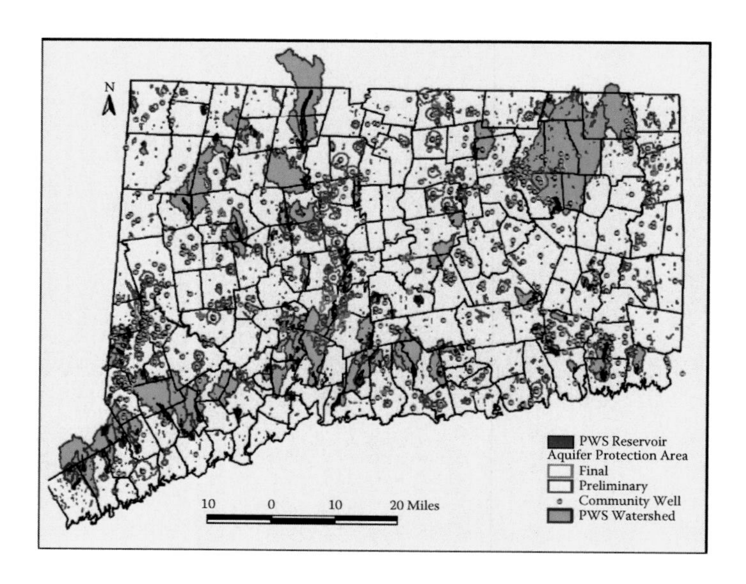

Figure 2.1 Map of Connecticut watersheds, illustrating surface watershed lands for cities and towns (green), community wells (gold) and preliminary (blue) and final (red) aquifer protection areas (From CT DPH, 2005).

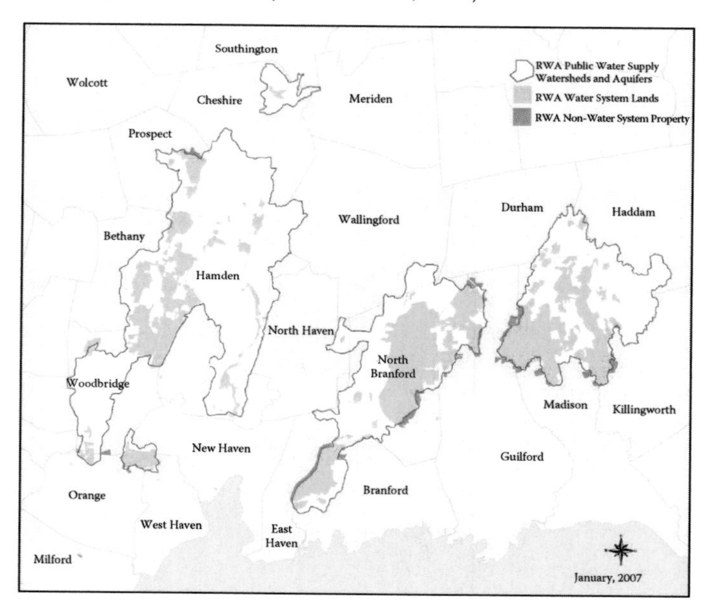

Figure 2.3 SCCRWA watershed lands. Green indicates land owned by SCCRWA. White indicates land within drinking water watersheds, about 3000 acres of which SCCRWA would like to purchase or protect. Orange indicates SCCRWA-owned land that is outside of their priority watershed protection areas (Class III), about 900 acres, that SCCRWA would like to sell (From SCCRWA, 2007b).

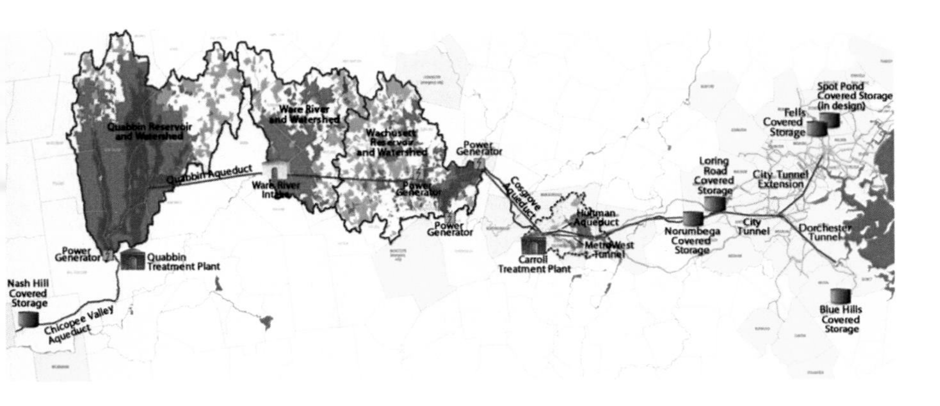

Figure 3.4 A schematic showing the general layout of the watersheds and infrastructure for the Boston water supply system.

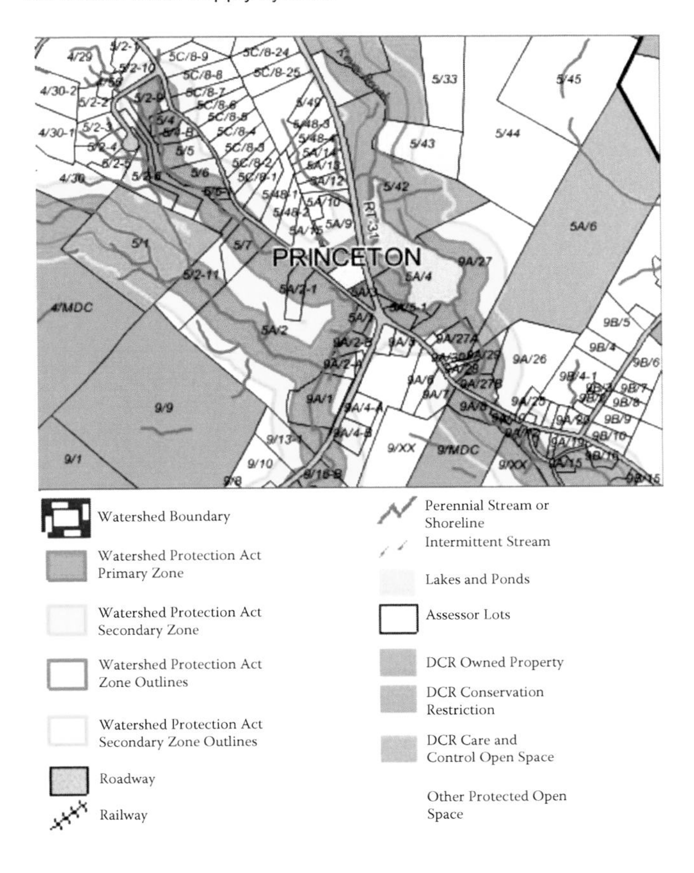

Figure 3.13 An example of the Watershed Protection Act zones taken from DEP interactive web resource for public access (From MassGIS, 2010).

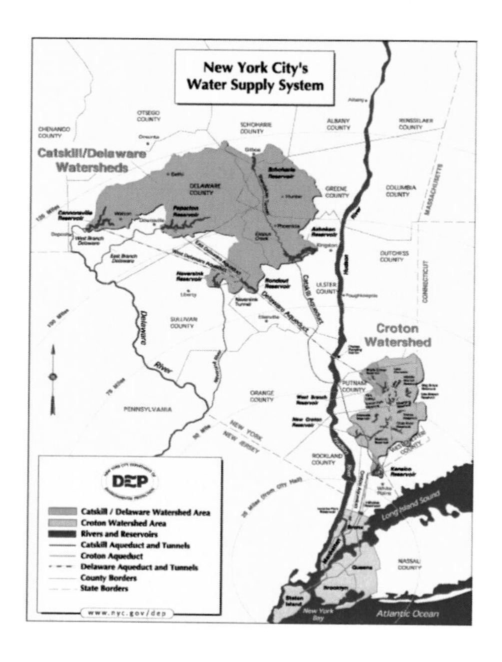

Figure 4.1 Map of the New York City water system (From DEP, 2007).

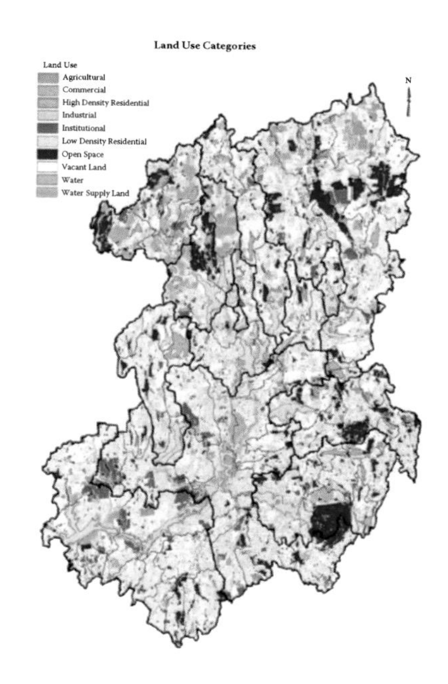

Figure 4.2 Croton land use (From Moffett et al., 2003).

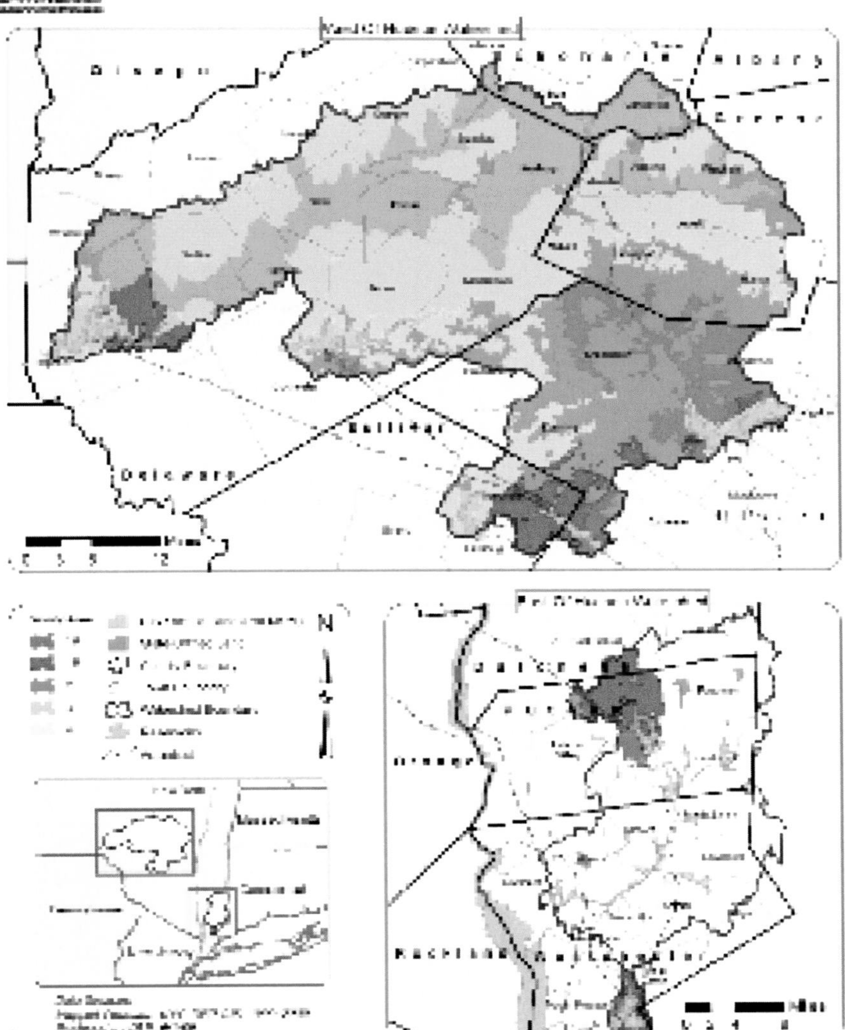

Figure 4.5 Priority areas for New York City Land Acquisition and Stewardship Program. NYC DEP has assigned each property a priority designation of 1–4 depending on its distance to terminal reservoirs and intakes, and the water's travel time to consumption (From DEP FEIS, 2010).

Table 4.2 Sample Matrix Layout

		Environmental Solutions							Engineered Solutions	
Categories	Issues	Agriculture Program	LAP	SMP	Forestry Program	WWTP	Septic Prog		Filtration	Disinfection
Public Health	Pathogens	*			*	*	*		*	*
	Nutrients	*			*	?	?		*	
	Turbidity	*			*	?	?		*	
	PPCPs					?			?	
	...									
Economic	Infrastructure co	*			*	?	?			
	Property values	*			*	?	?			
	Job creation	*			*	*	*		*	
	...									
Social	Aesthetics	*			*	?	?			
	Quality of life	*			?	?	?		*	
	Recreation				?	?	?			
	...									
Environmental	Air Quality	?			*					
	GHG mitigation	?			*	?				
	Ecosystem Health	*			*	*				
	...									

Note: Green indicates existing program coverage, while yellow indicates possible program coverage, red indicates a coverage gap, and a white or blank cell indicates issues are not applicable to the listed environmental or engineered solution, or that data is lacking. This example includes additional blank rows and columns to highlight the importance of flexible, adaptive program management. (Adapted from DEP WPPS, 2011).

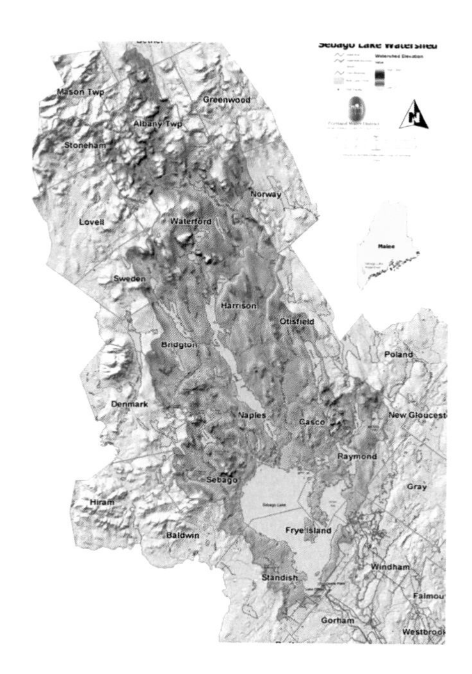

Figure 5.1 Sebago Lake watershed (From PWD, 2009).

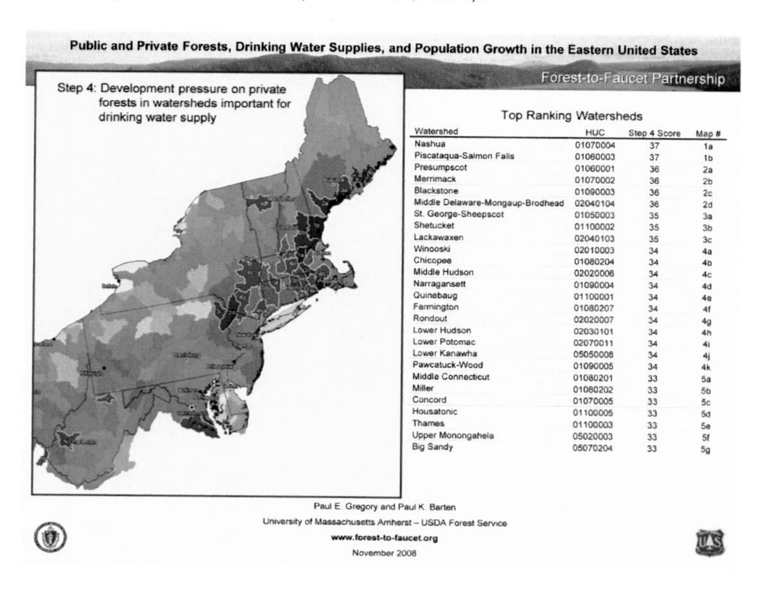

Watershed	HUC	Step 4 Score	Map #
Nashua	01070004	37	1a
Piscataqua-Salmon Falls	01060003	37	1b
Presumpscot	01060001	36	2a
Merrimack	01070002	36	2b
Blackstone	01090003	36	2c
Middle Delaware-Mongaup-Brodhead	02040104	36	2d
St. George-Sheepscot	01050003	35	3a
Shetucket	01100002	35	3b
Lackawaxen	02040103	35	3c
Winooski	02010003	34	4a
Chicopee	01080204	34	4b
Middle Hudson	02020006	34	4c
Narragansett	01090004	34	4d
Quinebaug	01100001	34	4e
Farmington	01080207	34	4f
Rondout	02020007	34	4g
Lower Hudson	02030101	34	4h
Lower Potomac	02070011	34	4i
Lower Kanawha	05050008	34	4j
Pawcatuck-Wood	01090005	34	4k
Middle Connecticut	01080201	33	5a
Miller	01080202	33	5b
Concord	01070005	33	5c
Housatonic	01100005	33	5d
Thames	01100003	33	5e
Upper Monongahela	05020003	33	5f
Big Sandy	05070204	33	5g

Figure 5.11 The Presumpscot River basin, which contains the Sebago Lake watershed and the Crooked River watershed, is ranked third among Eastern New England water supply watersheds at risk for development pressure on private forests (From Gregory and Barten, 2008).

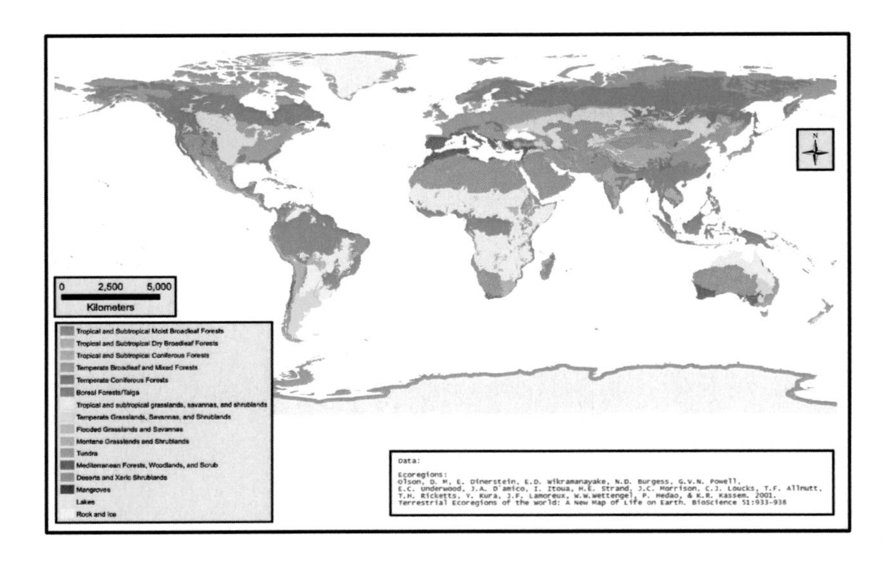

Figure 7.1 A worldwide map of biomes depicting the major vegetation types. The temperate broadleaf and mixed biome and the tropical and subtropical moist broadleaf biome represent only a small part of the earth's terrestrial systems but account for a disproportionate amount of the world's population and rainfall (Modified after Olson et al., 2001).

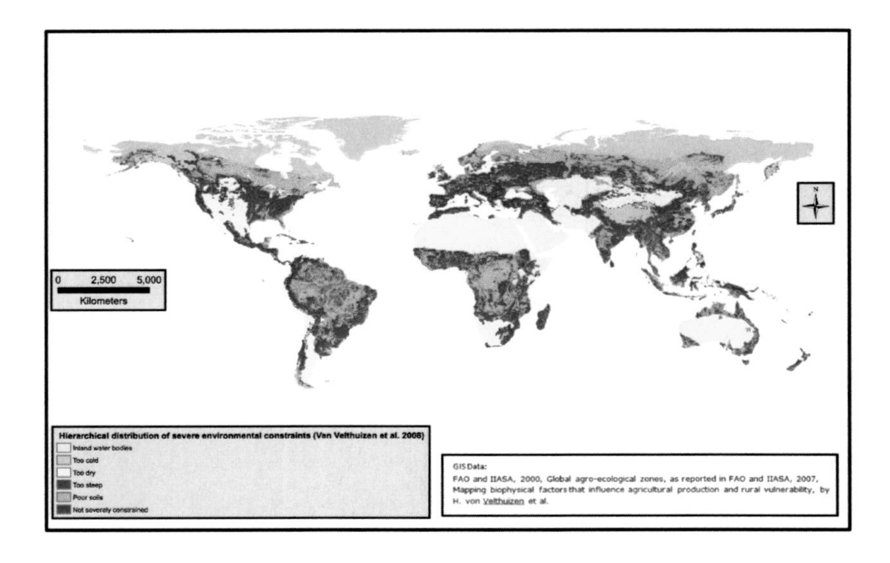

Figure 7.2 A worldwide map of climate and environmental constraints. The map depicts the results of a GIS analysis identifying areas of the world that are most favorable for growing plants. These areas would also be most favorable for developing resilient forested biofilters around drinking water systems (Modified after van Velthuizen et al., 2008, with permission).

CHAPTER 4

New York City Watershed Management

Past, Present, and Future

Justin Freiberg, Xiaoting Hou, Jason Nerenberg, Fauna Samuel, and Erin Derrington

CONTENTS

4.0 EXECUTIVE SUMMARY

The New York City (NYC) water supply system includes three upstate watersheds that span 1,972 square miles. The system's history is one of ongoing involvement and investment by the city in its upstate watersheds and the communities within them. The New York City Department of Environmental Protection (NYC DEP), the agency responsible for the operation, management, and maintenance of water supply infrastructure and protection, has engaged in an extensive land protection program to maintain high-quality water supplies. NYC DEP has worked to leverage common interests and build mutually beneficial relationships between upstate and downstate landowners and encourage source water protection efforts.

The NYC watershed management model demonstrates that source water protection is an important part of any drinking water supply strategy. The NYC DEP has maintained a filtration avoidance determination for the Catskill–Delaware system, which typically provides roughly 90 percent of New York City's drinking water; the Croton system, which remains offline until a finalization of the required filtration plan, provides the other 10 percent. The importance of these programs is emphasized by the high cost of building, operating, and maintaining a filtration plant for the Catskill–Delaware water supply. This chapter summarizes past and current drinking water management in NYC and highlights emerging challenges to watershed health as well as possible solutions. This model demonstrates the following:

- Land stewardship in source water areas can yield substantial benefits to water quality and supply systems.
- Incentive-laden partnership-based programs can support multiple resource management values and produce flexible conservation programs.
- Ongoing investment and comprehensive community engagement is critical to maintaining high-quality water systems.

The fact that the majority of New York City's water supply has remained unfiltered this long is a testament to both out-of-the-box thinking and strategic, continued investment. Nevertheless, as risks to sustainable water security and regulatory standards increase, proactive community-based planning will be essential to address future challenges.

4.1 INTRODUCTION

This chapter assesses the New York City water system to evaluate how, why, when, and where it makes environmental, economic, and social sense to protect and manage upland forests to maintain high-quality drinking water as a downstream service. To this end, we first highlight the history of land use, water extraction, and environmental regulation in the area to illuminate the decision-making process that led to extensive watershed protection efforts. Next, we provide a current profile of land protection, watershed management, and use in the NYC system. We then discuss details of current and future watershed management challenges and potential solutions and articulate lessons learned from this model that can be widely applicable to water system management. Finally we provide brief conclusions and recommendations.

4.2 HISTORY OF NEW YORK CITY WATERSHED

4.2.1 Land Use and Development History

To assess the New York City water supply and watershed protection efforts of NYC DEP and its partners, it is essential to understand the land use history and current land use patterns within the three upstate watersheds, known as the Catskill, Delaware, and Croton systems (Figure 4.1).

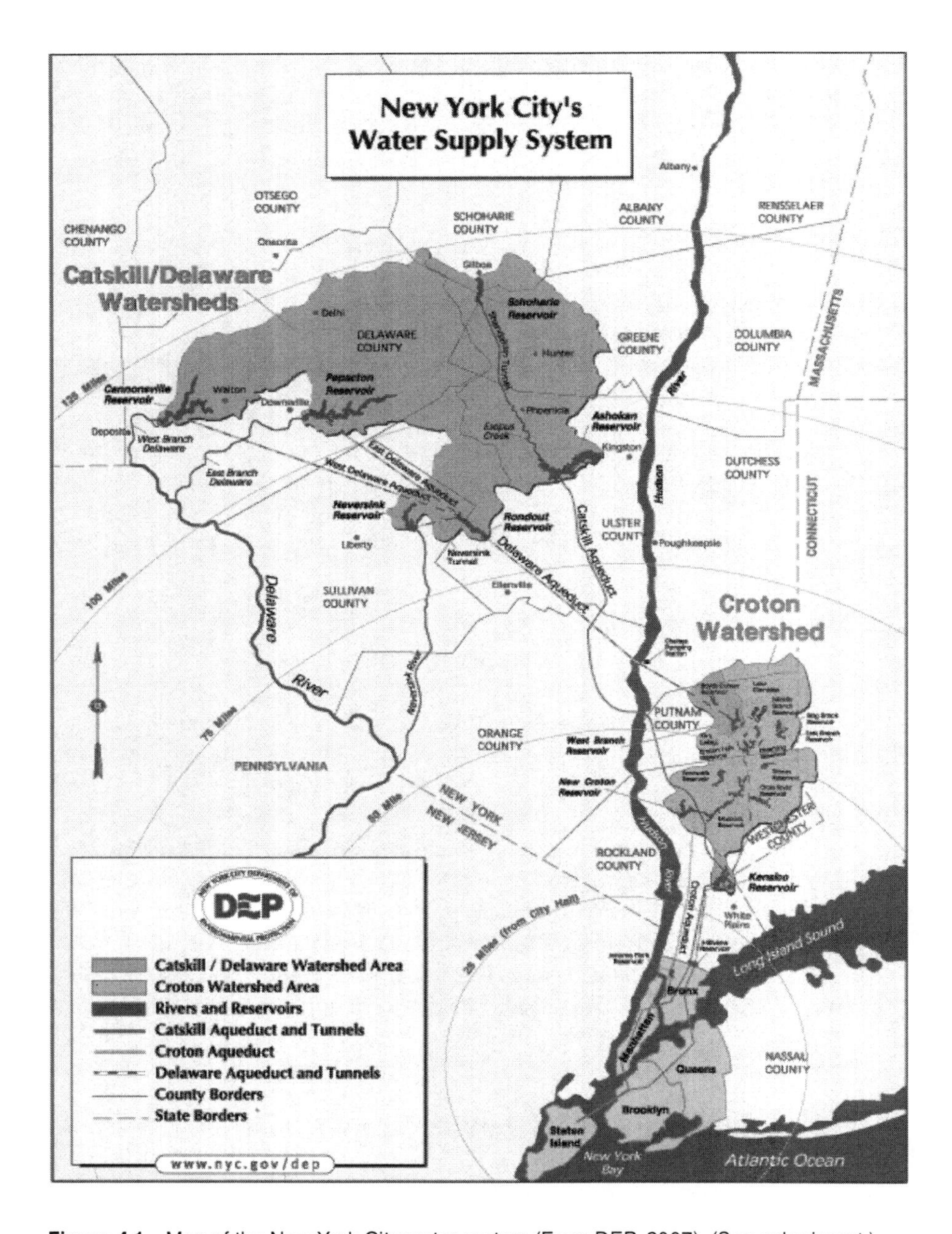

Figure 4.1 Map of the New York City water system (From DEP, 2007). (See color insert.)

4.2.1.1 The Croton Watershed

In response to water contamination and destructive fires, New York City started looking upstream for fresh water sources at the end of the eighteenth century (ASCE, 2005-11). After exploring supply options, the city focused supply expansion efforts on the nearby Croton River. The Croton watershed flows through Manhattan Hills, an area of hills and valleys ranging from sea level to nearly 900 feet above sea level (US FWS, 1997). It encompasses a network of 74 sub-basins, comprised of primarily mixed deciduous forests that begin in the more mountainous northern Hudson Highlands region and drain through the rolling hills of the Manhattan Prong in the southern region (Wilder and Kiviat, 2008). By 1842, the 40.5-mile gravity-fed aqueduct of the Croton water supply system delivered its first water (ASCE, 2005-11) (see Figure 4.1). In 1900, the population within the original 312-square-mile Croton watershed area was about 20,000 (Westchester County, 2009). By 2000, the permanent population within the expanded and modified 375-square-mile watershed was over 100,000, with a population density of about 352 people per square mile (NRC, 2000). Less than 15 percent of the area remains wooded and undeveloped (Burns, 2005).

Today, in part due to its proximity to New York City that makes the area desirable for commuters coupled with an absence of watershed protection during early expansion, nearly 80 percent of the Croton watershed is developed (Warne, 2010b). Development restrictions due to steep slopes, wetlands, and riparian zones apply to about 28 percent of currently developed land and 40 percent of undeveloped areas (Moffett et al., 2003). Although much of the region is composed of low-density residential development, land uses also include high-density town centers and lake communities; agricultural, commercial, and industrial zones; as well as open space, water protection areas, and undeveloped land (Moffett et al., 2003) (Figure 4.2). Substantial development within the Croton area has resulted in forest loss and high impervious surface coverage as well as associated runoff and water quality concerns (Wilder and Kiviat, 2008). The Catskill–Delaware watershed area offers a stark contrast in development history and current land use trends.

4.2.1.2 The Catskill–Delaware Watershed Systems

Although several development projects expanded the capacity of the Croton system throughout the nineteenth century, by the twentieth century city planners acknowledged the need for increased supplies. To accommodate growing demand, in the 1900s, the NYC water system began to expand to the Catskill and Delaware systems, over 100 miles northwest of the growing metropolis. The Catskill Water Supply System was completed in 1927 and the Delaware Water Supply System was completed in 1967 (NRC, 2000). The Catskill watershed is mountainous and drains to the Hudson River, while the rolling hills of the Delaware region drain to the Delaware River (US FWS, 1997). The Catskill and Delaware watershed systems, typically referred to as the Catskill–Delaware because of the commingling of their

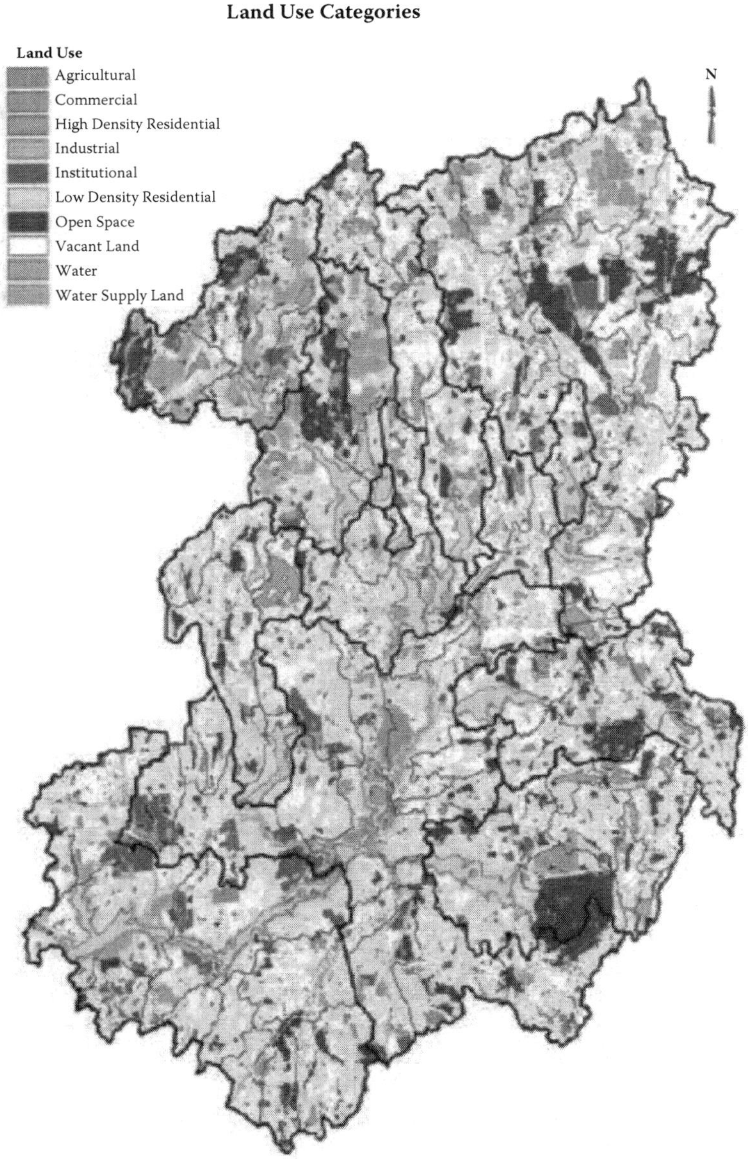

Land Use Categories

Land Use
Agricultural
Commercial
High Density Residential
Industrial
Institutional
Low Density Residential
Open Space
Vacant Land
Water
Water Supply Land

N

Figure 4.2 Croton land use (From Moffett et al., 2003). (See color insert.)

waters in the Kensico Reservoir, cover about 1,600 square miles of the Catskill Hills and the Hudson River Valley (NRC, 2000). Located on the west side of the Hudson River, this area does not offer a convenient commute to New York City and thus did not experience the same development pressures as the Croton region on the east side of the river (Figure 4.1).

A comprehensive National Research Council report details past and current land uses in the region (NRC, 2000). Early intensive land use in both watersheds consisted of trapping for fur, leather tanning, and logging. Dairy farming developed in both watersheds during the 1800s, but the post–World War II expansion of dairy farming in western New York and the Midwest led to large-scale abandonment of many of these farms as well as the opportunistic purchase of the land by the state to form the Catskill State Park. Today, the most intensive agricultural land uses are located in the valley bottoms of the Delaware watershed. Overall, agriculture accounts for roughly 5 percent of the total land use in the Catskill–Delaware watershed, ranging from 1 percent in the Ashokan and Neversink basin to 11 percent in the Cannonsville. High-density residential areas and commercial and industrial land use account for less than 1 percent of the total land area across all basins of this system. Low-density residential areas comprise anywhere from 4 percent of the area in the Neversink basin to 21 percent in the Cannonsville basin and 16 percent overall (NRC, 2000).

Although the region experienced high rates of development and construction of second homes in the 1960s–1980s, development has now slowed significantly (DEP, 2006). Population modeling shows that despite rapid growth in the West-of-Hudson counties comprising the Catskill–Delaware system, actual growth within the watershed boundaries has been minimal (Finnegan, 1997). In fact, the population of the watershed in 1990 was only 235 people greater than the estimated population in 1860 (NRC, 2000). The NYC DEP's 2006 Long-Term Watershed Protection Program reported that when the Land Acquisition Program began in 1996, New York City owned around 3.5 percent of the land in the Catskill–Delaware watershed (DEP, 2006). As of June 2011, the city has a controlling interest in and manages 14.8 percent of the land in the area, including conservation easements, with an additional 19.6 percent of the watershed protected through the state's Catskill Forest Preserve program (DEP, 2006). Deciduous forests remain the dominant land cover, ranging from 52 percent in the Cannonsville basin to 93 percent in the Ashokan, with an average of 68 percent overall (NRC, 2000). In 2006, 75 percent of the Catskill–Delaware watershed was classified as forested (DEP, 2006).

4.2.2 The History of the Engineered Water Supply System

The history of New York City water management demonstrates two trends: (1) a growing need for water, and (2) supply solutions obtaining water from increasingly faraway sources. The DEP reports that originally Manhattan was supplied with water from shallow wells from the 1600s to 1776, when the first reservoir was constructed on Broadway in the lower east side (DEP, 2006a). Water from this early reservoir was distributed throughout the city in hollow logs placed beneath the streets. In the early 1800s, additional reservoirs were constructed and water was dispensed through cast iron pipes to accommodate the growing population. In an effort to increase the supply of water to its residents, the downstream city decided to dam the Croton River to create the Old Croton Reservoir and construct an aqueduct to carry water from what is now Westchester County. Although expensive at the time, this infrastructure investment established a system providing city residents long-term access to the relatively pristine waters of upstate New York. Beginning in 1842, approximately 90 million gallons of water traveled through the Old Croton aqueduct daily to distribution reservoirs in Central Park and at 42nd Street (DEP, 2012; NRC, 2000).

As the Department of Environmental Protection recounts, when the population of New York City grew and demand increased, additional storage reservoirs near the Croton River were constructed, including Boyd's Corner in 1873 and Middle Branch in 1878 (DEP, 2006a). To increase supply, construction of the New Croton Aqueduct began in 1885 and was completed in 1893. Over the years the Croton water system was further expanded to span 375 square miles, including 12 reservoirs and three controlled lakes connected via open-channel streams and rivers, collecting water in the New Croton Reservoir before distributing it to the thirsty city downstream (DEP, 2006a) (Figure 4.3).

The Board of Water Supply was formed in 1905 and its first major project was to oversee the development of the Catskill region of upstate New York to further expand water supplies (DEP, 2006a). The board planned and constructed the Ashokan Reservoir through impounding Esopus Creek, as well as the 75-mile Catskill Aqueduct, which connects to the Kensico Reservoir in Westchester County (DEP, 2006a) (Figure 4.1). A roadway weir divides the Ashokan Reservoir into two basins: the upstream West Basin where the majority of settling occurs and the predominantly less turbid water spills over the weir into the downstream East Basin (DEP WPPS, 2011). Water then flows from the East Basin to the Kensico Reservoir through the Catskill Aqueduct. To address threats of high turbidity when extreme storms carry large loads of suspended sediment into the system that may be unable to settle before entering the Catskill Aqueduct, engineers designed and built a facility to hold chemical coagulants that could be added to enhance settling of particles entering the Kensico Reservoir (DEP, 2006). These projects were completed and the reservoirs placed into service in 1915. Less than 15 years later, the Catskill System was further expanded to include the Schoharie Reservoir and the 18-mile Shandaken Tunnel, which connects to a 12-mile stretch of Esopus Creek and then on to the West Basin of the Ashokan Reservoir (DEP, 2006) (See Figure 4.1). Ensuring increasing drinking water supplies remained a constant management concern.

Prior to the completion of the Catskill System in 1928, the Board of Water Supply proposed a plan to develop the upper section of the Rondout watershed and the Delaware River tributaries located within New York State lines. Construction of the Delaware System began in 1937 and was completed in five stages: the Delaware Aqueduct in 1944, the Rondout Reservoir in 1950, the Neversink Reservoir in 1954, the Pepacton Reservoir in 1955, and finally the Cannonsville Reservoir in 1964 (DEP, 2006). The Cannonsville, Pepacton, and Neversink Reservoirs supply water

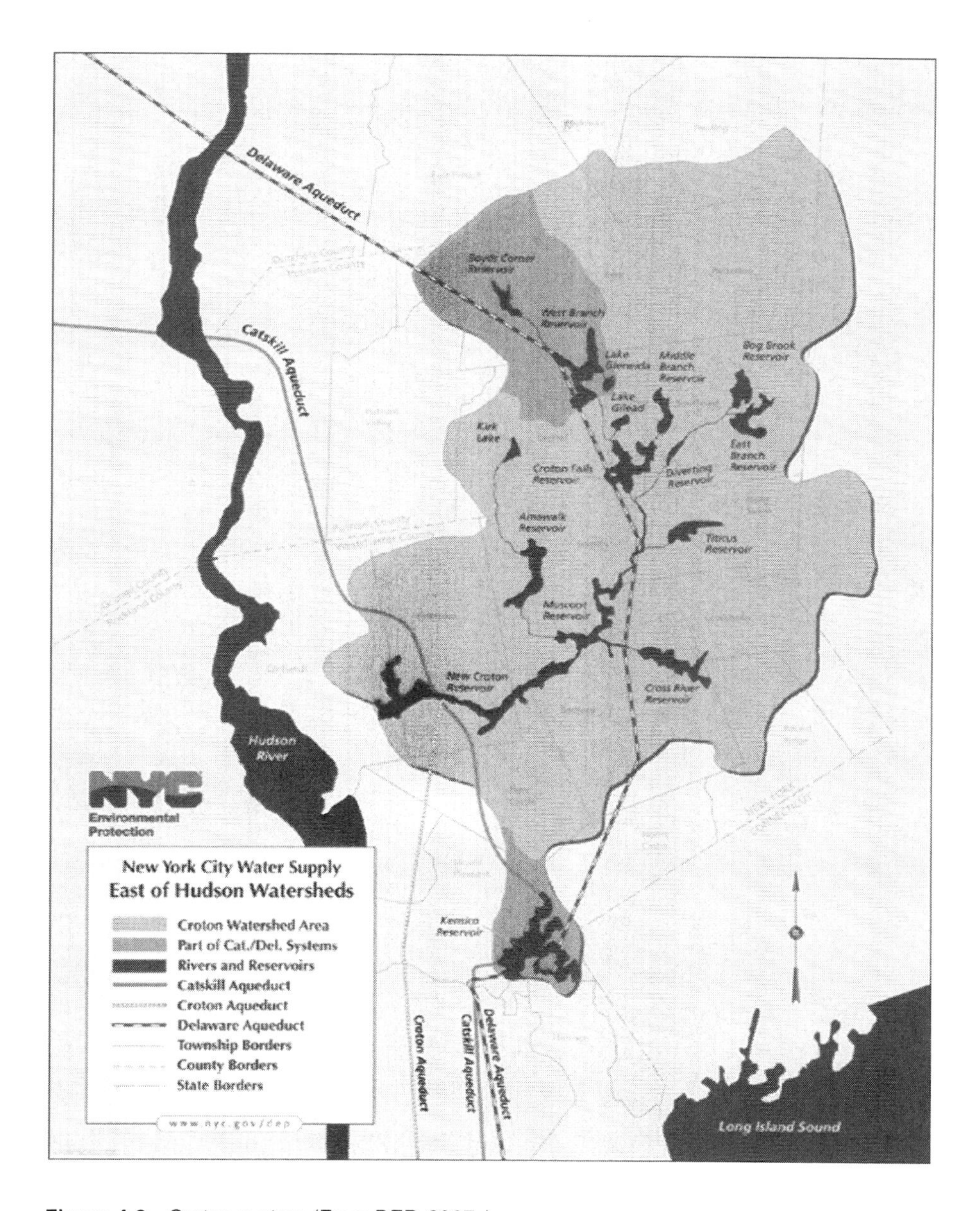

Figure 4.3 Croton system (From DEP, 2007a).

to the Rondout Reservoir through the West Delaware, East Delaware, and Neversink Tunnels (Figure 4.4).

After collection in the Rondout Reservoir, water flows through the 45-mile Delaware Aqueduct to the West Branch Reservoir in Putnam County and then on to the Kensico Reservoir to join with water from the Catskill source to supply drinking

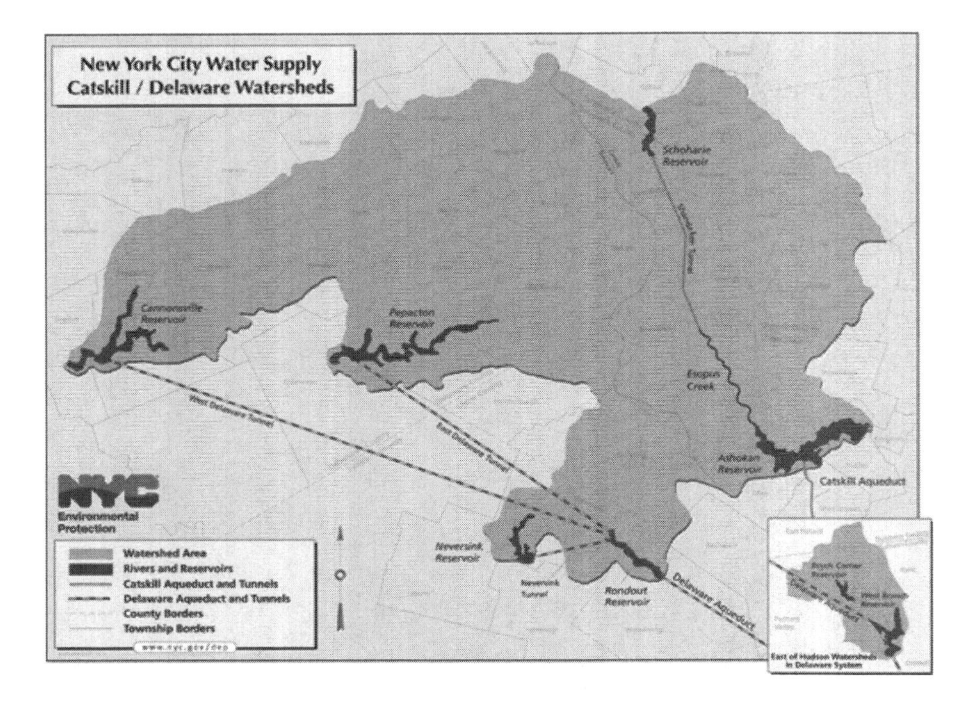

Figure 4.4 Catskill–Delaware systems (From DEP, 2007b).

water downstream (see Figure 4.1). In total, New York's water system spans three watersheds, includes 19 reservoirs and three controlled lakes (DEP, 2006). In 2008 the Catskill–Delaware and Croton systems provided 50, 40, and 10 percent, respectively, of the roughly 1.2 billion gallons of water consumed by over eight million New Yorkers and one million upstate residents every day (NYS DOH, 2008). In recent years with the construction of the Croton filtration facility this has been different.

4.2.3 Regulatory History

4.2.3.1 Regulatory Overview

New York City's drinking water system, though managed by the city, falls within a federal and state regulatory framework. The Safe Drinking Water Act (SDWA) is the federal law regulating both anthropogenic and naturally occurring contaminants in U.S. drinking water systems (EPA, 2010). Today, under the SDWA, the EPA establishes enforceable health-based standards called National Primary Drinking Water Regulations. These regulations establish enforceable maximum contaminant levels or treatment techniques for regulated drinking water contaminants (SDWA, 1974 et seq.; SDWA Amendments, 1989). They also provide guidelines called maximum contaminant level goals for nuisance contaminants under the National Secondary Drinking Water Regulations (SDWA, 1974 et seq.; Hecq et al., 2006; see EPA, 2011).

The SDWA applies to the more than 160,000 public water supply systems in the United States (EPA, 2010a). Program implementation is delegated to states, localities, and water suppliers, while the EPA oversees administration and compliance. New York state's Department of Health is charged with implementing the SDWA, yet the NYC DEP plays the primary role in structuring the programs that preserve New York City's watershed (Warne, 2010b).

While some national standards for drinking water quality were established in the United States as early as 1914, the SDWA, passed by Congress in 1974, expanded requirements to all public water systems (Hecq et al., 2006). Although originally focused on treatment interventions, major amendments in 1986 and 1996 changed the course of the law. The 1986 amendments focused on water treatment using specified "best available technology" and expanded EPA's regulatory mandates to include development of attainment of water quality standards at the point of consumption. EPA was required to develop maximum contaminant level goals (unenforceable health goals) and maximum contaminant levels simultaneously, and impose new monitoring requirements on public water systems for unregulated contaminants (EPA, 1986).

To implement the 1986 SDWA Amendments, on June 29, 1989, the EPA published drinking water regulations known as the Surface Water Treatment Rule (SWTR), which applies to all public systems using surface water or groundwater influenced by surface water (EPA, 1989; SDWA Amendments, 1989). The SWTR established criteria for filtration and disinfection requirements. To achieve a filtration avoidance determination (FAD), the regulations required water supply systems demonstrate at least 99.9 percent removal and/or inactivation of *Giardia lamblia* and at least 99.99 percent removal of waterborne viruses (SDWA Amendments, 1989).

The 1996 amendments recognized that source water protection is an integral part of the multiple barrier approach to ensuring safe drinking water, mandating and funding Source Water Assessment and Protection Programs (Hrudey et al., 2006). The 1996 SDWA Amendments further expanded water protection and management approaches by recognizing "source water protection, operator training, funding for water system improvements, and public information as important components of safe drinking water" (EPA, 2010, webpage). In fact, the Source Water Assessment and Protection Programs, one of the eight major elements of the 1996 amendments, made funding available for headwater protection projects[1] and required states to conduct source water assessments, where protection areas are delineated and inventoried for potential contaminants (SDWA, 1996). The New York State Department of Health completed the Final Assessment Program Plan in November 1999, noting that "[m]any aspects of source water protection have been practiced in New York State for decades" (NYS DOH, 1999). By focusing on risk reduction from source water to the tap, today's Safe Drinking Water Act and regulations employ a multiple barrier approach that aims to proactively ensure high quality of drinking water.

4.2.3.2 The Safe Drinking Water Act and New York City

The significance of the Safe Drinking Water Act to New York City and, in turn, New York's effect on SDWA merits consideration. In particular, how New York influenced SDWA filtration regulations through elements of the 1986 SDWA Amendments. These modifications in turn changed the way the NYC system operates today (Appleton, 2002). New York's influence on the 1986 amendments and the eventual achievement of a filtration waiver sets the stage for understanding the 1996 amendments and the current state of drinking water regulation.

In the mid-1980s, when the EPA asked Congress to pass an SDWA amendment that required filtration of all surface water sourcing systems, NYC resisted. Having historically invested in and relied on the consistently clean drinking water from its surface systems in the Croton and Catskill–Delaware watersheds, the cost of building new filtration plants seemed unreasonable (Appleton, 2002). Other large cities that still used unfiltered water—including Seattle, Portland (Oregon), San Francisco, Boston, Portland (Maine), and Syracuse—allied with New York to resist a uniform filtration requirement (Appleton, 2006). They advocated for a requirement to filter only when performance standards were not met. EPA supported uniform filtration standards, arguing that the construction of a filtration system could not be treated as a fallback upon failure to achieve standards put in place to avoid public health emergencies.

After much deliberation, the final SDWA surface water treatment rule included a provision allowing filtration avoidance if two conditions were met. First, water managers must demonstrate their water chemistry standards to be entirely compliant with water chemistry requirements. Second, a long-term plan for control and management of the surface drinking watershed was required (EPA, 1989; Appleton, 2009). The 1986 amendments allowed drinking water systems to apply for filtration avoidance until November 1991, while mandating that all construction of any new filtration works be completed by the end of 1993 (EPA, 2000). Because a large portion of the land in the upper watershed was privately held, with the majority of water pollution coming from nonpoint sources, which were not being successfully addressed through pollution regulations of the Clean Water Act, it was largely expected that New York City would have to filter to meet these new requirements (Appleton, 2009). Construction of a filtration plant for the Catskill–Delaware system was expected to cost between \$4 and \$8 billion (NYC IBO, 2002).

The city applied for filtration avoidance determinations (FADs) and received one-year waivers in 1991 and 1992 (NYC IBO, 2002). As EPA's *1997 Filtration Avoidance Determination Mid-Course Review for the Catskill/Delaware Water Supply Watershed* details, the city's November 1991 filtration avoidance application to the New York State Department of Health included provisions to spend up to \$47 million on land acquisition by 1992, with a goal to acquire 10,000 acres and develop a long-term land acquisition plan (EPA, 2000). In 1993, EPA issued a FAD for the Catskill–Delaware system, contingent on 150 conditions, including critical upstream conservation requirements (NYC IBO, 2002; EPA, 2000;

see 1993 FAD Conditions 13a, b, and c). In the June 1993 Long-Term Plan, the city committed up to $220 million to fund a seven-year acquisition program to add 70,000 acres to the watershed protection area (EPA, 2000). Unfortunately, it quickly became clear that meeting these land acquisition goals, and thus, the requirements of the FAD, was not as straightforward as simply appropriating funds. Complexity of land purchase contracts and water supply permits were primary stumbling blocks to the planned acquisition program, and by 1994, no land had been acquired (EPA, 2000).

The effects of the SDWA's implementation and NYC's desire to maintain filtration avoidance led to changes in the relationship between populations living within the watersheds and the city water users, resulting in contentious upstate–downstate land use issues (Schneeweiss, 1997; Appleton, 2009). Although achieving a filtration avoidance determination would require a long-term transboundary plan including land use regulations, some residents within the watershed areas complained that these restrictions would create significant burdens (Schneeweiss, 1997). This conflict necessitated negotiations between the upstream, often agricultural communities of the Catskills and downstream stakeholders. In 1995, the city, state, upstate communities, EPA, and environmental parties began negotiating what would become the New York City Watershed Memorandum of Agreement (MOA) (EPA MOA, 1997).

Signed on January 21, 1997, the MOA established land acquisition requirements, creating the NYC Watershed Protection and Partnership Council and corresponding watershed protection provisions and programs (EPA MOA, 1997).[2] The U.S. EPA issued a five-year FAD when the Watershed Rules and Regulations became effective in May of 1997 (EPA MOA, 1997). New York's past and present approaches to water system management further illuminate the challenges and successes of filtration avoidance for the Catskill–Delaware watersheds.

4.2.4 Decision-Making History

For over two centuries after the founding of New York City in 1625, drinking water management decision making was driven by public health concerns and the continuous need to locate reliable local drinking water resources. Not until the early nineteenth century did the city finally tap into the resources of Croton watershed and begin construction of aqueducts and dams to capture remote supplies. Connecting to this new source alleviated concerns over the ability to meet demand and pushed public health concern to the top of the agenda (Gandy, 1997).

In the late nineteenth century, with the growing status and involvement of new public service professions including engineers, watershed foresters, and urban planners, the decision-making process for the NYC water supply shifted from laissez-faire to a more rationalized extensive central planning process where public health and engineering feasibility were the key factors (Gandy, 1997). During the construction of the infrastructure related to the Catskill–Delaware watershed, state-of-the-art technologies were explored and adopted to increase water supplies. This further entrenched the role of engineers in the technical management of NYC's water system (NRC, 2000). By the time the Catskill–Delaware system was completed in 1967, more participants were involved in the planning process for watershed management, yet the operation and maintenance (O&M) of the water supply system remained the responsibility of the city and the engineers (NRC, 2000). While NYC has remained the key decision maker, today planning responsibilities are shared between the NYC DEP (for water quality and watershed protection) and NYC's Department of Health and Mental Hygiene (for water quality), the New York State Department of Environmental Conservation (NYS DEC) (for watershed protection), and the EPA (for setting water quality standards) (Gandy, 1997).

When the city agreed to federally mandated filtration for the Croton watershed in 1990 it also focused attention toward protecting the other 90 percent of the water supply from the Catskill–Delaware watersheds. Review of the Catskill–Delaware system revealed the key barrier to effective regulation of water quality was the lack of public ownership of the land in the watershed. Furthermore, the city also realized that compared to the expense of a filtration plant (capital cost estimates ranging from $6 billion to $8 billion, with annual O&M costs up to $350 million), a watershed protection program that could ensure sufficient water quality would be a bargain (NRC, 2000).

To avoid construction of a filtration plant for the Catskill–Delaware system, NYC proposed to acquire land as part of its watershed protection strategy in 1993. From 1993 to 1996 the city struggled to balance interests of different stakeholder groups and to develop a feasible watershed management plan in face of many political and fiscal obstacles (NRC, 2000). By proactively protecting NYC's water quality and enhancing the economic vitality of upstate watershed community, the city was able to satisfy EPA regulations and apply for a filtration avoidance determination (Appleton, 2009). The January 1997 Watershed Memorandum of Agreement began a new era focused on watershed management for New York City's Catskill and Delaware systems.

4.3 CURRENT PROFILE OF THE NEW YORK CITY WATERSHED

4.3.1 Current Source Water Protection

Source water protection is an important part of drinking water supply strategies, and the New York City watershed is a "nationally significant" success story (Porter, 2006). Source water protection is an especially important management tool in the case of NYC because of the high cost of building, operating, and maintaining a filtration plant for the Catskill–Delaware water supply (NRC, 2000; Warne, 2010). Protecting water sources requires significant cooperation and effort, and while there is no standard formula for success, source water protection planning techniques aim to identify and mitigate resource risks (Goss and Richards, 2008). NYC DEP assessment efforts identify potential water quality threats including but not limited to waterfowl on reservoirs, wastewater treatment plant discharges into waterways, leaky septic systems, nutrient runoff from farms, storm water runoff from developments, and sediment runoff from forest trails and roads (DEP, 2006).

The NYC DEP has developed and continues to implement three main source water protection programs: (1) the Land Acquisition Program (LAP); (2) Watershed Protection and Partnership Programs that include watershed forestry, wetlands protection, stream management, waterfowl management, and agricultural pollution prevention planning as well as public outreach and education; and (3) capital programs that include subprograms sewer extension, septic system rehabilitation and replacement, storm water retrofit, and wastewater treatment (DEP, WPPS, 2011).

The source water protection programs for the NYC water supply area are still evolving to changes in land use and feedback from watershed residents and partnering organizations. Because the Catskill–Delaware water system has a filtration waiver, protection programs are stronger in this system. However, the Croton watershed is also the target of many programs primarily aimed at reducing the costs of filtration (Warne, 2010a). As of 2010, these efforts have cost approximately $1.49 billion for the Catskills/Delaware watershed protection programs (Warne, 2010b).

4.3.1.1 Land Acquisition Program

Land use is inextricably linked to surface water quality. Surface runoff originating from forests, fields, and urban areas has very different characteristics relating to sediment load, nutrients, and pathogens (de la Cretaz and Barten, 2007; Tong and Chen, 2002). Given the significant, "unequivocal" relationship between land use and in-stream water quality (Tong and Chen, 2002), direct land acquisition and management is a primary objective for NYC DEP's watershed management program.

Since 1997, the goal of the Land Acquisition Program has been to acquire and manage the most sensitive areas in the Catskill–Delaware and Croton watersheds through fee simple purchase and conservation easements (DEP LTLAP, 2009). At the time of the program's inception, the city owned just 3.5 percent of the Catskill–Delaware watershed. The 1997 MOA required the city to solicit 355,050 acres within the watershed and allocate up to $300 million to acquire land and conservation easements from willing sellers (EPA MOA, 1997; DEP, 2006). The 2007 Filtration Avoidance Determination called for an additional $241 million to be spent on land acquisition (DEP LTLAP, 2009). The city additionally spent roughly $38.5 million on land acquisition in the Croton watershed (DEP, 2011; Freud, 2003). Since its inception through June 2011, the LAP program increased the city's ownership of real property interests in the Catskill–Delaware system by 113,375 acres, or 319 percent (DEP WPPS, 2011).

City officials tout the LAP as a continuing success. As of June 2011, the city had acquired 72,950 acres in fee simple ownership and protected an additional 44,732 acres through two types of conservation easements and in more than 1,350 separate transactions (DEP, 2011). These transactions bring NYC's ownership of the Catskill–Delaware watershed to almost 15 percent of the total area and the proportion of protected lands up to 36 percent (DEP, 2011). Since the Land Acquisition Program began in 1997, the City and its partner, the Watershed Agricultural Council

(WAC), have secured over 92,000 acres in fee simple or conservation easements. In all, the City now owns more than 137,000 acres, land which is now protected from development and managed pro-actively to protect water quality (DEP [NYC] 2011, p. 4). According to DEP, as of September 2011, approximately 721 acres of lands acquired in fee simple, or 1 percent of all acres acquired in fee, were outside the Catskill–Delaware or Croton watershed boundary[3] (DEP WPPS, 2011).

The city employs sophisticated land acquisition prioritization methods. The LAP first prioritizes properties based on location in the water supply system, and then considers site-specific characteristics (DEP FEIS, 2010). Within each subbasin, NYC DEP has assigned each property a priority designation of 1–4 depending on its distance to terminal reservoirs and intakes and the water's travel time to distribution consumption (Figure 4.5) (DEP FEIS, 2010). During the first phase of the program, NYC DEP officials solicited priority areas 1 and 2 more aggressively than 3 and 4, leading today to a 30 percent success rate (proportion of properties solicited that go into contracts, excluding farm easements) in areas 1 and 2 compared with a 24 percent success rate in areas 3 and 4 and a rate of 25 percent overall. NYC DEP officials now concentrate efforts on subbasins with less than 30 percent protected land and basins that are expected to be large contributors to the water supply in the future (DEP LTLAP, 2009). These principles are reflected in the Priority Area and Natural Features Criteria provisions of the MOA and the Long-Term Watershed Management Plan (DEP FEIS, 2010).[4]

Given concerns that the city's desire to own significant acreage in upstate New York could cause resentment and conflict between upstate residents and NYC DEP officials, it was important to design a LAP that was sensitive to the needs of upstate citizens and municipalities. The program is completely voluntary, and all interested landowners are offered fair market value based on an appraisal conducted by independent appraisers hired by the city (CWC, 2005). The LAP includes a community review process for property acquisitions and New York City pays all property taxes as assessed on acquired land (DEP FEIS, 2010; EPA MOA, 1997). For conservation easements, taxes are paid at a level equal to the proportion of the overall property value as if the property were unimproved (DEP, 2006; DEP LTLAP, 2009). The Water Supply Permit issued to NYC by NYS DEC in December 2010 for the 2012–2022 LAP (Extended LAP) includes several additional program changes, including expansion of town-designated Hamlet Expansion Areas, modification of Natural Feature Criteria, and development of a riparian buffer easement acquisition program (DEP FEIS, 2010).

In the 2010 Draft Environmental Impact Statement for the NYC LAP, DEP emphasized that land acquisition is an essential "anti-degradation strategy which seeks to avoid potential adverse water quality impacts associated with development and other land uses." It concluded that the future "Extended" LAP, which would run from 2012 to 2022 under a renewed Public Water Supply Permit,[5] is "needed to continue to support FAD requirements and to focus additional attention to basins and sub-basins with a low percentage of protected lands" (DEP FEIS, 2010, p. ES-1).

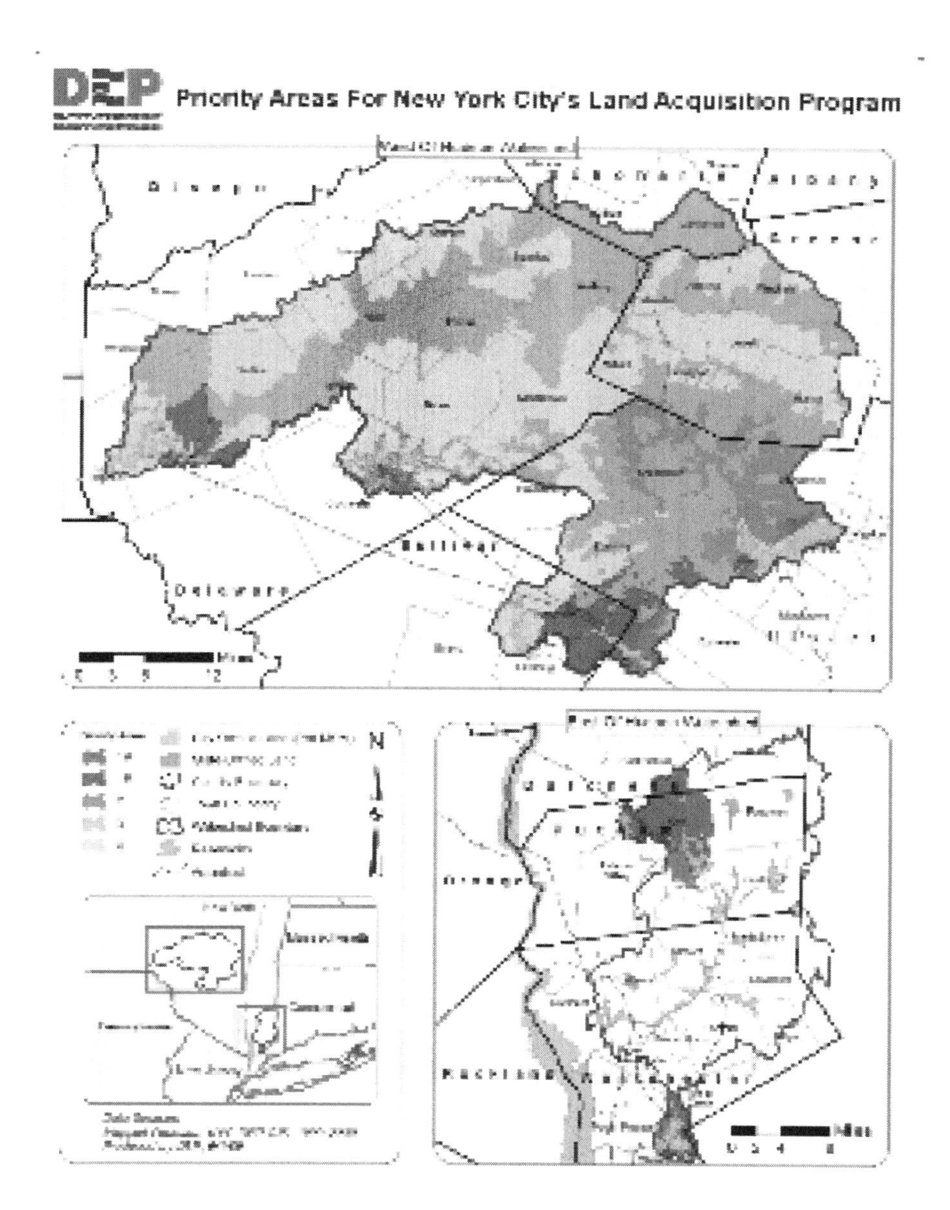

Figure 4.5 Priority areas for New York City Land Acquisition and Stewardship Program. NYC DEP has assigned each property a priority designation of 1–4 depending on its distance to terminal reservoirs and intakes, and the water's travel time to consumption (From DEP FEIS, 2010). (See color insert.)

4.3.1.2 Watershed Protection Partnership Programs

Complementing the LAP is the Watershed Protection Partnership Programs and their supporting program elements, which are designed to support comprehensive management of the city's property and easements in the watershed to meet the goals of source protection (DEP, LTAP 2009). These programs include reservoir waterfowl and wastewater management, stream and riparian buffer protection, as well as agricultural and forestry efforts.

Reservoir waterfowl management: In 1992, the NYC DEP enacted its Waterfowl Management Program to quantify and reduce the impact of bird populations in the vicinity of reservoirs (DEP, 2004). The NYC DEP acknowledges that waterfowl can be a significant source of fecal coliform, *cryptosporidium*, and *giardia*, so, in order to comply with the SWTR, NYC DEP engages in an aggressive avian management program that consists of bird harassment and control techniques. The dispersal and deterrent methods employed by DEP have been approved by the U.S. Department of Agriculture Wildlife Services (DEP, 2006).

Wastewater management: Human waste can also be a significant threat to water quality through the contamination of water by excess phosphorus, nitrogen, suspended solids, fecal coliform, bacteria, and viruses. To mitigate these risks the city has engaged in a comprehensive Wastewater Treatment Plant Upgrade Program. The city has devoted more than $140 million in funds toward upgrading wastewater treatment plants in the Catskill–Delaware watershed, as well as $96 million to construct seven new wastewater treatment plants (Warne, 2010a). As of November 2011 100 percent of the total flow of wastewater in the watershed in the Catskill–Delaware was being treated by upgraded plants (DEP, 2006).

In the Croton watershed, NYC DEP has devoted approximately $260 million to wastewater treatment plant upgrades and diversion (Freud, 2003). Additionally, in the Catskill–Delaware watershed, approximately $113 million has been allocated to the Community Wastewater Management Program, Septic Rehabilitation and Replacement Program, and Sewer Extension Program for the design, replacement, and maintenance of sewer systems and more than 3,400 septic systems to date (DEP, 2006; Warne, 2010). Storm water retrofits and future storm water controls have been the subject of an additional $55 million investment (Warne, 2010a).

Stream and riparian protection: Because the Catskill–Delaware region is mountainous and predominantly privately owned, upstream land management decisions that impact stream corridors can have significant downstream water quality implications. Due to this dynamic, the goal of the DEP's Stream Management Program is to preserve the stability and ecological function of riparian areas through stewardship. This voluntary program uses outreach, training, and demonstration to promote proper management of stream corridors (DEP, 2006). The program is credited with the creation of more than 13 stream restoration projects in the Catskill–Delaware watershed and nine stream management plans, costing $55 million (Warne, 2010a). The Riparian Buffer Protection Program is also devoted to protection of critical riparian buffers in the Catskill–Delaware watershed. As of 2008, the city owned 12 percent of land within 100 feet of streams, while an additional 18 percent is owned by the state (Rush, 2009).

Watershed agriculture and forestry programs: The Watershed Agricultural Program provides funding and assistance to farmers willing to create pollution prevention plans known as Whole Farm Plans. Operating since 1992 under the Watershed Agricultural Council (WAC), a nonprofit organization dedicated to promoting optimal agriculture and forestry activities within the New York City water supply area (WAC, 2012). Whole Farm Plans have been completed for more than 390 large and small farms in the Catskill–Delaware and Croton watersheds, enabling implementation of best management practices and prioritized water protection interventions (DEP, 2009). This figure constitutes about 94 percent of the large commercial farms in the West-of-Hudson watershed region. Of the 307 known large farms in the West of Hudson Watershed, 85.7 percent achieved the "substantially implemented" milestone (Rush and Holloway, 2009, p. 46). In the Croton watershed 50 farms had approved Whole Farm Plans at the end of 2009 (DEP, 2009). The program is designed to reduce or eliminate pollutant sources, prevent movement of runoff across landscapes, and promote the use of riparian buffers that serve to filter any polluted runoff that moves across the landscape.

The WAC also administers the Watershed Forestry Program, which aims to promote well-managed working forests to preserve beneficial land cover for the watersheds, in conjunction with the NYC DEP and the U.S. Forest Service (USFS) (DEP, 2006). This is achieved by assisting landowners to create Forest Management Plans, conducting training for loggers and foresters, and providing direct technical assistance. NYC DEP reports that WAC Forest Management Plans cover more than 100,000 acres of private forestland. Because approximately 75 percent of the watershed is forested, stewardship practices continue to play an important role in protecting water quality (DEP, 2006).

4.3.1.3 Capital Program

The capital expenditures DEP incurred are financed through debt that is to be repaid over the next 30-year period. The debt service is the single largest driver of the water and wastewater rate increases. Currently, the Water Finance Administration's total debt is $26.6 billion (NYC Water Board, 2011). From FY2002 to FY2010, DEP committed $20.8 billion to its capital program, and DEP planned to spend an additional $2.4 billion on its capital program in FY2011 (NYC Water Board, 2011).

Current capital improvement projects include the Croton Water Filtration Plant, the Catskill and Delaware Ultraviolet Disinfection Facility, City Water Tunnel No. 3, and the Newtown Creek Wastewater Treatment Plant. When operational in 2013, the Croton Water Filtration Plant will have the capacity to treat 290 million gallons of water per day, 30 percent of New York's daily demand (NYC Water Board, 2011). The plant is being constructed beneath Van Cortlandt Park in the Bronx.

Planned construction includes repair of the Delaware Aqueduct, one of New York City's primary drinking water supply tunnels, and repair and renewal of City Water Tunnels 1 and 2, which have been in constant operation since they first went into service in 1917 and 1936, respectively. Other planned work includes conducting a dependability study for the city's water supply and demand reduction and building the Cross River and Croton Falls Pumping Stations, which will be able to transfer water from the Croton system to the Delaware system during emergencies, planned service outages, and periods of drought (NYC Water Board, 2011).

Finally, every day, about 1 billion gallons of water are delivered to New York City, and every day the City's 14 wastewater treatment plants collectively treat about 1.3 billion gallons of wastewater. As with most city infrastructure, the 14 plants are aging, and DEP must invest a significant amount of its resources to maintain them in a state of good repair and to modernize them to meet constantly evolving state and federal standards (NYC Water Board, 2011).

4.3.2 Current Engineering and Filtration Technology

The NYC water supply system infrastructure remains highly interconnected today. One interesting feature of this system is that when the water leaves the Catskill–Delaware region, it does not flow directly to New York City. Instead, the waters of the Catskill–Delaware system are collected in the Kensico reservoir in Westchester County (DEP, 2006). This interconnected design allows for redundancy in the system, which helps ensure water security. For example, NYC has the ability to reroute or completely shut off water coming from one of the reservoirs that has higher turbidity levels caused by a storm or other disturbance (NYS DOH, 2008).

The NYC water system has also been cost-effective in the long term because unlike numerous other major cities, New York has been able to avoid the expense of constructing filtration facilities to treat the majority of its drinking water. However, due to a 1998 consent order, filtration is required for the Croton plant, and site preparation for construction began in 2005 (DEP, 2005). The Croton supply system itself has been offline since 2008 and will remain offline until construction of the Croton Water Filtration Plant is completed (DEP Croton, 2009).

In the late 1990s, planning for the $2.1 billion Croton filtration plant began (NYC DEP, FSEIS, 2004). Initially the 12-acre plant was slated for completion in 2009, but now is scheduled to enter service in 2013. The Croton filtration plant will use a stacked dissolved air flotation system to remove remaining contaminants suspended in the water after pretreatment of mixing, coagulation, and flocculation (Crossley and Valade, 2006). After the dissolved air flotation stage, the water is then filtered through 60-cm anthracite and 30-cm silica sand filter and disinfected with a dual system of UV and chlorine. The Croton filtration plant is rated to treat 320 million gallons of water per day with resulting postfiltration water quality measurements of <0.1 NTU turbidity, <2 mg/l total organic carbon, >99.9 percent removal and inactivation of *Giardia* and viruses, and 99.9 percent removal and inactivation of *Cryptosporidium* (NRI, 2011).

Pollutants that are of primary concern to New York City's surface drinking water supplies include microbial pathogens, organic carbon, nitrogen, phosphorus, and sediment. Microbial pathogens (bacteria, viruses, and protozoa) are common contaminants of surface water systems (NRC, 2000). While many bacteria and virus strains can be treated with chlorine disinfection, two microorganisms, the protozoa *Cryptosporidium parvum* and *Giardia lamblia*, are resistant to chlorine, which has posed a problem for the currently unfiltered water of the Catskill–Delaware system (Smith et al., 2011). To ensure the highest water quality for this system, EPA's 2002 Filtration Avoidance Determination stipulated that a UV treatment plant for the Catskill–Delaware water supply be completed by August 2012 (EPA, 2007b). The Catskill and Delaware UV Disinfection Facility will provide an additional barrier against these types of microorganisms that cannot be addressed solely by a combination of forest filtration and chlorine disinfection techniques (EPA, 2007b).

Unlike biological contaminants, other pollutants of concern including nitrogen, phosphorus, biodegradable organic carbons, and toxins must be addressed using coagulation, flocculation and sedimentation, or filtration processes (EPA, 2007a). The EPA is particularly concerned with nutrient loading due to increased risks of microbes and the potentially "toxic and/or carcinogenic disinfection by-products" produced by reactions between certain disinfectants and organic carbon (EPA, 2007a). The organic carbon is a legacy of algae that have fed off the nutrients. As the Natural Research Council's watershed management assessment describes, nutrient imbalances from nitrogen and phosphorus have various sources and can cause significant water quality impairment (NRC, 2000). Nitrogen can be introduced into the watershed from wet and dry atmospheric deposition of NOx in both the Croton and Catskill–Delaware systems. In addition, nitrogen as well as phosphorus can be introduced via septic tank leakage (predominantly in the Croton system), runoff from agricultural lands, and urban storm water runoff (NRC, 2000). Phosphorus has been one of the highest profile pollutants in the NYC water supply system because of its ability to cause algal blooms, which can lead to eutrophication.

Eutrophication can have negative impacts on water supply including increased turbidity from algal material and byproducts, increased organic carbon that can form disinfection by-products (DBPs), negative impacts on fish habitat due to decreased dissolved oxygen levels, and potential toxic by-products formed by the algae blooms (NRC, 2000). DBPs may be the most serious concern in terms of human health effects. These by-products, including trihalomethane and haloacetic acid, which are regulated by the EPA, are considered to be carcinogenic (Pereira, 2009; EPA, 2001; EPA DBP Rule, 2010). In addition to DBPs, the EPA also regulates the quantity of other toxic compounds in the water supply. Because the Catskill–Delaware watersheds are predominately undisturbed forestland, there are far fewer occurrences of toxic compounds and hazardous wastes than the more urbanized and industrial Croton watershed. However, pesticides are used in both watersheds on a regular basis and may also pose human and environmental health concerns (NRC, 2000).

4.3.3 Regulatory Effects on the New York Water Supply

Public water supply regulation in New York predates the Federal Safe Drinking Water Act by decades. As in California, New York has over the years modified its sanitary code to implement the rules in the federal code.

4.3.3.1 The Surface Water Treatment Rule

To implement the SDWA, the Surface Water Treatment Rule (SWTR) was codified in 1989 to address the potential danger of microorganisms in unfiltered surface water (EPA SDWAA, 2010). The SWTR mandates filtration of public surface water supplies unless a system "minimizes the potential for contamination of *Giardia* cysts and viruses in the source water" (Platt et al., 2000) and meets the following criteria:

- Source water must meet turbidity and fecal coliform standards of the SWTR.
- There must be no source-related violations of the coliform rule.
- There must be no waterborne disease outbreaks in the supplied population.
- The system must demonstrate compliance with certain disinfection requirements, maintain a minimum chlorine residual entering and throughout the distribution system, and undergo an annual on-site inspection to review the condition of the disinfection equipment (SWTR, 1989).

4.3.3.2 The Memorandum of Agreement and Filtration Avoidance Determinations

On January 21, 1997, the NYC Watershed Memorandum of Agreement (MOA) was signed, forming a new and historic partnership among the many stakeholders to (1) preserve high-quality drinking water and (2) guarantee the economic health and vitality of communities located within the watershed. The 1997 MOA establishes the rules and parameters for the programs and partnerships that will meet the requirements of the SDWA for long-term watershed protection of New York (EPA MOA, 1997). By meeting the requirements of the MOA, the City of New York was granted an interim Filtration Avoidance Determination in both 2002 and 2007, allowing continued filtration avoidance for the Catskill–Delaware system for 5 to 10 years (2007–2017) (EPA, 2007b). Under the periodic FADs, the city continues to develop programs that attempt to balance between water quality preservation and community development interests.

The current programs under both the MOA and the FAD fall into two categories: protective (antidegradation) and remedial (specific actions taken to reduce pollution generation from identified sources). These approaches differ in the sense that most monitoring effects are long-term and big scale for programs focused on protection, while remedial programs are short term and small. Further, there are different expectations and measures of success for long-term protection programs, while remedial programs are focused on quickly mitigating specific problems and evaluating the degree to which they can reduce pollutant loadings from entering the water supply (Principe et al., 2000).

The collection and maintenance of a comprehensive water quality data set is fundamental to the evaluation of both categories of the program. Two fundamental monitoring activities exist to inform decision makers. First, the quality of source water is closely monitored. Samples are collected and tests are conducted throughout the watershed including sites at aqueducts, reservoirs, streams, and watershed wastewater treatment plants. Each year more than 35,000 samples are collected from

300 sites and more than 300,000 laboratory analyses are performed (Principe et al., 2000). A geographic information system has been developed and adopted to better track and evaluate the watershed inventory. Second, the city conducts active epidemiological disease surveillance within New York City and microbial risk assessments on the source water. Current activities that are evaluated include the detection of waterborne disease outbreaks and epidemiological studies for determining the proportion of illness attributable to drinking water (DEP, 2009a).

Currently, benchmarks for these evaluations are based on the objective criteria contained in the SWTR.[6] The data baseline is 1987 when data collection was extended out into the watershed to provide substantial spatial and temporal coverage. The DEP NY provides progress reports annually on its various watershed protection and remedial programs and conducts an evaluation every five years (Principe et al., 2000). NYC DEP's programs are also informally evaluated by third parties. For example, Riverkeeper, one of the signatories of MOA, creates a report card on NYC DEP's programs. The following criteria are used to give a grade for each program whether or not NYC DEP had complied with the program's requirements as stipulated in the MOA or FADs, if compliance was timely, if funding for the program was adequate, if DEP worked well in partnerships where applicable, and if DEP was putting its best efforts into the program (Riverkeeper, 2008).

4.3.3.3 Current Decision-Making Context

Within the constraints of the budget, the main focus of current decision making has clearly prioritized the public health issues of watershed management. Some environmental and social benefits are treated as "side benefits" and have not necessarily been institutionalized into the decision-making process of NYC DEP. For example, although one of the main objectives of the MOA is to ensure an "adequate supply of clean and healthful drinking water" which is "vital to the health and social and economic well being of the People of the State of New York," no formal evaluation is conducted to compare how programs are faring in relation this objective (EPA MOA, 1997, p. 1). There is also no standardized optimization process for managing the watershed. The decision on how much to invest in watershed protection programs could benefit from cost-benefit analysis and other quantitative financial and environmental techniques to evaluate strategic options.

4.4 FUTURE THREATS AND SUGGESTED ACTIONS

The source water protection efforts implemented by NYC DEP and its partnering organizations offer impressive examples of systemwide water supply management. New York City is fortunate in its abundant water resources and should continue to aggressively fund its source water protection programs in the Catskill–Delaware watershed to maintain its filtration avoidance determination. In this way, NYC can avoid the high costs of a filtration plant and also strengthen the environmental and social benefits that the current programs bring. For the Croton system, NYC DEP should continue to implement a multiple barrier approach to water supply by continuing its source water protection programs even after the filtration plant is completed. Doing so will minimize environmental risks of pollutants and sediments and maximize the social benefits that watershed protection can provide through open space recreation, wildlife habitat, increased property values, and climate mitigation.

However, potential problems in NYC's watershed are numerous, and these challenges require varied analyses and approaches. Some of the most pressing threats to drinking water security include declines in forest ecosystem health from invasive species and climate change, contamination risks from natural gas exploration, pharmaceuticals and personal care products (PPCPs), disinfection by-products (DBPs), and infrastructure and operational interruptions due to decreasing water demand and the fragile nature of many partnerships.

Each of these challenges has different implications for the potential filtration of the Catskill–Delaware system. One such example is the potential increased frequency and severity of storms related to climate change. Such a scenario could result in traditional problems such as increased runoff and turbidity that could be remedied through filtration and other engineered interventions. However, other threats, for example the water quality impairment from PPCPs, DBPs, and the many chemicals associated with the slurry used in hydrological fracturing for natural gas mining, have the potential to pollute the watershed in ways that existing regulations and technology do not currently address (Tiemann, 2010). These problems support an enhanced, alternative calculus to measure the severity of water security threats and interventions, as well as an expanded toolbox of stewardship interventions. Some of these challenges can be approached through the traditional stakeholder-engagement approach that NYC has relied upon in maintaining its filtration avoidance to this point. Others may demand new approaches altogether.

The following sections address some of the pressing threats to water quality from the NYC watershed and make some recommendations to NYC DEP. Many of these challenges and recommendations may be broadly applicable to a wide range of water supply systems. Because the NYC DEP has already taken comprehensive steps to monitor, model, and proactively implement climate change adaptation, potential implications of climate change on watershed health and drinking water security are not addressed here.

4.4.1 Forest Insects and Pathogens

4.4.1.1 Threat

A healthy intact forest has the potential to yield quality surface drinking water for downstream residents. However, not all forests are healthy and intact; some are at risk from disturbances that alter what is often considered the status quo or a stable state (Foster et al., 1998). Common disturbances in the Northeast include convectional windstorms (Woods, 2004), hurricanes (Foster et al., 1998), ice storms (Rhoads et al., 2002), and insect defoliators (Orwig and Foster, 1998). It is important that forest managers actively administer their properties to mitigate these risks on a landscape scale. Such efforts are underway, for example, in the Quabbin watershed in Massachusetts, where managers focus on maintaining a diverse mix of species and age classes so as to increase the resiliency of the forest against a disturbance that could affect the land's ability to act as a filter (MA DCR, 2007).

Insects are perhaps the most visible threat to the New York watershed forests. New York State lies near or within the expanding distribution range of several introduced and invasive insects that are now dramatically altering the composition and future health of the northeastern forest. The emerald ash borer (*Agrilus planipennis*) is a beetle native to Asia that was discovered in Michigan in 2002 (USDA, 2011). The New York State Department of Environmental Conservation reports that the Forest Service attributes the death and decline of tens of millions of ash trees in the United States to the emerald ash borer, and efforts to control and eradicate this pest have not been successful (NYS DEC EAB, 2011). These insects have been discovered in several counties in western New York and, most recently, in Greene and Ulster counties, north of Westchester and the Croton watershed and within the easternmost portion of the Catskill–Delaware Watershed. Ash, a tree that occupies critical watershed riparian and wetland areas, is particularly at risk of emerald ash borer infestation (Crocker et al., 2007; USFS, 2009). In January of 2011, the New York State Department of Environmental Conservation's Bureau of Private Land Services published the Emerald Ash Borer Management Response Plan to address this statewide threat using both "tiered" and "threat-based" risk management approaches (NYS DEC EAB Plan, 2011).

Other invasive insects, such as the hemlock woolly adelgid (*Adelges tsugae*) and the Asian long-horned beetle (*Anoplophora glabripennis*), also threaten forest health and stand dynamics and the water quality these ecosystems protect. The first time hemlock woolly adelgid was recorded in the United States was in 1951 in Virginia; it is now present across 21 percent of the eastern hemlock's range, leading to widespread decline in the health and vigor in these stands (Evans and Gregoire, 2006). The dieback of hemlock in the Croton and Catskill–Delaware watersheds has been ongoing for the past 25 years. Hemlock provides critical thermal cover to cold water stream ecosystems which changes forest stand dynamics and could have water quality impacts on watershed catchments (see Stradler et al., 2006; Gordon and Pugh, 2011).

Asian long-horned beetle infestations, of which the first U.S. documentation was in Brooklyn, New York, in 1996, also pose a significant threat to forest health (Haack et al., 1997; DPR, 2009). Asian long-horned beetles attack at least 29 tree species including maple, elm, willow, birch, and ash, and infestations eventually lead to tree deaths (Bancroft and Smith, 2001). A recent outbreak of Asian long-horned beetle in Worcester, Massachusetts, where over 25,000 street trees had to be felled, provides evidence of the insect's potential impact (Alsop, 2009).

In addition, throughout the Northeast high deer populations can limit seedling establishment, threatening the regeneration of forests and their future resiliency to climate change, storms, and unpredictable disturbances (Cote et al., 2004). Any combination of these factors could threaten the forest's function to act as an effective filter for incoming precipitation. Threats to forest health have the potential to reduce species diversity, alter age class distribution and limit the resiliency of New York

City's watershed forests. For forested watersheds to continue to provide clean drinking water, watershed managers must continue to manage these areas and promote healthy forests.

4.4.1.2 Action

To address threats to forest health the NYC DEP should engage in public outreach programs. Public education efforts should continue to teach invasive species detection and control to minimize risks of outbreaks in New York State. Because the only effective means to control infestations of Asian long-horned beetle is to remove and destroy infested trees, proactive eradication methods must continue to be a management priority. Programs to address invasive pests should be continued and expanded.

Current efforts by the NYC DEP seem to have been centered on NYC itself. In the Catskills, a training program organized by a local conservation nonprofit in the fall of 2009 reports training 13 individuals through a day-long training session, which does not seem to match the scale of the problem (Catskill Center for Conservation & Development, 2009). The New York Department of Parks and Recreation has already established a hotline to report insect disturbance, as well as a map of sightings and quarantine areas, which can be easily accessed online. To support these efforts, larger scale, more easily accessible training sessions with Catskills–Delaware residents should be funded and facilitated by NYC DEP. Participatory planning to enhance invasive pest monitoring and management should be a priority in addressing threats to forest health in important watershed areas.

4.4.2 Natural Gas Exploration

4.4.2.1 Threat

Abundant gas reserves are present and being aggressively developed in 31 states, including New York (Lustgarten, 2009).The NYC DEP has stated that the potential development of natural gas wells in the Catskill–Delaware watershed is of concern to the watershed's health (Kane, 2010). The Marcellus shale formation, the focus of much current natural gas development interest, is speculated to contain 20 years of gas reserves and extends through all or part of 29 counties in southern and central New York, including the entire West-of-Hudson watershed (NYS DEC, 2011b; DEP RIAR, 2009; DEP FIAR, 2009).

Hydraulic fracturing, or *fracking*, is a process where pressurized fluid is injected into deep horizontally bored holes to facilitate extraction of natural gas, a process that DEP's 2009 Impact Assessment of Natural Gas Production found to have been "associated with the movement of natural gas and contaminants into aquifers or surface water bodies" (DEP RIAR, 2009, p. ES-4). DEP acknowledges that it is difficult to assess the impact that these chemicals might have. These activities have been exempted from Clean Water Act permitted discharge requirements, and due to trade law secrets, oil and gas companies are not currently required to disclose all of the chemicals used in the drilling fluids.[7] Despite this limitation, the DEP report identifies numerous ways that natural gas extraction could pose a risk to the integrity of the water quality, quantity, and infrastructure of the NYC drinking water supply system. The DEP's rapid impact assessment examined the failures of environmental quality assurances from similar operations in other states and found reasons to be seriously concerned with the impact such drilling technologies would have on the Catskill–Delaware watershed. Similar operations in the Marcellus shale formation in Pennsylvania have led to reports of methane-contaminated groundwater (Kobell, 2009). The chemicals in hydraulic fracturing could pose water security risks if they impact ecosystems or contain health-adverse contaminants that enter NYC's drinking water.

DEP's assessment found that the activities during natural gas development could contaminate groundwater or surface water supplies, cause reliability problems from water withdrawals, or damage critical NYC DEP infrastructure (DEP RIAR, 2009). The NYC DEP also suggests the threat of leaks could lead to a negative change in public perception about the future protection of the watershed, which in turn could add to the difficulty of maintaining filtration avoidance. Despite DEP's jurisdiction over many aspects of New York City's drinking watershed management, NYC DEP has little regulatory control regarding hydraulic fracturing activities (Kane, 2010). Instead, the New York State Department of Environmental Conservation (NYS DEC) is the agency with regulatory jurisdiction over proposed gas developments. To address perceived risks to water security, NYC DEP requested that NYS DEC establish a protective perimeter around the city's six major Catskill reservoirs and connecting infrastructure (Lloyd, 2008). In April 2010, the NYS DEC announced that the supplemental environmental analysis would not apply to the watershed area (Navarro, 2010). On July 1, 2011, the NYS DEC released its recommendations on mitigating the environmental impacts of high-volume hydraulic fracturing, and went a step further by stating that fracking would be prohibited in the New York City and Syracuse watersheds and buffer zones for at least the foreseeable future (NYS DEC, 2011a). Some immediate concerns left out of the proposed regulations relate to ensuring sizable buffers along aqueducts and tunnels and around watersheds to prevent impacts from horizontal drilling, ensuring that the chemicals used in fracking are declared, and waste residues treated as hazardous, with the eventual requirement of promoting chemical alternatives.

Despite the uncertainty and risks associated with hydraulic fracturing, economic and energy security considerations lead some to support natural gas extraction. Former New York Governor Paterson stated that shale gas extraction could provide up to $1 billion a year of revenue for economically depressed rural towns (Lisberg, 2010). Current DEC commissioner Joe Martens has expressed the belief that hydrofracking "can be done safely" and the Cuomo administration has been eager to strike a balance between the economic benefits of natural gas extraction and environmental concerns (Robin, 2011). Conversely, some environmental groups continue to assert that nothing but an outright ban will keep gas companies out (Navarro, 2010). These issues remain a land and water management challenge.

4.4.2.2 Action

Due to jurisdictional restrictions, the NYC DEP has limited options to address threats from hydraulic fracturing in the Catskill–Delaware watershed. This issue may exacerbate upstate and downstate sociopolitical divides given the uncertain risks of gas extraction and the certainty of economic development potential. Nevertheless, this is not so unlike many of the divisive watershed management challenges that have been successfully overcome in the past. Tried and true methods to deal with this threat that should continue to be employed include targeted regulation, precautionary planning, and interagency cooperation. There are many coalitions of upstate groups that can and do advocate against the drilling, but after accurate valuation of the potential of the wells is complete, it may also be worth developing an extraction rights payment system for the landowners near the watersheds who might otherwise lease their land for drilling. Recently, however, the state imposed strict regulations all but banning drilling within the drinking water supply watershed for New York City and Syracuse, though lawsuits from private landowners are likely to occur (NYS DEC, 2011b).

The pace of development of these potential wells will play a significant role in how well NYC DEP maintains safeguards against extraction activities in sensitive watershed areas (DEP RIAR, 2009). Constant interagency oversight and precautionary management practices should be a primary objective in addressing hydraulic fracturing activities in and near watershed areas. If land and watershed managers are able to design a successful voluntary easement program with landowners who would have leased or sold their land for natural gas drilling, their ability to reduce associated water quality risks may be expanded.

NYC DEP should continue to proactively identify and minimize potential risks to water security from natural gas extraction by providing the best available information of risks to policy makers and the public. Scaling up public outreach programs by holding workshops and publishing educational materials is an important step in continuing regulatory dialogs about fracking. Public support will be critical to ensure precautionary regulations that are protective of New York City's invaluable water supply. NYC DEP can also work with DEC to encourage disclosure or limited disclosure of fracturing chemicals by oil and gas companies to better assess risks. Additionally, to maximize watershed protection and balance environmental and economic development values, developing a payment for extraction rights for landowners in and around sensitive watershed regions may be desirable.

4.4.3 Pharmaceuticals and Personal Care Products

4.4.3.1 Threat

Pharmaceutical and personal care products[8] have been cited by the NYC DEP as a potential threat to the safety of the New York City watershed (Warne, 2010b). They are a pervasive issue wherever humans who use PPCPs are present and have been detected in receiving waters influenced by raw or treated sewage throughout the world (Lipsky et al., 2009; Ellis, 2006). Some reports suggest there are as many as six million PPCP substances available worldwide and that conventional water treatment processes are insufficient to remove these contaminants (Ellis, 2006). In their 2010 report on the occurrence of PPCPs in NYC source waters, DEP detected PPCPs at extremely low levels in every sample (DEP PPCPs, 2010).

As the EPA and NYC DEP report, PPCPs are still being categorized and the technology to measure them is still being refined (EPA PPCPs, 2010; DEP PPCPs, 2010). Thus far, in New York, the Department of Health has concluded that detected concentrations are below levels expected to have any effect on human health, but research suggests some of these chemicals may present ecological concerns. In particularly, the EPA reports that PPCPs can include "endocrine disrupting chemicals, xenobiotics, hormonally active chemicals, pesticides and newly emerging contaminants" and that additional research is necessary to determine the cumulative effects of long-term exposure (EPA PPCPs, 2010, webpage). To identify contaminants, assess risks, and maintain SDWA compliance, NYC DEP already collects PPCP data in the East-of-Hudson watersheds (Figure 4.6; Table 4.1) (Lipsky et al., 2009).

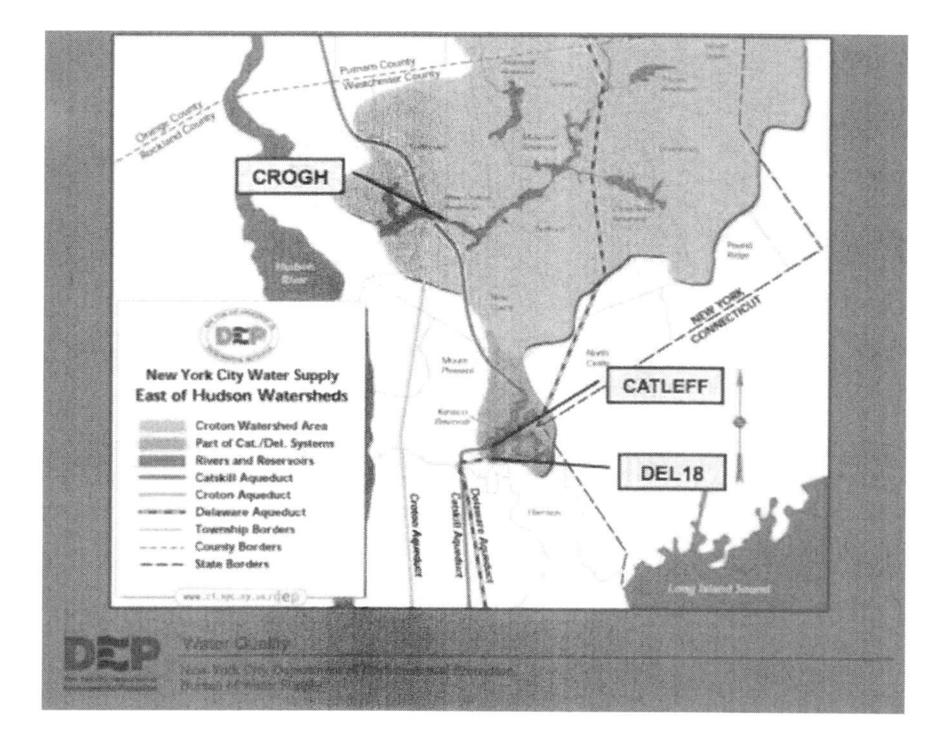

Figure 4.6 Monitoring points for PPCPs. The NYC DEP has begun collection of data related to PPCPs in the East-of-Hudson Watersheds (From Lipsky et al., 2009).

Table 4.1 Monitoring Site Descriptions for PPCPs; These Sites Monitor Prechlorination Water Quality

PPCP Monitoring Sites		
Site Code	**Site Description**	**Reason for Site Selection**
CROGH	Croton Gatehouse, untreated effluent from New Croton Reservoir	Keypoint sampling location. Prechlorination.
DEL18	Delaware Aqueduct, Shaft 18 untreated effluent from Kensico Reservoir	
CATLEFF	Catskill Aqueduct, lower effluent chamber, untreated Kensico Reservoir effluent	

Source: Lipsky et al., 2009.

4.4.3.2 Action

The combination of still-emerging technology for detection of lower levels of contaminants, as well as the lack of a defined methodology for collection or analysis (Lipsky et al., 2009), could certainly lead to a prolonged period of uncertainty that may delay action regarding PPCP detection and regulation. NYC DEP will need to continue to refer to the EPA for new regulations while also proactively communicating with and engaging watershed stakeholders to implement programs to address PPCP contamination. Partnering with "big consumers" of PPCPs in the watershed (e.g., health centers, as well as agriculture and aquaculture operations) can help target major nondomestic sources of PPCPs (see Ellis, 2006). Working to educate the public of potential risks of PPCPs may be the only way to reduce domestic sources. Besides educating consumers how to prevent PPCPs from entering the water system, DEP can also help create some incentives for them to change their behavior. For example, DEP can create PPCP "recycling centers" to collect unused PPCPs. Because PPCPs are an emerging concern worldwide, NYC DEP could also engage watershed managers in other cities or nations to combine efforts to understand and address PPCPs in drinking water supplies.

4.4.4 Finding the New York Watershed Icon

The benefits of building public awareness of and appreciation for an unfiltered water supply could be a strategic move for the NYC DEP. This could help build public understanding for any potential rate increases used for upstate watershed programming or to counter falling demand. To build funding needed for programmatic changes, NYC DEP could imitate the Seattle Public Utility by designing a public outreach program based on (1) finding a watershed-related icon that the public embraced and (2) aligning the potential rate increase with a daily item that minimized the impact of the rate increase in the mind of the public. In Seattle, the water utility picked the salmon as an icon and a single latte as its rate increase equivalent (Modie, 2002; SPU, 1995–2011). NYC DEP should investigate the potential of a parallel strategy—perhaps using a beloved northeastern fish species such as trout and the cost of a subway ride or a slice of pizza—to appeal to the ecological and economic sensibilities of their constituents. Furthermore, New Yorkers rightfully have pride in their tap water—this pride alone, when developed through an appropriate marketing campaign, could be leveraged to build a citywide movement emphasizing the importance of continued investments to ensure healthy drinking water and ecosystems through watershed protection programs.

4.4.5 Building Public Outreach and Decision-Making Tools

NYC DEP has been conducting a variety of watershed protection programs but has not been able to communicate the values of these programs effectively to the public, either in the upstream watershed community or in the city. NYC DEP has the potential to adopt a value matrix to better communicate to the public in an easily understandable way the benefits its different programs can deliver. Each program under NYC DEP could be evaluated against each key issue to ascertain efficacy, costs, and benefits. Such a value matrix can also be used to argue for implementing both watershed management protection programs and engineered solutions, as engineered solutions can only help achieve a limited set of benefits (Table 4.2). A multibarrier approach can not only help reduce risks using green and gray infrastructure and organizational redundancies, but can also help create a set of resilient public benefits such as for public health and also for economic, environmental, and social values. This matrix can also be used to manage stakeholder relationships in the watershed. By inviting stakeholders to help map out the key issues, NYC DEP can match actions to local priorities. Hosting workshops could create a sense of ownership among stakeholders by engaging them in the decision-making process. By highlighting issues that concern different individuals, DEP can identify priority issues.

A stakeholder value matrix could also be developed representing the different groups and those important resource issues highlighted that affect or impact each of these groups. Such an analysis may help to clarify challenges and suggest criteria for watershed protection using Triple Bottom Line analysis, an approach already in use in cities such as Seattle (SPU, 2007). Additionally, by soliciting and incorporating stakeholder-identified criteria into the adopted matrix, DEP can secure more public support for expanding benefit assessment and valuation.

Given the current public-health-focused monitoring data (see Box 4.1), NYC DEP can only perform qualitative evaluations for most parts of the matrix, except for engineered solutions and public health issues where numbers can be more readily defined. If NYC DEP finds the use of a matrix to be a valuable decision-making tool, with proper technology and a broadened monitoring program (see Box 4.2 for some data sets that could be considered to expand such assessment), the matrix can be further quantified to serve as an optimization tool for social values beyond engineering and public health.

Table 4.2 Sample Matrix Layout

		Environmental Solutions							Engineered Solutions	
Categories	Issues	Agriculture Program	LAP	SMP	Forestry Program	WWTP	Septic Prog		Filtration	Disinfection
Public Health	Pathogens	*			*	*	*		*	*
	Nutrients	*			*	?	?		*	
	Turbidity	*			*	?	?		*	
	PPCPs					?			?	
	...									
Economic	Infrastructure co	*			*	?	?			
	Property values	*			*	?	?			
	Job creation	*			*	*	*		*	
	...									
Social	Aesthetics	*			*	?	?			
	Quality of life	*			?	?	?		*	
	Recreation				?	?	?			
	...									
Environmental	Air Quality	?			*					
	GHG mitigation	?			*	?				
	Ecosystem Health	*			*	*				
	...									

Note: Green indicates existing program coverage, while yellow indicates possible program coverage, red indicates a coverage gap, and a white or blank cell indicates issues are not applicable to the listed environmental or engineered solution, or that data is lacking. This example includes additional blank rows and columns to highlight the importance of flexible, adaptive program management. (Adapted from DEP WPPS, 2011). (See color insert.)

BOX 4.1 2008 MONITORING DATA IN THE NEW YORK CITY WATERSHED

WATER QUALITY

A. Physical

(1) Temperature (2) pH (3) Alkalinity (4) Conductivity (5) Hardness (6) Color (7) Turbidity (8) Secchi Disk Depth (9) UV 254 Absorbency

B. Chemical

(1) Dissolved Organic Carbon (2) Total Phosphorus (3) Total Nitrogen (4) Nitrate+Nitrite-N (5) Total Ammonia-N (6) Iron (7) Manganese (8) Lead (9) Copper (10) Calcium (11) Sodium (12) Chloride (13) Barium (14) Corrosivity (15) Fluoride (16) Magnesium (17) Potassium (18) Silica (19) Specific Conductance (20) Strontium (21) Sulfate (22) Total Dissolved Solids (23) Zinc (24) Disinfection Byproducts

C. Biological

(1) Fecal Coliform (2) Chlorophyll a (3) Total Phytoplankton (4) *Cryptosporidium* (5) *Giardia* (6) HEV (7) Benthic Macroinvertebrate (8) Zebra Mussels

LAND USE CHANGES

(1) Wetland Acreage and Biodiversity (West-of-Hudson) (2) Forest Ecosystem (Forest Inventory)

WEATHER DATA

Precipitation, air temperature, relative humidity, rainfall, snow depth (snow water), solar radiation, wind speed, and wind direction.

(*Source*: DEP New York City Drinking Water Supply and Quality Report, 2008).

4.4.6 Tapping into Market Power: Water Quality Trading

From 1997 to 2007 NYC DEP conducted a 10-year phosphorus trading pilot program in the Croton system (Kane, 2007). The program was not carried beyond the trial session due mainly to the following reasons:

- Demand for the program was far less than expected.
- The administrative burden was significantly greater than expected due to a prolonged initial selection process and the level of guidance participants needed.

BOX 4.2 KEY DATA NEEDED TO QUANTIFY THE PROPOSED MATRIX

A. Economics in upstream community: (not an exhaustive list)

- Change in property values near the land-use related watershed protection programs
- Number of jobs created
- Average wage for watershed protection program employed personnel
- Emerging industry: services, tourism, art, recreation etc.

B. Social benefits and environmental impacts in upstream community: (not an exhaustive list)

- Survey on how the upstream community feel about the impacts that watershed protection programs on their life quality: data on whether there is an increased recreation use and by how much
- Air quality monitoring data where there is watershed protection program
- The amount of greenhouse gas reduction created by all the watershed protection programs (especially the Land Acquisition Program and the Forestry Program)
- Impact on biodiversity from the watershed management programs in the watersheds

- Applicants found it difficult and time-consuming to identify approvable offsets.
- There was a lack of capacity for monitoring and verification of reliable offsets at the time (Kane, 2007).

Although this pilot program was not successful, water quality trading for other measures of water health still holds some potential for future watershed protection as a relatively cost-effective tool in specific geographic areas if it can address the issues that led to its initial lack of success. Learning from the experiences of the phosphorus trading program, NYC DEP could investigate the establishment of a water quality trading program where strong monitoring programs exist or where they can be created, potentially cutting down on added administrative costs of monitoring and verification. For instance, within the soon-to-be-filtered Croton watershed, such monitoring additions to the engineering infrastructure may help quantify benefits of technical and land management interventions. Such comparisons may also help maintain stakeholder support of watershed stewardship efforts. To encourage greater trading program participation, NYC DEP could provide incentives, for example, by providing a revolving fund for low-interest loans to lower the pressure for up-front investments before offset payments begin. NYC DEP could also continue to utilize the *Water Quality Trading Toolkit* (EPA, 2007c) and other sources to take advantage of new trading schemes. For example, the EPA has published a guide for sediment and nutrient trades between forestry and a drinking water treatment facility. The guide clearly lays out how to generate sediment credits with vegetative planting, including riparian buffer and vegetated filter strips, and how to build a trading program (EPA WQT, 2010).

4.5 FUTURE DECISION-MAKING OUTLOOK

Given projections of lower water consumption rates and decreased revenues, the NYC DEP may be operating under a more constrained budget in the near future. It is important for the agency to have the ability to defend decisions to spend money within the Catskill–Delaware and Croton watersheds. In addition to strengthening political will and public support for watershed protection programs, they could take steps to evaluate its protection efforts in a more quantitative fashion. Up to this point, NYC DEP has been able to use the often quoted \$6–\$8 billion filtration plant avoidance as justification for its many source water protection programs. At the very least with the construction of the Croton filtration plant, NYC DEP will have to work harder to defend its land management and acquisition programs in the Croton watershed. If the Catskill–Delaware supply is ever subject to filtration, it will still be important to be able to quantify the financial benefits of source water protection programs.

Quantifying benefits of source water protection offers a variety of challenges. With an expectation that a lot of technology and methods will be perfected in the future, a quantitative decision-making model may be very useful (Figure 4.7). Such models can use one engineered solution as a baseline to look at the total value (including public health, economic, environmental, and social) that each program can create to inform decision making (Linvok et al., 2006). The two key methods for quantifying different values are (1) evaluating comparative improvement criteria and (2) assessing triple bottom line criteria that include economic, social, and environmental values.

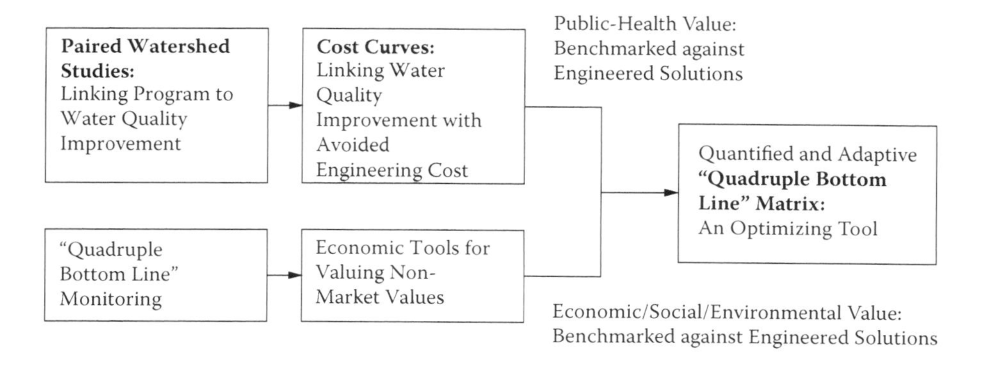

Figure 4.7 An example of a quantitative decision-making model.

1. *Evaluating comparative improvement criteria.* To evaluate comparative improvement criteria, it is important to know how much water quality improvement programs can bring. Improvements to public health or water quality can be quantified, and then the engineering cost-curve can be used to calculate how much operation and maintenance cost (C_E) would be needed to achieve the same level of engineered water quality improvements. If the cost of the evaluated program is Cp, then the public health value of the program benchmarked against the engineered solution will be: Cp – C_E. For example, to reduce turbidity by 10 mg/L needs 100 acres of new forest that will cost $300 as compared to an investment of $680 more in engineering.
2. *Assessing the triple bottom line.* To evaluate economic, social, and environmental criteria, market values for any of the criteria can be used after they have been established. If social or ecological enhancements do not have a defined market value, economic nonmarket-value tools can estimate the benefit of these services. Ideally, every program's value can be benchmarked against the chosen baseline program.

Theory and methods of these evaluation approaches will be discussed in more detail in the following subsections.

4.5.1 Evaluating across Criteria

4.5.1.1 Linking Source Water Protection Programs with Water Quality Changes

One way to quantify the benefits of source water protection is to calculate the avoided water treatment costs land management programs can provide. To do this, the first step is to establish relationships between programs and the actual water quality improvements they create. The goal of many of NYC DEP's source water protection efforts is to reduce pollutant loading to its streams and reservoirs. Quantifying the change in water quality after the implementation of a land management program, as currently done for the FAD five-year assessments (DEP WPPS, 2011), would allow the description of the difference in cost of treating the cleaner water compared to water that was not improved by the program.

Although quantifying the effects of source water protection programs can be difficult, one approach is to evaluate changes in water quality using paired watershed experiments. A paired watershed experiment is designed to measure the impact of an experimental treatment on stream flow and water quality (Brooks et al., 2003). These treatments can include clear-cutting forests, applying herbicides or fertilizers, and cutting riparian buffers. Such a study allows researchers to separate the effect of a treatment from natural variation in stream flow by comparing a treated watershed to a control (untreated) watershed. The process involves several steps:

1. Choosing two or more watersheds that are as similar as possible
2. Installing gauges to measure stream flow and water quality parameters
3. Collecting baseline data during the calibration period to develop regression models that can later be used to compare future measurements with the baseline

4. Applying some sort of treatment to the experimental watershed while the control is left undisturbed
5. Collecting data on the change in water characteristics to evaluate the impact of the treatment (Brooks et al., 2003)

Paired watershed experimental design has been a successful method of controlling for environmental variability. While it is easier and tempting to simply measure water quality before and after a treatment in just one watershed, such study designs are unlikely to document statistically significant water quality effects at the watershed level (Bishop et al., 2005). Paired watershed experiments are widely used throughout the United States including at the Hubbard Brook Experimental Forest in New Hampshire, the Coweeta Hydrologic Laboratory in North Carolina, and the H.J. Andrews Experimental Forest in Oregon (de la Cretaz and Barten, 2007).

NYC DEP could employ a paired watershed experiment model to document changes in water quality in the Catskill–Delaware and Croton watersheds. NYC DEP already conducts extensive watershed monitoring that covers the entire watershed and includes meteorological stations, snow surveys, stream sites, reservoir sites, aqueducts, and wastewater treatment plants (EPA, 2000). The current monitoring program was designed to meet legally binding mandates and is used to identify potential sources of pollution and quantification of pollutant loading so that appropriate measures can be taken to protect drinking water quality (DEP, 2009). This monitoring program is sufficient to measure water quality changes throughout the entire New York City watershed and distribution system. To conduct paired watershed studies, it may be beneficial to expand the monitoring system to watersheds adjacent to the Catskill–Delaware and Croton regions. These watersheds could act as controls for the "experimental" watersheds within the NYC drinking water supply area.

Many paired watershed studies have taken place in areas that were highly controlled and easily manipulated by researchers (de la Cretaz and Barten, 2007). Conducting paired studies on New York City watersheds is complicated by the presence of residents, as well as modeling complications including the variable fate and effects of contaminants in relation to watershed geology, soils, and climate. Nevertheless, paired studies have been conducted in residential watersheds (Groffman et al., 2004). NYC DEP could model their approach after studies that have measured water quality impacts in watersheds that have been impacted by development (Pickett et al., 2007) to expand their focus on water quality and growth management.

Conducting paired studies to evaluate the effectiveness of source water protection programs in the Catskill–Delaware and Croton areas would take long-term commitment to experimentation. Alternatively, the NYC DEP could attempt to estimate improvements from many different watershed protection programs by conducting several trials for each program. NYC DEP could pick several control areas that would remain untreated. Other watersheds could each be used to test a land management program in isolation (see Table 4.3). After a best management practice (BMP) has been tested and an estimate of water quality improvements is made, NYC DEP officials would be able to extrapolate benefits of future implementation of similar

Table 4.3 Potential Design of Paired Watershed Studies

Watershed	Program Trial	Specific Treatment	Expected Water Quality Changes
Control A	None	None	None
Control B	None	None	None
Control C	None	None	None
Trial A	Watershed agricultural program	Installing riparian buffer	Nitrogen, phosphorus, pathogens
Trial B	Watershed agricultural program	Nutrient management plan	Nitrogen, phosphorus, pathogens
Trial C	Wastewater treatment plant upgrade	Upgrade wastewater treatment plant to reduce nutrient loading to streams	Phosphorus, pathogens
Trial D	Waterfowl management program	Bird harassment	Pathogens

BMPs. For instance, the implementation of a program to address manure management, rotational grazing, and improved infrastructure on a farm in the Delaware system (Cannonsville Reservoir basin) was shown to reduce stream loading of total dissolved phosphorus by 43 percent and particulate phosphorus by 29 percent (Bishop et al., 2005). These metrics can add to the perceived value of land management interventions.

There are several difficulties to using this method. First, such studies could slow the implementation of certain protection programs in the test watersheds, which may violate MOA or FAD regulations. Second, such studies may involve long-term investments, as it takes varying amounts of time for water quality improvements to be detected in a watershed after treatment testing (Boesch et al., 2001; Wang et al., 2002). Third, there are challenges with funding and project coordination, as source water protection funds are disbursed by NYC DEP in the Catskill–Delaware system, but by Westchester and Putnam counties in the Croton. This gives NYC DEP less control over implementation of BMPs and infrastructure upgrades in the Croton watershed in order to use it as a study area, despite the desirability of obtaining such data.

Paired water quality studies could yield particularly valuable information that can help quantify real benefits of watershed stewardship. Given the challenges of determining benefits of source water protection efforts, long-term comparative water quality data is desirable to develop metrics for designed water quality improvements. Nevertheless, it can often be difficult to attribute watershed interventions with water quality improvements, a challenge that is highlighted by the example of attempts to reduce phosphorus loading in the New York City watersheds in the 1990s. The NYC DEP has documented water quality improvements since the start of its watershed protection programs in the early 1990s (Table 4.4).

Several trends in phosphorus can be detected (Table 4.4). Although there were marked declines in three out of four Delaware reservoirs and several Croton

Table 4.4 Phosphorus Concentrations in NYC Drinking Water Reservoirs 1991–1998

	Phosphorus Concentration (μg/l; geometric mean, May–Oct)							
	1991	**1992**	**1993**	**1994**	**1995**	**1996**	**1997**	**1998**
				Delaware District				
Cannonsville	26.8	18.6	22.9	15.9	24.1	16.9	21.0	17.1
Pepacton	11.2	9.0	10.5	9.0	8.2	9.9	8.2	7.9
Neversink	5.6	4.3	8.2	5.4	5.8	5.3	5.1	3.3
Rondout	7.3	7.9	9.1	8.1	8.8	8.3	6.3	7.6
				Catskill District				
Schoharie	11.0	10.9	13.3	13.9	14.5	34.5	18.4	18.7
Ashokan East	8.5	11.0	13.4	9.0	13.0	16.2	13.7	12.7
Ashokan West	8.4	11.4	7.0	11.4	12.3	22.6	15.1	14.2
				Croton District				
Amawalk	*	21.7	*	19.0	17.3	20.5	21.1	23.5
Bog Brook	*	*	17.1	17.5	21.0	14.5	14.1	19.8
Boyd Corners	18.7	14.2	13.3	16.3	11.9	14.6	5.1	8.7
Croton Falls	32.7	34.4	24.5	21.3	22.2	25.8	19.8	19.6
Cross River	*	14.8	7.1	10.4	13.3	*	*	16.8
Diverting	*	33.4	32.7	25.5	23.5	23.9	23.1	33.4
East Branch	*	*	27.0	25.3	27.0	20.0	25.1	31.6
Middle Branch	*	*	20.7	19.0	29.5	18.7	18.9	26
Muscoot	24.5	24.4	24.4	24.2	25.7	24.4	23.3	29.3
Titicus		25.7	12.2	21.3	18.5	21.3	*	38.1
West Branch	12.8	15.7	10.3	14.0	10.1	9.5	5.6	6.6
Lake Gleneida	*	*	*	*	30.2	26.4	24.0	21.3
Lake Gilead	*	*	*	*	22.7	27.6	21.5	15.1
				Source Water				
New Croton	22.8	17.4	17.3	18.7	17.3	16.4	15.0	15.8
Kensico	13.6	8.6	13.0	8.5	7.1	7.4	5.4	5.3

Source: Principe et al., 2000.

reservoirs, during this period phosphorus concentrations increased in all Catskill reservoirs and several Croton reservoirs. Despite increases in some reservoirs, the concentrations decreased in the terminal reservoirs for the Croton and Catskill–Delaware systems. While DEP can claim to have reduced overall phosphorus loading, it is difficult to determine which program should receive credit for the reductions. For example, there were several programs that might be responsible, most notably wastewater treatment plant upgrades and the implementation of BMPs on farms funded by the Watershed Agricultural Program (DEP, 2009). Additionally, should attempts at reducing phosphorus loading in the Catskill system be deemed a failure because Catskill reservoirs increased in phosphorus concentrations during this

Table 4.5 Phosphorus Export Coefficients for a Variety of Land Uses

Range of Unadjusted Export Coefficient (EC) Values for Land-Cover Classes, West Branch Delaware River Watershed

	Range of EC Values		
Land-Cover Class	·······kilograms per hectare per year·······		
Deciduous forest	0.0350	0.1400	0.2325
Coniferous forest	0.0600	0.2000	0.2750
Grass, shrub	0.0305	0.2075	0.2538
Pasture	0.1900	0.2500	0.3775
Cornfield	0.6700	0.9500	2.6750
Winter spread manure, cornfield	3.0500	8.7000	15.1500
Alfalfa	0.6400	0.7600	1.2400
Bare soil	0.1000	0.1500	0.2000
Urban area	0.4875	0.9250	2.4500

Source: Reckhow et al., 1980.

period? Does it matter, as long as the terminal reservoirs experienced a decrease in phosphorus? These are the types of questions that NYC DEP's scientists and policymakers will need to answer if they wish to attribute water quality improvements to a specific program, a goal that can be important in assessing the effectiveness of source water protection programs. This objective may be achieved through assessing experimental work, literature, and careful spatial and temporal analyses using a geographic information system to quantify benefits of source water protection.

Another important and perhaps more straightforward calculation of benefits may be achieved for prevention programs such as the Land Acquisition Program by quantifying the amount of avoided future pollutants from these lands by comparing the yearly nutrient budget for forests, farms, and residential areas. Based on Reckhow et al.'s export coefficient modeling in the Delaware watershed (Table 4.5), and assuming that 30,000 of the 41,558 hectares (ha) of land protected by NYC DEP is deciduous forest, high-end estimates in total phosphorus runoff from this land would be 6,975 kg/yr (30,000 ha × 0.2325 kg/ha/yr) (Reckhow et al., 1980).

If all 30,000 hectares were converted to urban land use, total phosphorus would be 73,500 kg/yr (30,000 kg × 2.45 kg/ha/yr). Therefore preserving this land as forest prevents 66,525 kg of phosphorus from entering New York City reservoirs annually. This estimate, however, is limited for several reasons. First, it is unlikely that this land, if not purchased, would be converted to urban area—it is more likely that it would be developed for vacation homes or subdivisions. The phosphorus budget for single-family homes on large lots would be markedly different from the urban phosphorus budget (Reckhow et al., 1980). Second, it is unlikely that this land would have immediately been converted to another land use. Rather, this transition would have taken place over a long period of time. Given these challenges, to estimate avoided pollution from prevention programs like the Land Acquisition Program, NYC DEP would need to

1. Create reliable nutrient budgets for the cross-section of land uses within its watersheds (e.g., deciduous forest, coniferous forest, dairy production, corn production, low-density residential, high-density residential, etc.)
2. Determine which parcels would have eventually been converted to each different land use if they had not been purchased
3. Decide when each parcel would have been converted to each different land use so as to determine when to start considering the benefit of avoided contamination

Quantitatively linking different watershed protection programs to the water quality improvements alone is not enough to attach a dollar value to a specific program. Nevertheless, it is important that watershed managers quantify program achievements and costs avoided—for example, by creating a cost curve for water treatment in relation to water quality improvements gained through land management programs—to communicate the benefits of watershed stewardship to decision makers.

4.5.1.2 Creating Cost Curves for Water Treatment

While water quality cost curves are a useful tool to quantify actual savings from water quality improvements, they can be challenging to construct. Estimates of the average cost of drinking water treatment facilities are largely based on the sheer size of the plant and the capacity of water it can treat. A treatment facility for the Catskill–Delaware water supply has been estimated to cost anywhere between $6 billion and $8 billion (NRC, 2000). In reality, actual drinking water costs vary significantly, even among systems of the same size (EPA, 2005).

Municipal water facilities often use a number of treatment chemicals for filtration and disinfection. The expense and extent of treatment needed increases as water quality decreases (Daily, 1997; EPA, 2005). High turbidity levels further complicate the disinfection process, thus making it more costly. The U.S. EPA conducted a study to examine the cost of chemicals used in water treatment facilities to predict the increase in O&M costs required to comply with stage 2 disinfectant requirements for *Cryptosporidium* removal. The EPA reported average cost for treatment chemicals per ton (Table 4.6) as well as comparative costs of disinfection (Figure 4.8a) and removal technologies (Figure 4.8b) (EPA, 2005). Despite rising treatment costs associated with declining water quality, changes in influent water quality may have little to no effect on the price, quality, or quantity of drinking water seen by consumers. Instead, the decrease in quality affects the cost of maintaining acceptable water supply, which is primarily seen only by the drinking water suppliers. To accurately predict these costs, it is essential to monitor and collect data to examine the intricate relationship between cost (capital and O&M) and influent water quality. For example, one study of a group of municipal water treatment facilities in Texas concluded that the chemical costs increased as a function of influent surface water turbidity (Dearmont, 1998).[9]

While it may be easier to relate the increased cost of treatment of source water of lower quality, it can be more difficult to show which specific policies relate to a direct change in water quality. With the scheduled opening of the Croton filtration

Table 4.6 Chemical Costs for Water Treatment Facilities in 2000

Chemical	Cost	Units
Alum, dry stock	$300	per ton
Alum, liquid stock	$230	per ton
Carbon dioxide, liquid	$340	per ton
Chlorine, 1 ton cylinder	$280	per ton
Chlorine. 150-pound cylinder	$600	per ton
Chlorine, bulk	$280	per ton
Ferric chloride	$400	per ton
Hexametaphosphate	$1300	per ton
Lime, hydrated	$110	per ton
Lime, quick lime	$100	per ton
Phosphoric acid	$650	per ton
Polymer	$1.00	per lb
Potassium permanganate	$2900	per ton
Sodium hydroxide, 50%	$350	per ton
Sodium hypochlorite, 12%	$1100	per ton
Sodium chlorite	$325	per ton
Sodium chloride	$100	per ton
Sulfuric acid	$100	per ton
Surfactant, 5%	$0.15	per gal

Sources: EPA, 2005; vendor quotes, 2000.

plant in 2013, NYC could use this facility to monitor the costs associated with variable surface water quality. Creating such cost curves requires data including raw water turbidity, color, algae, iron, and manganese and the relative chemical costs used to treat them (Faust and Osman, 1998). In the Croton system, these curves could be used to identify the costs of treatment options and to assist in development of source water protection policies to address area-specific water quality issues. Since the water in the Croton system will be filtered by 2013, it may be difficult to make a case for enhanced source water protection in the region, despite the fact that watershed stewardship provides numerous other benefits in addition to improved water quality and decreased treatment costs. Cost curves relating influent water quality to filtration plant treatment costs could support arguments for continued source protection in the Croton system.

4.5.1.3 Evaluating across Economic, Social, and Environmental Criteria

To assess triple bottom line benefits, NYC DEP can link more monitoring data for economic, social, and environmental values (Box 4.2) to valuing models for nonmarketable goods (Box 4.3) to benchmark management programs against engineered solutions and quantify benefits of these programs. To keep the values in the matrix consistent, all solutions should be benchmarked against the engineered option. For example, if the jobs and the average salaries created by the Forest Management Program and a drinking water treatment plant are known, the job creation value for

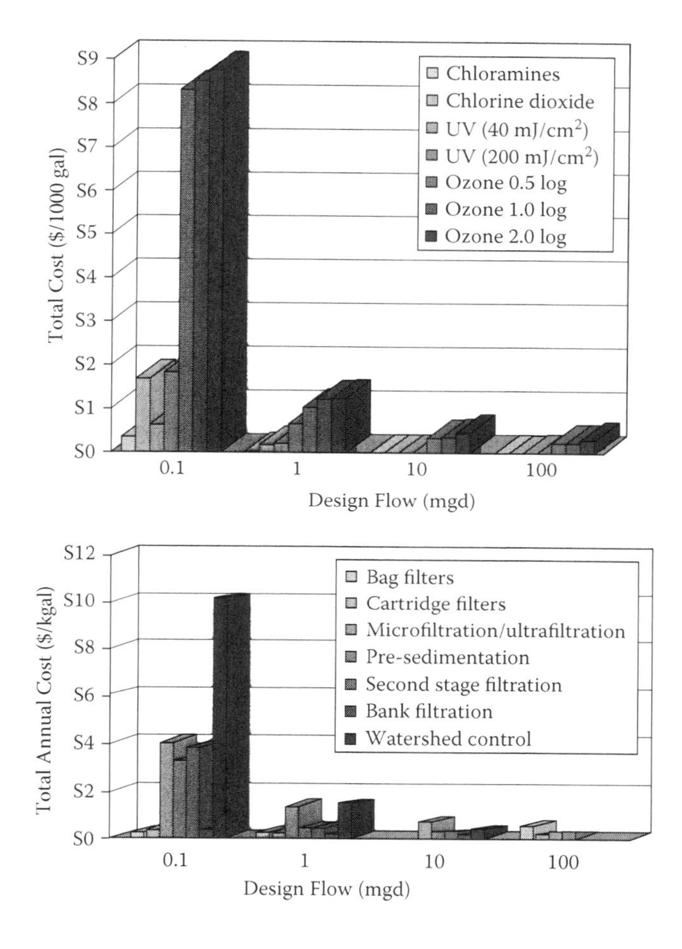

Figure 4.8 (a) Cost comparison of disinfection costs (From EPA, 2005); (b) cost comparison of removal technologies (From EPA, 2005).

the forest program in comparison to the water treatment option can be determined. Similarly, using data on the increased number of recreational users of designated watershed areas and computing travel costs (see Box 4.3) to give recreational values for different land uses. Again, this data can be benchmarked against the engineered solution, which, in this case, would be zero. Such comparisons help to demonstrate the costs and values of both engineered and land management water security solutions.

4.5.2 A Quantified and Adaptive Matrix for Optimizing Decision Making

With improved land modeling programs, perfected O&M cost curves, and better monitoring data, a better understanding can be gained over how much O&M

BOX 4.3 VALUING NON-MARKETED GOODS

Hedonic Price Model: Non-marketed goods sometimes affect the prices of goods that are marketed. For example, a housing price not only reflects the characteristics of the house, but also such location factors as the quality of forests, water, and associated recreational activities nearby. Statistical techniques can often be used to identify the independent contribution of the characteristic of interest on price. However, unless data describing all the major characteristics affecting price are available, a reliable estimate of the independent contribution of the characteristic of interest cannot be made.

Opinion Survey: Directly survey public's willingness to pay.

Travel Cost Method: Instead of surveying people's behavior is valued directly. The most well-known example of this method is the evaluation of the distance to recreational sites. It will not be a good measure for people who view travel itself as desirable.

(*Source*: David Leo Weimer, *Policy Analysis: Concepts and Practice*, Upper Saddle River, N.J.: Pearson Prentice Hall, 2005.)

can cost each land-use-related program. Such information can be used to develop a benchmarking program using values to fill the matrix (Table 4.3) for the different programs across different economic, social, and environmental values to facilitate decision making based in more quantitatively comparative costs and benefits. By calculating economic costs and benefits of O&M interventions, a quantified matrix for different environmental programs with engineered solutions as a baseline can be developed. Such a quantified matrix tool should also be adaptive to integrate better data sets, better evaluation methods, and evolving criteria. Combining the benefits with the cost for each program and analyzing the marginal benefit for investing more

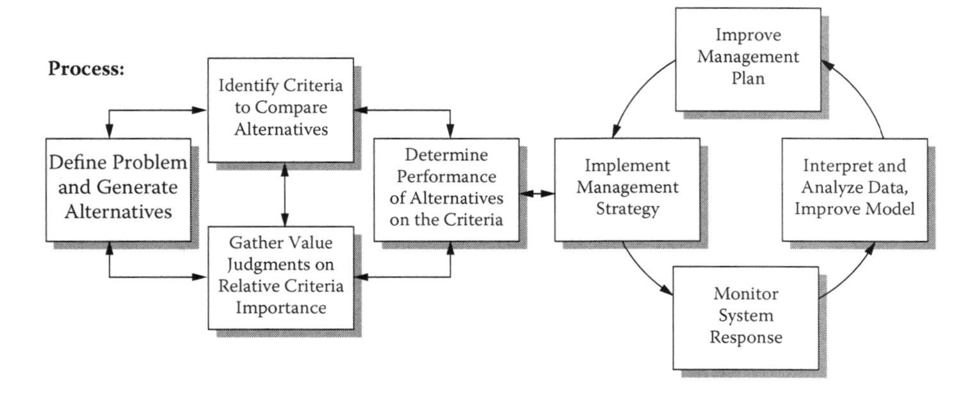

Figure 4.9 Adaptive decision-making process. An adaptive decision-making model is a flexible management approach that ensures criteria for decision making can be revisited and rebuilt to incorporate lessons learned (from Linvok et al., 2006).

in a certain program, the matrix may help the NYC DEP to better understand where to keep investing, where to expand, and where to reduce the spending.

Each step of this "ideal" decision-making model will include inherent limitations and uncertainties, which are important to acknowledge. In addition to the limitations described in the above sessions for land use models and cost curves, quantifying social and environmental benefits can be difficult even with transferred value techniques. For example, there is still no generally endorsed method to capture the value of increased biodiversity that a land conservation program can bring, and there is always going to be a debate on how to give a value to community's improved quality of life.

It is also important to note the danger in trying to solely quantify source water protection programs through monetary valuation, even though quantitative methods may be convincing to some stakeholders. To this point, these authors emphasize, and data support, that source water protection remains a significant and essential part of water supply management, even when filtration is in place. The difficulty of fully quantifying benefits of source water protection programs reinforces the importance of supporting source water protection on two fronts. First, the economic benefits of protection should be calculated through models that are improved when possible; and second, the "inner salmon" approach should be undertaken as well, appealing to the intangible aspects of people's values through extensive educational programs. By quantifying program benefits and maintaining public buy-in, land management programs can continue to provide crucial water quality protection benefits to the New York City drinking water supply system.

4.6 SUMMARY TRENDS AND RECOMMENDATIONS

4.6.1 Trends, Status and History of the New York City Watershed

- In 1900, the population within the 375-square-mile Croton watershed area was about 20,000; but by 2000, the population within the watershed was more than 100,000. Today, less than 15 percent of the area remains wooded and undeveloped.
- The Catskill watershed is mountainous and drains to the Hudson River, while the rolling hills of the Delaware region drain to the Delaware River. Population models show actual growth within the watershed boundaries has been minimal. Today, the city owns 10.9 percent of the land, including conservation easements, with an additional 20 percent of the watershed protected through the state's Catskill Forest Preserve program with 75 percent classified as forested.
- The 1996 amendments to the Safe Drinking water Act recognized that source water protection is an integral part of the multiple barrier approach to ensuring safe drinking water, mandating and funding Source Water Assessment and Protection Programs.
- The New York City Watershed Memorandum of Agreement, signed on January 21, 1997, established land acquisition requirements, creating the NYC Watershed Protection and Partnership Council and corresponding watershed protection provisions and programs.

- The NYC DEP has developed and continues to implement three main source water protection programs: (1) the Land Acquisition Program; (2) Watershed Protection Partnership Programs that include the subprograms watershed forestry, wetlands protection, stream management, waterfowl management, and agricultural pollution prevention planning, as well as public outreach and education; and (3) capital programs that include subprograms sewer extension, septic system rehabilitation and replacement, storm water retrofit, and wastewater treatment.

4.6.2 Future Threats and Suggested Actions

- Outside natural forces of insects, disease and ungulates are the most visible threat to the New York watershed forests, threatening the regeneration of forests and their future resiliency to climate change, storms, and unpredictable disturbances. To address threats to forest health the NYC DEP must engage in public outreach programs.
- Abundant gas reserves are present and being aggressively developed in 31 states, including New York in the Marcellus shale formation. Due to jurisdictional restrictions, the NYC DEP has limited options to address threats from hydraulic fracturing in the Catskill–Delaware watershed. This issue may exacerbate upstate and downstate sociopolitical divides, given the uncertain risks of gas extraction and the certainty of economic development potential.
- Pharmaceutical and personal care products have been cited by the NYC DEP as a potential threat to the safety of the New York City watershed. NYC DEP will need to continue to refer to the EPA for new regulations while also proactively communicating with and engaging watershed stakeholders to implement programs to address PPCP contamination.
- NYC DEP could initiate a public outreach program based on (1) finding a watershed-related icon that the public embraced and (2) aligning the potential rate increase with a daily item that minimized the impact of the rate increase in the mind of the public. In addition, by inviting stakeholders to help map out the key issues, NYC DEP can match actions to local priorities, creating a sense of ownership among stakeholders by engaging them in the decision-making process.

4.6.3 Summary Recommendations

The New York City watershed protection efforts demonstrate that it is possible to meet both downstream water quality objectives and upstream economic development goals through voluntary partnerships and implementation of watershed-based stewardship. Key elements that contributed to the success of NYC's efforts to maintain filtration avoidance for the Catskill–Delaware water supply systems include the following:

- Supporting early stakeholder involvement in participatory planning processes guided by local leadership
- Establishing watershed-wide development policies that are connected to incentives that support implementation of best management practices and encourage long-term planning to ensure system sustainability

Both interpersonal relationships and proactive watershed protection initiatives will continue to play important roles in the success of New York City's drinking water management efforts.

New York City will always have to work with landowners within a larger watershed area that don't necessarily see watershed protection as a primary goal. The ability of city officials to work with these landowners and impress upon them the importance of compatible land stewardship will continue to be of the utmost importance. Up to this point, the partnerships created by the DEP and other watershed organizations have proved capable of managing conflict between upstate and downstate expectations of the NYC water supply watersheds, leading to a watershed that can still supply unfiltered water. It is important that these relationships remain intact.

Additionally, it will be essential to continue and expand proactive systemwide management planning and implementation to address current and future water security threats. Recommendations developed from examination of water supply trends in the New York City system include the following:

- Continuing and expanding forest management planning in and around sensitive watershed areas
- Implementing risk management education campaigns to address natural and anthropogenic water quality threats

To support these efforts, the DEP can continue its robust, incentive-laden, partnership-based programs, as it has in the past. Additional, NYC DEP can further communicate the value of source water protection programs by

- Calculating economic benefits when possible
- Incorporating appeals to intangible values through continued and expanded education and outreach

It is also important that the DEP develop decision-making tools that acknowledge the complexity of oncoming threats and the potential need to filter because of them. Such tools should take many shapes initially during their test phases and should be adaptive to address changes in water quality threats as well as management goals and program outcomes. Some tools, such as a quantified and adaptive matrix for optimizing decision making, strategic cost curves, and paired watershed studies, have been discussed in this chapter. Moreover, it will remain essential to continue to engage stakeholders in participatory watershed-wide planning to address current and future threats. That New York's water has remained unfiltered this long is a testament to both out-of-the-box thinking and strategic, continued investment by the city. Looking forward, the city will have to work to continue both of these trends.

ENDNOTES

1. However, the Association of State Drinking Water Administrators laments "the Federal funds dedicated to source water assessments are no longer available, and in many states, no regulatory mechanisms exist to compel water systems to use the building blocks of source water assessment to implement a source water protection plan" (ASDWA/GWPC, 2006).
2. The 1997 MOA is discussed in more detail in Section 4.3.3.
3. In the Croton watershed there are a variety of other programs in addition to watershed regulations, land acquisition, WAP and WWTP upgrades, operating at a smaller scale than in the Catskill–Delaware watershed. The largest among these are the East-of-Hudson Nonpoint Source Program, which focuses on mitigating common sources of nonpoint source pollution like storm and sanitary sewers. A variety of other programs to monitor and protect Croton water quality are reviewed in Freud (2003).
4. The Final Environmental Impact Statement for the 2012 LAP contemplates resolicitation of solicited land not yet acquired as well as additional solicitation criteria, including prioritization of parcels adjoining previously acquired land (DEP FEIS, 2010).
5. The existing Public Water Supply Permit for the current LAP will expire in January 2012, and NYC DEC submitted an application for a new permit in January 2010 with permit approval requested prior to January 2012 in order to continue LAP efforts (DEP FEIS, 2010).
6. That is, source water fecal coliform bacteria and turbidity levels, distribution total coliform bacteria and disinfection by-product levels, pathogen and virus inactivation efficacy, and maintenance of entry point and distribution system and disinfectant concentrations.
7. In 2009 congressional representatives introduced the Fracturing Responsibility and Awareness of Chemicals Act, FRAC Act, before the House and Senate to regulate the practice under the SDWA and require disclosure of the chemicals; howeverl these bills have not moved beyond subcommittee consideration (see H.R. 2766, S. 1215, 111th Congress, 2009–2010).
8. NYC DEP reported that PPCPs that have been detected include atenolol, caffeine, carbamazepine, ibuprofen, estrogen, and trimethoprim, and relates that "[w]hile PPCPs can originate from numerous sources, effluents from wastewater treatment plants (WWTPs) have been identified as a significant source to surface waters" (DEP PPCPs, 2010).
9. The study was conducted over a three-year period and data were collected from 12 separate water treatment facilities in Texas with turbidity ranges from 5.85 to 89.16 NTU. Results showed that a 1 percent increase in surface water turbidity resulted in a 0.25 percent increase in chemical costs (Dearmont et al., 1998).

REFERENCES

Alsop, P., (2009, November). Invasion of the Longhorn Beetles: In Worcester, Massachusetts, Authorities Are Battling an Invasive Insect That Is Poised to Devastate the Forests of New England. *Smithsonian Magazine.* http://www.smithsonianmag.com/science-nature/Invasion-of-the-Longhorns.html

American Society of Civil Engineers (ASCE), 2005-11. Croton Water Supply System. http://www.ascemetsection.org/content/view/341/875/ Accessed 01/30/12.

Appleton, A., (2002). How New York City Used an Ecosystem Services Strategy Carried Out through an Urban-Rural Partnership to Preserve the Pristine Quality of Its Drinking Water and Save Billions of Dollars. Forest Trends, Tokyo.

Appleton, A., (2006). Regional Landscape Preservation and the New York City Watershed Protection Program: Some Reflections, in: Goldfeld, K. S. (Ed.), *The Race for Space: The Politics and Economics of State Open Space Programs*. Princeton University. 39–72.

Appleton, A., (2009). Regional Landscape Preservation and the New York City Watershed Protection Program: Some Reflections. Presentation at Princeton University.

ASDWA/GWPC (Association of State Drinking Water Administrators / Ground Water Protection Council), (2006). Elements of an Effective State Source Water Protection Program. Available at http://www.gwpc.org/e-library/documents/general/Elements%20of%20an%20effective%20state%20source%20water%20protection%20program.pdf (Accessed 07/07/11).

Bancroft, J.S., and Smith, M.T., (2001). Modeling Dispersal of the Asian Longhorned Beetle. USDA Agricultural Research Service. Available at http://www.uvm.edu/albeetle/research/DispModl.pdf (Accessed 07/15/11).

Bishop, P.L., Hively, W.D., Stedinger, J.R., Rafferty, M.R., Lojpersberger, J.L., and Bloomfield, J.A., (2005). Multivariate Analysis of Paired Watershed Data to Evaluate Agricultural Best Management Practice Effects on Stream Water Phosphorus. *Journal of Environmental Quality*, 34: 1087–1101.

Boesch, D.F., Brinsfield, R.B., and Magnien, R.E., (2001). Chesapeake Bay Eutrophication: Scientific Understanding, Ecosystem Restoration, and Challenges for Agriculture. *Journal of Environmental Quality*, 30: 303–320.

Brooks, K.N., Ffolliott, P.F., Gregerson, H.M., and DeBano, L.F., (2003). *Hydrology and the Management of Watersheds*, Iowa State University Press, Ames, IA.

Burns, D., T. Vitvar, et al., (2005). Effects of suburban development on runoff generation in the Croton River basin, New York, USA. *Journal of Hydrology* 311, 266-281. http://www.cof.orst.edu/cof/fe/watershd/pdf/2005/Burns_et_al_2005.pdf. Accessed 01/30/12.

Catskill Center for Conservation and Development. (2009). Educational programs. http://www.catskillcenter.org/index.php?option=com_content&view=article&id=79&Itemid=111

Cote, S.D., Rooney, T.P., Tremblay, J.P., Dussault, C., and Waller, D.M., (2004). Ecological Impacts of Deer Overabundance. *Annual Review of Ecol. Evol. Syst.*, 35: 113–147.

Crocker, S. J., Moser, W. K., Hansen, M. H., and Nelson, M. D., (2007). The spatial distribution of riparian ash: Implications for the dispersal of the emerald ash borer. In McRoberts, R. E., Reams, G. A., Van Deusen, P. C., and McWilliams, W. H., (Eds.) Proceedings of the seventh annual forest inventory and analysis symposium; October 3-6, 2005; Portland, ME. *Gen. Tech. Rep.* WO-77. Washington, DC: U.S. Department of Agriculture, Forest Service: 155–160.

CWC (Catskill Watershed Corporation) (2005). MOA Summary Guide: Land Acquisition Program. http://www.cwconline.org/pubs/moa/moaland.html (Accessed 07/15/11).

Daily, G. (1997). *Nature's Services: Societal Dependence on Natural Ecosystems*. Washington DC: Island Press.

de la Cretaz, A.L. and Barten P.K., (2007). *Land Use Effects on Stream Flow and Water Quality in the Northeastern United States*. Boca Raton, FL: CRC Press.

Dearmont, D., McCarl, B.A., and Tolman, D.A., (1998). Costs of Water Treatment Due to Diminished Water Quality: A Case Study in Texas, *Water Resource. Res.*, 34(4), p. 849–853.

DEP (Department of Environmental Protection, New York City), (2004, June 30) Final Environmental Impact Statement: The Extended New York City Watershed Land Acquisition Program.

DEP, (2006). Long Term Watershed Protection Plan.

DEP, (2006a). Water Conservation Plan. Available at http://www.awwa.org/files/resources/waterwiser/references/pdfs/GENERAL_NYC_Water_Conservation_Plan_2006.pdf (Accessed 07/17/11).

DEP, (2007). New York City's Water Supply System. http://www.nyc.gov/html/dep/html/drinking_water/wsmaps_wide.shtml (Accessed 07/17/11).

DEP, (2007a). Croton Watershed Map. http://www.nyc.gov/html/dep/html/dep_projects/croton_wide.shtml (Accessed 07/17/11).

DEP, (2007b). Catskill/Delaware Watershed Map. http://www.nyc.gov/html/dep/html/dep_projects/catdel_wide.shtml (Accessed 07/17/11).

DEP, (2008). 2008 Drinking Water Supply and Quality Report. http://www.nyc.gov/html/dep/pdf/wsstate08.pdf

DEP, (2009). 2008 Watershed Water Quality Annual Report.

DEP, (2011). Ashokan. http://www.nyc.gov/html/dep/html/watershed_protection/ashokan.shtml (Accessed 07/17/11).

DEP (NYC). (July, 2011). Semi-Annual Report: East of Hudson Nonpoint Source Pollution Control Program. http://www.nyc.gov/html/dep/pdf/reports/fad_4.9_eoh_non-point_source_pollution_control_program_semi-annual_report.pdf. Accessed 01/30/12.

DEP, (2012). History of New York City's Water Supply System. http://www.nyc.gov/html/dep/html/drinking_water/history.shtml. Accessed 01/30/12.

DEP FEIS, (2010, December 10). Final Environmental Impact Statement: The Extended New York City Watershed Land Acquisition Program. Available at http://www.nyc.gov/html/dep/pdf/reviews/land_acquisition_program/extended_lap_feis_.pdf (Accessed 05/20/11).

DEP FIAR, (2009, December. Final Impact Assessment Report: Impact Assessment of Natural Gas Production in the New York City Water Supply Watershed. Available at http://www.nyc.gov/html/dep/pdf/natural_gas_drilling/12_23_2009_final_assessment_report.pdf (Accessed 4/04/11).

DEP LTLAP, (2009). Long Term Land Acquisition Plan 2012–2022. Bureau of Water Supply, Division of Watershed Lands and Community Planning,

DEP PPCPs, (2010, May 26). Occurrence of Pharmaceutical and Personal Care Products (PPCPs) in Source Water of the New York City Water Supply. Available at http://www.nyc.gov/html/dep/pdf/quality/nyc_dep_2009_ppcp_report.pdf (Accessed 07/07/11).

DEP RIAR, (2009, September). Rapid Impact Assessment Report: Impact Assessment of Natural Gas Production in the New York City Water Supply Watershed. Available at http://www.nyc.gov/html/dep/pdf/natural_gas_drilling/rapid_impact_assessment_091609.pdf (Accessed 4/07/11)

DEP WPPS, (2011, March). 2011 Watershed Protection Program Summary. http://www.nyc.gov/html/dep/pdf/reports/fad51_wmp_2011_fad_assessment_report_03-11.pdf (Accessed 07/15/11).

DPR (Department of Parks and Recreation, New York City), (2009). Asian Long-horned Beetle Alert. http://www.nycgovparks.org/sub_your_park/trees_greenstreets/beetle_alert/beetle_alert.html (Accessed 4/04/11).

Ellis, J. B. (2006, November) Pharmaceutical and Personal Care Products (PPCPs) in Urban Receiving Waters. Environmental Pollution, Soil and Sediment Remediation (SSR) 144:1, 184–189.

EPA (U.S. Environment Protection Agency), (1986, June 20). Press Release: President Signs Safe Drinking Water Act Amendments.

EPA, (1989, June 29). Surface Water Treatment Rule. 54 Fed. Reg. 27486, 40 C.F.R. §§ 141 et seq.

EPA, (2000, May). The 1997 Filtration Avoidance Determination Mid-Course Review for the Catskill/Delaware Water Supply Watershed.

EPA, (2001, June). Drinking Water Priority Rulemaking: Microbial and Disinfection Byproduct Rules, 816-F-01-012.

EPA, (2005). Technologies and Costs Document for the Final Long Term 2 Enhanced Surface Water Treatment Rule and Final Stage 2 Disinfectants and Disinfection Byproducts Rule, 815-R-05-013.

EPA, (2007a). Effect of Treatment on Nutrient Availability. Total Coliform Rule Issue Paper.

EPA, (July, 2007b). Surface Water Treatment Rule Determination for New York City's Catskill/Delaware Water Supply System. http://www.epa.gov/region2/water/nycshed/2007finalfad.pdf. Accessed 01/30/12.

EPA, (August, 2007c). *Water Quality Trading Toolkit for Permit Writers*. http://water.epa.gov/type/watersheds/trading/WQTToolkit.cfm. Accessed 01/30/12.

EPA, (2010). Safe Drinking Water Act. Available at http://www.epa.gov/safewater/sdwa/. (Accessed 04/01/11)

EPA, (2011). National Primary Drinking Water Regulations. http://water.epa.gov/drink/contaminants/index.cfm (Accessed 07/07/11).

EPA DBP Rule, (2010, August). Comprehensive Disinfectants and Disinfection Byproducts Rule (Stage 1 and Stage 2): Quick Reference Guide. Available at http://www.epa.gov/ogwdw000/mdbp/qrg_st1.pdf (Accessed 07/07/11).

EPA MOA, (1997, January 21). New York City Watershed Memorandum of Agreement. Available at http://www.epa.gov/region02/water/nycshed/nycmoa.htm (Accessed 4/04/11).

EPA PPCPs, (2010). Pharmaceuticals and Personal Care Products as Pollutants (PPCPs): Frequent Questions. Available at http://www.epa.gov/ppcp/faq.html (Accessed 4/03/11)

EPA SDWAA, (2010). Safe Drinking Water Act Amendments of 1996. http://water.epa.gov/lawsregs/guidance/sdwa/theme.cfm (Accessed 04/01/11).

EPA WQT, (2010). Water Quality Trading Appendix G, Sediment and Nutrient Trades with Forestry and Drinking Water Treatment Facility (Added May 2009) http://www.epa.gov/npdes/pubs/wqtradingtoolkit_app_g_sediment_nutrient.pdf (Accessed 4/19/11).

Evans, A.M., and T.G. Gregoire, (2006). A geographically variable model of hemlock woolly adelgid spread. *Biol. Invasions*. 9:369–382.

Faust, S.D., and Osman, A.M., (1998). *Chemistry of Water Treatment*. Boca Raton, FL: CRC Press.

Finnegan, M.C., (1997). New York City's watershed agreement: A lesson in sharing responsibility. *Pace Environmental Law Review* 14(2): 577–644.

Foster, D.R., Motzkin, G., and B. Slater (1998). Land-Use History as Long-Term Broad-Scale Disturbance: Regional Forest Dynamics in Central New England. *Ecosystems*. 1: 96–119.

Freud, S., et al., (2003). Why New York City Needs a Filtered Croton Supply. Department of Environmental Protection, NYC.

Gandy, M., (1997). The Making of a Regulatory Crisis: Restructuring New York City's Water Supply. *Transactions of the Institute of British Geographers, New Series*, 22(3): 338–358.

Gordon, E., and Pugh, E., (2011). A Conceptual Model of Water Quality Impacts from Insect-Induced Tree Death in Snow-Dominated Coniferous Forests. Western Water Assessment. University of Colorado, Boulder. Presentation. Available at http://wwa.colorado.edu/ecology/beetle/docs/2011presentations/2011mpb_gordon.pdf (Accessed 07/07/11).

Goss, M., and Richards, C., (2008, June). Development of a Risk-Based Index for Source Water Protection Planning, Which Supports the Reduction of Pathogens from Agricultural Activity Entering Water Resources. *Journal of Environmental Management,* 87:4 623–632.

Groffman, P.M., Law, N.L., Belt, K.T., Band, L.E., and Fisher, G.T., (2004). Nitrogen Fluxes and Retention in Urban Watershed Ecosystems. *Ecosystems*, 7: 393–403.

Haack, R.A., Law, K.R., Mastro, V.C., Ossenbrugen, H.S., and Raimo, B.J., (1997). New York's Battle with the Asian Long-Horned Beetle. *Journal of Forestry*, 95(12): 11–15.

Hecq, P., Hulsmann, A., Hauchman, F.S., McLain, J.L., and Schmitz, F., (2006). Drinking Water Regulations. in: *Analytical Methods for Drinking Water: Advances in Sampling and Analysis,* P.P. Quevauviller and C. Thompson, (Eds.) John Wiley and Sons Ltd., Chichester, West Sussex, England.

Hrudey, S.E., Hrudey, E.J., and Pollard, S.J.T., (2006). Risk Management for Assuring Safe Drinking Water. *Environmental International*, 32(8): 948–957.

Kane, K., (2007). *Evaluation of the Phosphorus Offset Pilot Program*. Unpublished manuscript. New York City Department of Environmental Protection.

Kane, K., (2010, April). Personal communication. Director of Watershed Management Studies, New Yprk City, Department of Environmental Protection.

Kobell, R., (2009, December). Marcellus Shale: Pipe Dreams in Pennsylvania? *Chesapeake Bay Journal*. Available at http://www.bayjournal.com/article.cfm?article=3715 (Accessed 04/02/11).

Linvok I., Satterstrom, F.K., Kiker, G., Batchelor, C., Bridges, T., and Ferguson, E., (2006). From Comparative Risk Assessment to Multi-Criteria Decision Analysis and Adaptive Management: Recent Developments and Applications. *Environment International*, 32(8): 1072–1093.

Lipsky, D., Martin, T., Hurley, J., and Glasser, C., (2009). Pharmaceutical and Personal Care Products in the NYC Watershed. New York City Department of Environmental Protection Presentation. Available at http://www.dos.state.ny.us/watershed/pptlist2009.html (Accessed 4/04/11).

Lisberg, A., (2010, January 31). Gov. Paterson in a Rush to Begin 'Hydrofracking' Drilling Near City's Upstate Reservoirs. *New York Daily News*. Available at http://www.nydailynews.com/ny_local/2010/01/31/2010-01-31_daves_rush_to_drill_upstate.html (Accessed 4/04/11).

Lloyd, E., (2008, July 18) Letter to Alexander B. Grannis. Re: S. 8169- A/A Oil and Gas Spacing Bill.

Lustgarten, A., (2009, August 25). EPA: Chemicals Found in Wyo. Drinking Water Might Be from Fracking. Pro Publica. http://www.propublica.org/article/epa-chemicals-found-in-wyo.-drinking-water-might-be-from-fracking-825 (Accessed 07/17/11).

MA DCR (Massachusetts Department of Conservation and Recreation, Division of Water Supply Protection), (2007). Quabbin Reservoir Watershed System: Land Management Plan 2007–2017.

Modie, N., (2002, August 12). Seattle Water Rate Increase Passed, but Is Trimmed a Bit. *Seattle Post-Intelligencer Reporter*.

Moffett, K., Atamian, A., How, C., Wordsman, L., and Kane, K., (2003). *Evaluating Management Scenarios in the Crotin Watershed.* Malcolm Pirnie Inc., White Plains, NY.

Navarro, M., (2010, April 23). State Decision Blocks Drilling for Gas in Catskills. *The New York Times*. Available at http://www.nytimes.com/2010/04/24/science/earth/24drill.html?hp (Accessed 07/15/11).

NRC (National Research Council), (2000). *Watershed Management for Potable Water Supply: Assessing the New York City Strategy. Water Science and Technology Board, Commission on Geosciences, Environment and Resources*. National Academy Press.

NRI (Net Resources International), (2011). Croton Water Filtration Plant, New York, US, Specifications. Water-Technoloty.Net. http://www.water-technology.net/projects/crotonfiltration/specs.html (Accessed 07/07/11).

NYC IBO (New York City Independent Budget Office), (2002, March 18). Letter from Merrill Pond, IBO Senior Budget and Policy Analyst to Ms. Breen of the New York Public Interest Research Group. Available at http://www.ibo.nyc.ny.us/iboreports/watershedlet.pdf (Accessed 07/07/11).

NYC Water Board. 2011. Public information regarding water and waste water rates. http://www.nyc.gov/html/nycwaterboard/pdf/blue_book/bluebook_2012.pdf

NYS DEC (New York State Department of Environmental Conservation), (2011a, June 30). Press Release: New Recommendations Issued in Hydraulic Fracturing Review.

NYS DEC, (2011b). Marcellus Shale: Gas well drilling in the Marcellus Shale. http://www.dec.ny.gov/energy/46288.html (Accessed 07/07/11).

NYS DEC EAB, (2011, June 16). Press Release: Emerald Ash Borer Detected in Buffalo. http://www.dec.ny.gov/press/75097.html (Accessed 07/20/11).

NYS DEC EAB Plan, (2011, January 12). Emerald Ash Borer Management Response Plan. Bureau of Public Land Services Available at http://www.dec.ny.gov/docs/lands_forests_pdf/eabresponseplan.pdf (Accessed 07/20/11).

NYS DOH (New York State Department of Health), (November, 1999). Final New York State Source Water Assessment Program Plan. http://www.health.ny.gov/environmental/water/drinking/swapp.pdf. Accessed 01/30/12.

NYS DOH (New York State Department of Health), (2008, December). Drinking Water Infrastructure Needs of New York State. Available at http://www.health.state.ny.us/environmental/water/drinking/docs/infrastructure_needs.pdf (Accessed 07/07/11).

Orwig, D.A., and D.R. Foster, (1998). Forest response to the introduced hemlock woolly adelgid in Southern New England, *USA. J. Torrey Bot. Soc.* 125(1): 60–73.

Pereira, M. A., (2009). Health Risk of the Trihalomethanes Found in Drinking Water Carcinogenic Activity and Interactions: Final Report. United States Environmental Protection Agency. Available at http://cfpub.epa.gov/ncer_abstracts/index.cfm/fuseaction/display.abstractDetail/abstract/22/report/F (Accessed 07/07/11).

Pickett, S.T.A., Belt, K.T., Galvin, M.F., Groffman, P.M., and J.M. Grove, (2007, June). Watersheds in Baltimore, Maryland: Understanding and Application of Integrated Ecological and Social Processes. Universities Council on Water Resources. *Journal of Contemporary Water Research and Education*, 136(1): 44–55.

Platt, R.H., Barten, P.K., and Pfeffer, M. J., (2000). A Full, Clean Glass? Managing New York City's Watersheds. *Environment: Science and Policy for Sustainable Development*, 42: 5.

Porter, K.S. (2006). Fixing Our Drinking Water: From Field and Forest to Faucet. *Pace Environmental Law Review*, 23(2): 389–422.

Principe, M.A., Janus, L., and Warne, D., (2000). Assessing the Effectiveness of New York City's Watershed Protection Program. New York City Department of Environmental Protection, Valhalla, NY.

Reckhow, K.H., Beaulac, M.N., and Simpson, J.T., (1980). *Modeling Phosphorus Loading and Lake Response under Uncertainty: A Manual and Compilation of Export Coefficients*. Report 440/5-80-11. Washington, DC: US Environmental Protection Agency.

Riverkeeper, (2008). Multi-Barrier Approach to Protect NYC Drinking Water in Croton Watershed. http://www.riverkeeper.org

Robin, J., (2011, July 1). Despite Risks, Top Official Calls Hydrofracking Safe. *New York 1 News*. Time Warner Cable Inc. Available at http://www.ny1.com/content/news_beats/politics/142114/despite-risks—top-environmental-official-calls-hydrofracking-safe (Accessed 07/07/11).

Rush, P., and Holloway, C., (2009). Filtration Avoidance Annual Report. New York City, Department of Environmental Protection.

Schneeweiss, J. (1997). Watershed Protection Strategies: A Case Study of the New York City Watershed in Light of the 1996 Amendments to the Safe Drinking Water Act. *Villanova Environmental Law Journal*, 9: 77–119.

SDWA (Safe Drinking Water Act of 1974), (1974, December 16). 42 USC § 300(f).

SDWA, (1986, June 19). SDWA Amendments of 1986. Pub. L. § 99-339.

SDWA, (1996, August 6). SDWA Amendments of 1996. Pub. L. § 104-182.

SDWA Amendments, (1989). Ohanian, E. J., Gilbert, C. E. and H. Pastides (Eds.). Library of Congress.

Smith, P. D., Keesler, D., Valade, M., Schaefer, J., and G. Kroll (2011). New York City Embraces UV Disinfection of its Water Supplies. Horizons, Hazen and Sawyer.

SPU (Seattle Public Utilities), (2007). Drinking Water Quality Report 2007: A Report to the Community. http://www.seattle.gov/util/stellent/groups/public/@spu/@ssw/documents/webcontent/spu01_003730.pdf (Accessed 07/11/11).

SPU (1995–2011). Salmon Friendly Seattle. http://www.seattle.gov/util/About_SPU/Management/SPU_&_the_Environment/SalmonFriendlySeattle/index.htm (Accessed 07/11/11).

Stradler, B., Muller, T. and Orwig, D., (2006, July). The Ecology of Energy and Nutrient Fluxes in Hemlock Forests Invaded by Hemlock Woolly Adelgid. Ecological Society of America. *Ecology*, 87(7): 1792–1804.

SWTR (Surface Water Treatment Rule of 1989), (1989). 40 CFR § 141.70.

Tiemann, M. (2010, July 27). Safe Drinking Water Act: Selected Regulatory and Legislative Issues. Congressional Research Service. Available at http://www.fas.org/sgp/crs/misc/RL34201.pdf (Accessed 07/07/11).

Tong, S.T.Y., and Chen, W., (2002). Modeling the Relationship between Land Use and Surface Water Quality. *Journal of Environmental Management*, 66: 377–393.

USDA (United States Department of Agriculture), Animal and Plant Health Inspection Service, (2011). Emerald Ash Borer. http://www.aphis.usda.gov/plant_health/plant_pest_info/emerald_ash_b/index.shtml. (Accessed 04/21/11).

USFS (United States Forest Service), (2009). Emerald Ash Borer. Report to Congress. http://www.fs.fed.us/aboutus/budget/requests/DDBR428_Emerald_Ash_Borer_report_to_Congress.pdf. Accessed 01/30/12.

USFWS (United States Fish and Wildlife Service), (1997). Geomorphic Provinces and Sections of the New York Bight Watershed. Significant Habitats and Habitat Complexes of the New York Bight Watershed. US Department of the Interior.

WAC (Watershed Agricultural Council), (2012). http://www.nycwatershed.org. Accessed 01/30/12.

Wang, L., Lyons, J., and Kanehl, P., (2002). Effects of Watershed Best Management Practices on Habitat and Fish in Wisconsin Streams. *Journal of American Water Resources Association*, 38: 663–680.

Warne, D., (2010a, March). Optimizing Green and Gray Drinking Water Infrastructure: Policy Issues. Presentation.

Warne, D., (2010b, March–April). Personal communication. Assistant Commissioner, New York City Department of Environmental Protection.

Weimer, D.L., (2005). *Policy Analysis: Concepts and Practice*. Upper Saddle River, N.J.: Pearson Prentice Hall.

Westchester, County of, (2010). Land Use in Westchester: A Detailed Look at Existing Conditions and Development Trends. County Board of Legislators Department of Planning. http://www.westchestergov.com/planningdocs/Reports/LandUseReport1.pdf. Accessed 01/30/12.

Wilder, A., and Kiviat, E., (2008). The Functions and Importance of Forests, with Applications to the Croton and Catskill/Delaware Watersheds of New York. Report to the Croton Watershed Clean Water Coalition. http://www.ellahhh.net/uploads/resourcelibrary/Forests.pdf_.pdf. Accessed 01/30/12.

Woods, K.D., (2004). Intermediate disturbance in a late-successional hemlock-northern hardwood forest. *Journal of Ecology* 92:464–476.

CHAPTER 5

The Crooked River Watershed, Sebago Lake, and the Drinking Water Supply for the City of Portland, Maine

Jennifer Hoyle

CONTENTS

5.0 EXECUTIVE SUMMARY

The Crooked River watershed is unique among rural watersheds in New England given the wide range of values placed on the watershed land and on the Crooked River itself. The river supplies more than 40 percent of the surface water inflow to Sebago Lake. Sebago Lake is not only a recreation destination for southern Maine but also the municipal water supply source for 11 communities and approximately 200,000 customers in and around Portland, Maine, the largest and fastest growing urban center in the state. Recognition of development pressures on the watershed land have led to projected high future risk to surface water supply. Consequently, management of Crooked River watershed land for drinking water quality and supply protection has been identified as a special interest of critical importance to the wider water supply delivery and watershed land conservation community. This chapter describes the following:

- The biophysical characteristics of Sebago Lake as a water supply reservoir
- A history of watershed development within the region
- The Watershed Control Program of the Portland Water District
- Water supply trends for Portland
- Development pressures on watershed lands for the region and Maine in general
- Recommendations for the Portland Water District for improved watershed management

In summary, it is important to recognize how trends in surface water management for drinking water quality in the Sebago Lake watershed support the creation of a payment for ecosystem service program in the Crooked River watershed. These trends include decreasing demand for drinking water, increasing stringency of regulations on drinking water quality, development pressures on watershed properties, the need throughout the water supply industry to quantify the benefits of watershed protection, and the facilitation of source water protection in the state of Maine. Because the watershed comprises mostly private, forested properties, it is necessary to create a program that would enable landowners to manage land for surface water quality protection. This program should ideally include a portfolio of market-based incentives to facilitate water quality protection and enhancement through private land management. This chapter highlights necessary components for a complete payment for watershed services process that will achieve this objective.

5.1 INTRODUCTION

The Crooked River watershed has many unique characteristics and uses that make it a river system of high social, economic, and recreational value to both residents of the watershed and stakeholders throughout Maine and New England. Under the State of Maine's Water Classification Program initiated in 1950, the Crooked River is rated entirely class AA. In 2006 only 7 percent of Maine's rivers and streams achieved this classification as an outstanding natural resource that should be preserved for its unique ecological, social, scenic, or recreational importance (Brakeley and Ezor, 2009). No direct discharge of pollutants is allowed into the river without approval from Maine Department of Environmental Protection (Brakeley and Ezor, 2009). This is because aquatic life, dissolved oxygen, and bacterial content in the river are at naturally occurring levels (Brakeley and Ezor, 2009).

The Crooked River provides spawning habitat for approximately 70 percent of the wild fish in Sebago Lake (Maine Rivers, 2009), including one of only four known indigenous populations of landlocked Atlantic salmon (*Salmo salar Sebago*) (Gartner, 2009). According to Maine Rivers, "The Crooked River was identified in the 1982 Maine Rivers Study as one of only seven rivers in Maine that are 'the state's most significant inland fishery rivers'" (Maine Rivers, 2009, p. 1). It is the only one in the southern part of the state. The 1983 Maine State Rivers Act designated the Crooked River as worthy of special protection because of its value as a fishery resource (Maine Rivers, 2009).

Significantly, the Crooked River supplies over 40 percent of the surface water inflow to Sebago Lake, which is the municipal supply reservoir for 200,000 people in 10 communities, including Portland (PWD, 2009). According to the research team of the Forest-to-Faucet Partnership at the University of Massachusetts Amherst and U.S. Department of Agriculture (USDA) Forest Service, the larger Presumscot River basin surrounding the Crooked River watershed ranks very high among eastern New England river basins for development pressure (Gregory and Barten, 2008a). The report notes that this is a common trend, where, in general, watersheds important for drinking water supply have strong pressures on private landowners for development and forestland conversion (Gregory and Barten, 2008b).

Given that the Crooked River watershed currently supplies high-quality water yet is at significant risk for future development pressures, the challenge facing water suppliers and water consumers is how to incentivize private landowners in the Crooked River watershed to manage their land to maintain high surface water quality for both the local ecosystem and the downstream water users. The purpose of this paper is to describe the land management circumstances in the Crooked River watershed, and recommend guidelines for market-based incentive program that would enable landowners to manage watershed land for water quality protection.

5.2 WATER SUPPLY PROTECTION AND THE PORTLAND WATER DISTRICT

5.2.1 Sebago Lake as a Water Supply Reservoir

The Sebago Lake has a surface area of approximately 30,500 acres and contains about 995 billion gallons of water (Table 5.1) (Colby, 2010). Inflow is estimated at 544 million gallons per day (218 million gallons per day from the Crooked River) and outflow at 498 million gallons per day, with 24 million gallons per day used by the Portland Water District (PWD) (Colby, 2010). At current water use rates, Sebago Lake stores over 100 years' worth of drinking water for PWD customers. Sebago Lake is the deepest lake in Maine, with an average water depth of 101 feet and a maximum water depth of 316 feet, 49 feet below sea level (Johnston, 2005). The

Table 5.1 Sebago Lake Water Supply Profile

	Sebago Lake Supply
Category	**Description**
Water System Name:	Portland Water District
Surface Water Source:	Sebago Lake
Water System Type:	Nontransient, community system
Watershed Location:	Windham, Standish, Sebago, Naples, Casco, Frye Island, Raymond, Harrison, Otisfield, Norway, Waterford, Albany Township, Greenwood, and Bethel
Source Surface Water Area:	30,512 acres
Source Watershed Area:	142,009 acres
Maximum & Average Depths	316 ft maximum, 101 ft average
Water Volume in Storage	995 billion gallons
Population Served (customers):	Services: 47,000 +/– Population: 190,000 +/–
Type of Treatment:	Ozonation, chloramination, fluoridation, corrosion control
Filtration:	Filtration waiver
Estimated Daily Water Use:	24 MGD average, 42 MGD peak

Source: SWAP, 2003.

depth of the lake is due to the fact that the underlying nutrient-poor granitic bedrock is more easily eroded than the metamorphic rock found throughout the rest of Maine. Before glacial activity in New England, the Androscoggin River carved the deep valley in the bedrock from the Crooked River to the southern end of Sebago (Johnston, 2005). Glacial activity further eroded and enlarged the basin (Johnston, 2005). The depth of Sebago Lake allows water storage residence time ranging from five to six years. This is a long reservoir residence time compared with that of other unfiltered surface water supply systems throughout New England. For example, according to the Environmental Protection Agency (EPA), the residence time is approximately less than 60 days for New York City's Catskill–Delaware reservoir system (USEPA, 2006).

The 1996 amendments to the Safe Drinking Water Act required states to create Source Water Assessment Programs (SWAPs) to identify the status and future sustainability of drinking water supplies (SDWA, 1996). According to the SWAP report on the Sebago Lake water supply completed in 2003 by the Maine Drinking Water Program, "water quality data collected by the Portland Water District classify Sebago Lake as oligotrophic, which suggests minimal nutrient enrichment from the watershed area. The data show that the lake has low phosphorus and bacteria counts, high dissolved oxygen concentrations and good clarity" (SWAP, 2003). The large size and volume of the lake, combined with the 45-mile long watershed, provides ample opportunity for natural processes to attenuate and mitigate potential contaminants from entering the PWD water supply (SWAP, 2003).

5.2.2 Land Use and Threats to Surface Water Quality in the Sebago Lake Watershed

Sebago Lake is located 10 miles northwest of the City of Portland, Maine. The Sebago Lake watershed (Figure 5.1) is approximately 300,000 acres, stretching more than 50 miles from north to south and 13 miles from east to west at the widest point. Approximately 86 percent of the watershed is undeveloped woodland, 6 to 7 percent is agricultural land, and 7 to 8 percent is residential, commercial, and industrial development (PWD, 2009). While the majority of the watershed land is privately owned, areas of significant conservation ownership or management include the 2,500 acres of shoreline land owned by PWD; the Steep Falls Wildlife Management Area in the Town of Sebago; the Sebago Lake State Park, which surrounds the drainage of the Songo and Crooked Rivers leading into Sebago Lake; the Jug-Town Easement, which is a conservation easement on 3,200 acres owned by the Hancock Lumber Company; and a portion of the White Mountain National Forest located in the Albany township (PWD, 2009).

The SWAP process identified a few nonresidential land uses that could negatively impact surface water quality including gravel pits (located in the Crooked River valley), waterfront recreation activities, human and animal waste storage, and handling and storage of petroleum and related chemicals (both storage and transportation on major roadways in the watershed) (SWAP, 2003). However, because the majority of the land in the watershed is privately owned, the main threat to surface water quality

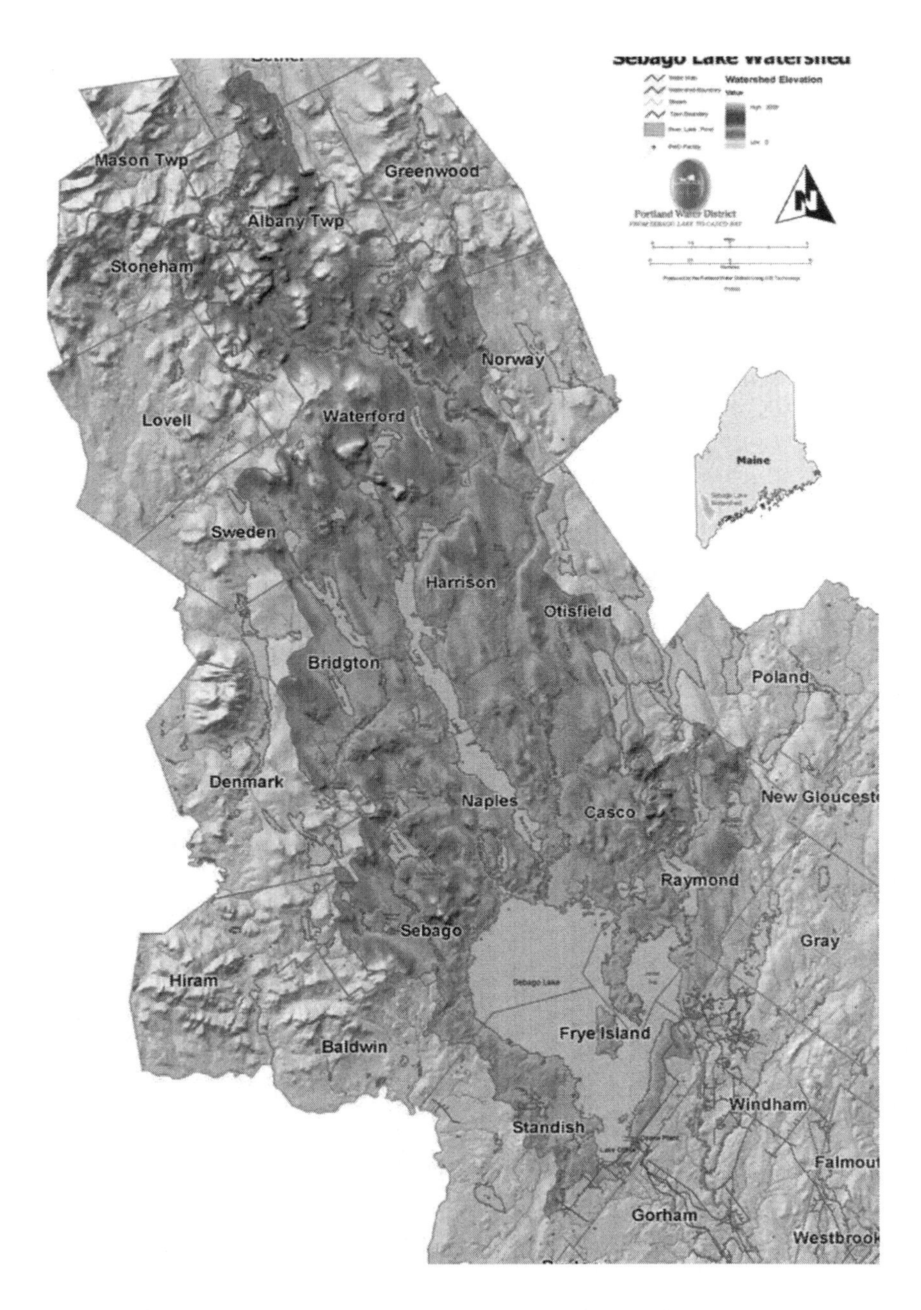

Figure 5. Sebago Lake watershed (From PWD, 2009). (See color insert.)

is development. There are approximately 4,000 parcels partly or entirely within 200 feet of the lakeshore, which are developed or could be developed in the future. As stated in the SWAP report, "shorefront development poses a threat to the quality of the lake from subsurface wastewater disposal, lawn fertilizers, gravel roads and driveways, and increased recreational access" (SWAP, 2003). This assessment does not consider development pressure in the upper watershed within the Route 302 corridor and within the Crooked River watershed.

5.2.3 Development of the Portland Water Supply System

The establishment and expansion of the PWD water supply infrastructure paralleled development of water systems throughout New England (Figure 5.2). On July 4, 1866, a fire destroyed downtown Portland, leaving 12,000 residents homeless (Stickney, 1942). This disaster galvanized the population around the need for a public water system, despite costs (Stickney, 1942). Subsequently, the Portland Water Company was formed as a private utility and on November 18, 1869, construction of a 20-inch-diameter water main was completed connecting Sebago Lake to downtown Portland (Stickney, 1942). Throughout the late nineteenth century the Portland Water Company expanded capacity to meet growing demand, until Portland residents became upset with high rates and a lack of transparency around financial management of the water supply. On March 27, 1907, the legislature of Maine passed laws to incorporate the Portland Water District as a public municipal water supply district (McCarthy, 2008). In 1942, Fred Stickney wrote *History of the Water District,* a paper that details the financial struggle inherent in acquiring the water systems that now make up the PWD, as well as the equipment and infrastructure of the system at the time. In summary, throughout the first half of the 20th century, PWD consistently undertook major capital improvement projects in an effort to ensure water supply capacity that met drinking water demand as well as emergency storage needs. These projects included construction of a 42-inch main to triple the capacity of PWD to deliver water from Sebago to customers in Portland in 1912 (Figures 5.3 and 5.4) and construction of a 60-inch-diameter intake pipe to again increase intake capacity in 1952 (Stickney, 1942). During this time PWD recognized the value of delivery of safe water to customers downstream from Sebago, as manifested in PWD efforts both to protect source water quality and to disinfect and treat water before distribution.

5.2.4 Disinfection and Treatment of Water at the Portland Water District

In 1914, as water systems around the country found that bacteria, parasites, and viruses in water caused and spread disease, the Portland Water District began using chlorine as a disinfectant (Stickney, 1942). In the mid-1920s the PWD board of trustees commissioned a team of experts to evaluate the raw water quality of Sebago Lake and to advise PWD on whether or not to invest in further filtration technologies. The specialists determined that the raw water did not require filtration, as

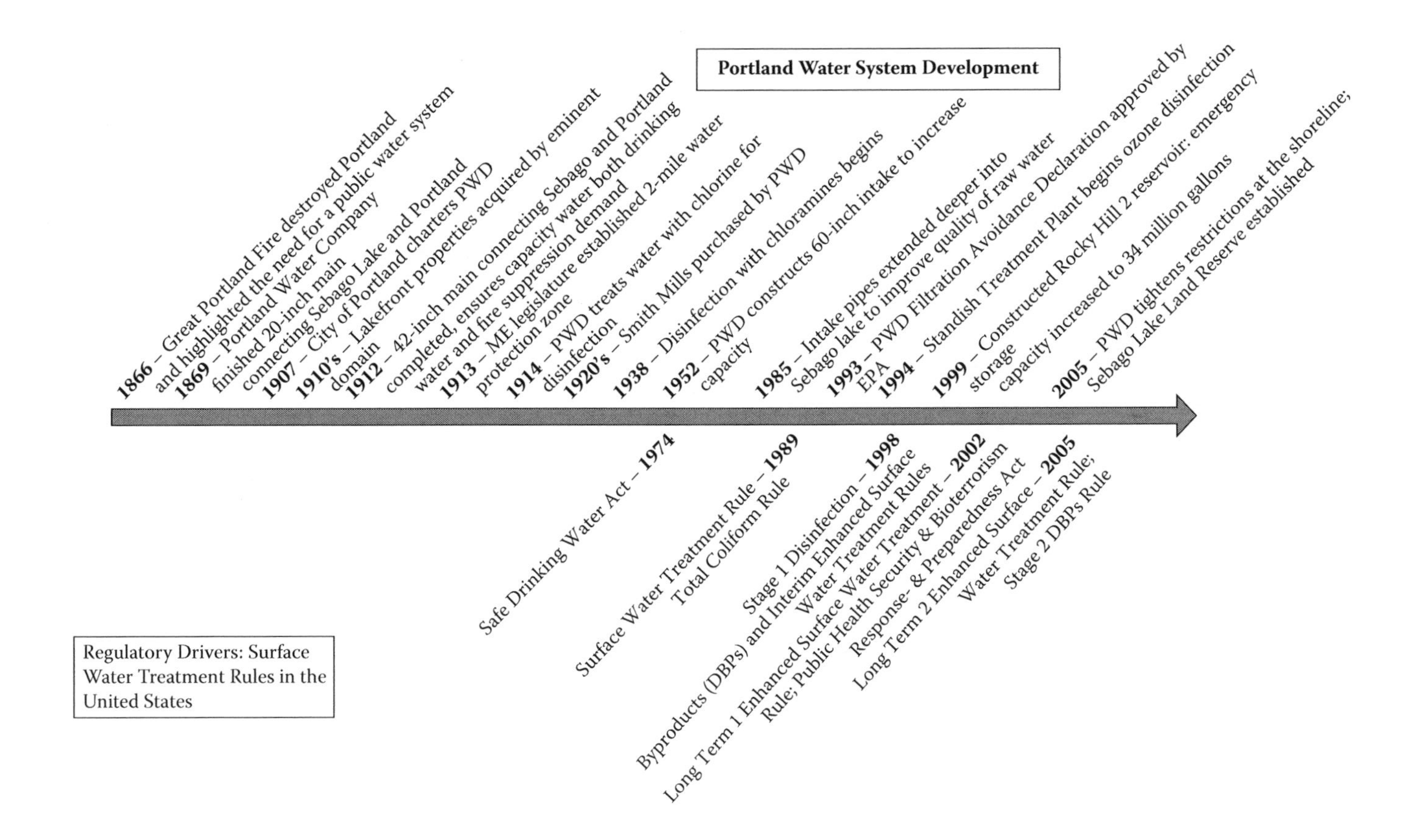

Figure 5.2 Timeline of PWD development (From EPA, 2010; Stickney, 1942; McCarthy, 2008).

Figure 5.3 Photograph of the 1912 installation of the 42-inch main designed to deliver water from Sebago Lake to PWD customers in Portland, Maine (From Stickney, 1942).

explained in the PWD's 1926 Annual Report (see Box 5.1). However, in 1938 PWD began disinfection with chloramines for increased stability and residence time in the growing distribution system (Stickney, 1942).

With the passage of the Surface Water Treatment Rule (SWTR) (SDWAR, 1989), PWD applied for and obtained a waiver from new surface water filtration requirements. This waiver was largely due to the high quality of raw water from Sebago Lake and ongoing watershed protection efforts. A compliance agreement and order setting forth a schedule to meet avoidance criteria was issued to PWD on September 21, 1993, by the Maine Department of Health and Human Services (DHHS) Drinking Water Program (SWAP, 2003), which has primacy to enforce drinking water regulations in Maine (SWAP, 2003). The agreement summarizes the monitoring and reporting PWD completed to prove the consistently high quality of water in Sebago Lake, reviews the regulatory framework for the filtration avoidance, and outlines the PWD-specific and general compliance requirements for unfiltered water systems to remain unfiltered.

The 1986 amendments to the Safe Drinking Water Act instructed EPA to establish criteria under which filtration is required as a treatment technique for public water systems supplied by a surface water source (SDWA, 1986; DHHS, 1993). The 1989 SWTR requires public water systems to provide treatment of source water for

Figure 5.4 Photograph of a major junction of the 42-inch main installed in 1912 (From Stickney, 1942).

3-log (99.9 percent) removal of *giardia* cysts and 4-log (99.99 percent) removal of viruses (SDWAR, 1989). From December 30, 1990, through December 31, 1991, PWD submitted at least six months of data to the DHHS demonstrating that it met these requirements (DHHS, 1993). The conditions for the PWD system to remain unfiltered also include detailed reporting and inspection requirements with water quality parameter concentrations that demanded that (1) fecal coliform concentrations be less than or equal to 20/100 milliliters in at least 90 percent of the measurements made for the previous 6 months; (2) turbidity measurements be made on representative grab samples of raw water prior to disinfection at least once every four hours and that the turbidity level cannot exceed 5.5 NTU; and (3) residual disinfectant concentration must be continuously monitored at specific locations and frequencies (DHHS, 1993; PWD, 2009). The rule also requires maintenance of both appropriately high residual disinfectant concentrations and appropriately low disinfection by-product levels (SDWAR, 1989; PWD, 2009).

The other major requirement from DHHS for a filtration waiver was the construction and operation of an ozone disinfection facility by September 1993 (DHHS, 1993). In 1994 PWD finished construction of a new water treatment plant in Standish that uses ozone as a primary disinfectant (PWD, 2009). Since 1993, both turbidity and fecal coliform levels from Sebago Lake remained well below the compliance levels specified in the compliance agreement (Figures 5.5 and 5.6). In the future PWD must address regulatory requirements for concentrations of *cryptosporidium* included in

BOX 5.1 THE REPORT OF LEONARD METCALF IN THE 1926 ANNUAL REPORT OF THE PORTLAND WATER DISTRICT ON THE QUALITY OF SEBAGO LAKE WATER

"The District has also retained the services of Leonard Metcalf, of the firm of Metcalf & Eddy, and he was consulting engineer of the District until his death in the fall of 1925. Mr. Metcalf was also consulting engineer of the Metropolitan Water Supply of Massachusetts, and has testified in nearly all of the important cases in late years on these matters. These eminent specialists made exhaustive examinations of Lake Sebago and its waters, and unanimously agreed that Sebago water in its natural condition was a first class drinking water and did not require filtration. At the same time the Portland Water Company engaged the services of other eminent sanitary engineers, among whom was Prof. William P. Mason, of the Rennselaer Polytechnic Institutes, author of several standard textbooks and an acknowledged sanitary expert. At the hearing on the valuation of the Portland Water Company system, it was unanimously agreed by the experts on both sides and made a part of the record in the case that Sebago water did not require filtration. Dr. Mason on the stand testified as follows:

"I have considered the water as at least as good and in many cases better than filtered water for this reason, that bacteriologically and chemically it compares favorably, shows equally as good, and a pure water as furnished by nature is always better than a equally pure furnished by art, because art might possibly fail in its operation sometime." (PWD, 1926)

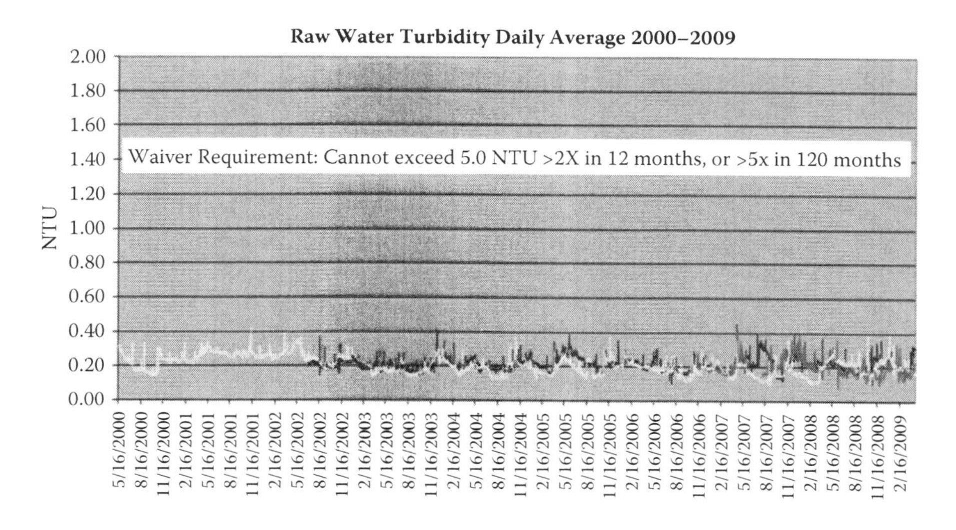

Figure 5.5 Raw water turbidity daily average at PWD from 2000 through 2009 (From Hunt, 2010).

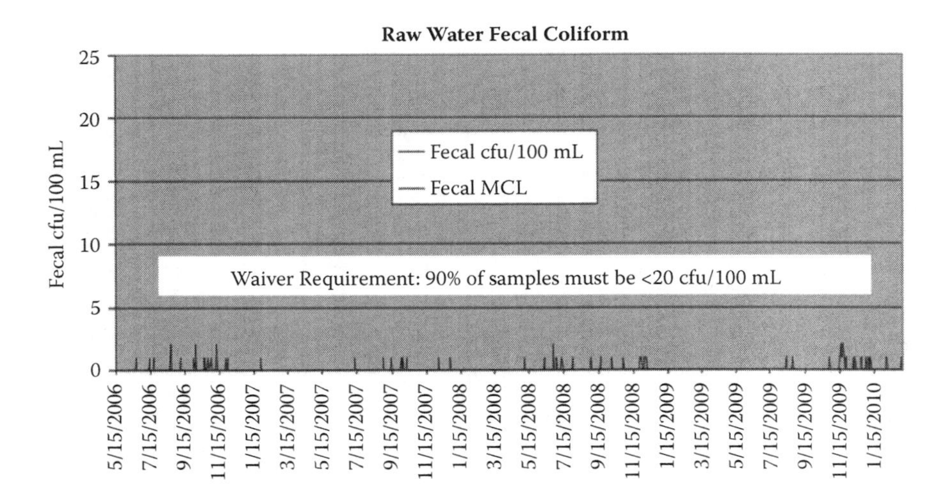

Figure 5.6 Raw water fecal coliform concentrations from May 2006 through January 2010 (From Hunt, 2010).

the Long-Term 2 Surface Water Treatment Rule by modifying the existing plant to include ultraviolet disinfection as the first treatment process (PWD, 2009).

5.2.5 History of Watershed Protection at the Portland Water District

Concurrently with the effort to protect water quality in the distribution system using disinfection, PWD also invested heavily in watershed protection. From 1910 to 1930 the Sebago Lake area was strongly impacted by the creation and oversight of PWD and the realization that human activity around the lake was degrading drinking water quality (Stickney, 1942). Prior to the PWD presence, steamer ships on the lake and shoreline properties would actively discharge wastewater directly into the lake (Stickney, 1942). These practices were halted by PWD using local ordinances and later by the federal Clean Water Act and Safe Drinking Water Act, which aimed to, respectively, protect water quality and prevent spread of waterborne disease into drinking water supplies (SDWA, 1986). Furthermore, PWD purchased much of the shorefront properties around the intake pipes from willing customers where possible and by condemnation when necessary. In 1913, the Maine legislature established a two-mile water protection zone around the water supply intake pipes in Sebago Lake (Stickney, 1942). In the 1920s PWD went so far as to purchase the entire village of Smith Mills, a part of Standish that included the Dupont Saw Mill (Stickney, 1942). PWD's 1926 Annual Report details watershed protection recommendations as well as some of the conflict around PWD control of lakefront properties (see Box 5.2).

Aside from acquisition of lakefront properties from willing sellers whenever possible, PWD's watershed management efforts remained steady until the filtration waiver effort of the early 1990s, prompted by the SWTR as discussed above as well as other amendments to the federal SDWA, which required each state to complete a

BOX 5.2 ADVICE FROM CONSULTANTS ON LAND ACQUISITION IN THE 1926 ANNUAL REPORT OF THE PORTLAND WATER DISTRICT

"The experts did recommend, however, that the District acquire lands around the shore of the lower bay, and that bathing near the intake be prohibited, and that the watershed be policed to prevent violation of the law, all of which has been done. In accordance with these recommendations, the District has expended more than $200,000 in purchasing land in the lower bay, until they now control practically its entire water front, and the Legislature has prohibited bathing within two miles of the intake.

During the past year the Trustees were informed that the law against bathing was being broken, and they ordered that personal notice be given all within the restricted area that the law must be observed. This aroused the resentment of some of the cottage owners, who claimed they should be permitted to bathe in the lake and that the District should remedy any resulting pollution by the construction of a filtration plant...

... The Trustees know ... that no necessity exists for a filtration plant provided the present policy of protecting the supply is continued ...

... The Trustees have, however, caused an investigation to be made to determine the cost of a filtration plant, should the same at any time be found necessary. Such a plant would cost $1,000,000, and the expense of operating it would amount to $50,000 per year, which, with interest on cost, would make an expense of $100,000 per year. As the total water income of the District is only $500,000 this would require a 20 percent increase in water rates throughout the entire District and impose an entirely unnecessary burden upon the water takers.... The Trustees wish to state unequivocally that the water supply of Portland is being constantly guarded and is perfectly safe, and if the Trustees can be upheld in their enforcement of present regulations, filtration will probably never be necessary." (PWD, 1926)

source water assessment to evaluate vulnerability of each public water supply source (see SDWA, 1986; SDWAR, 1989; and SDWA, 1996). These regulatory drivers led to renewed commitment and effort around watershed protection in the late 1990s and into the next decade, which continue today. As part of the application process for the waiver from filtration requirements, Maine DHHS outlined a few requirements for PWD focused on watershed protection, including, but not limited to, the following:

1. The removal of a boat-launching facility within the two-mile water protection zone in the Town of Standish. This requirement was later removed by DHHS.
2. The commissioning of a study to evaluate the effects of ice fishing, specifically ice-fishing shacks, upon water quality.
3. Continued monitoring of and attempts to mitigate poor water quality from Standish Brook.

4. Continued public education activities, especially at points of public access to Sebago Lake, detailing the potential impacts of humans on the public water supply via recreation on the lake (DHHS, 1991).

In addition, in 2003 the SWAP review of threats to public water supply in the Sebago Lake watershed ranked the susceptibility of the water supplied by Sebago Lake (Table 5.2) and provided recommendations for added protection of surface water supply in the future based on an evaluation of the watershed, the shoreline, and the intake area (SWAP, 2003). Overall, the susceptibility of the Sebago Lake water supply to degradation of water quality was evaluated as moderate as compared to low or significant (SWAP, 2003).

The SWAP report credited the PWD Watershed Control Program already in place as a "strong base for protecting the quality of the lake" (SWAP, 2003, p. 17). It also included four additional actions to increase water quality protection given the threats identified throughout the assessment process:

1. Portland Water District should foster more land conservation in the watershed—PWD should facilitate improved communication and coordination among landowners and land trusts in the watershed.
2. Shoreline towns should have development and land use zoning that incorporates protection for the lake. As illustrated in Table 5.3, the towns of Sebago and Frye Island have no land use zoning; this leavesthem vulnerable to badly planned, high-density growth and development.
3. PWD and the Town of Standish should collaborate on an alternative boat launch location. The boat launch located at the intersection of Routes 35 and 114 is currently a risk to water quality due to its proximity to the PWD intake pipes.
4. PWD should advocate for land use practices that are protective of high water quality in the Upper Watershed, especially Crooked River and its tributaries (SWAP, 2003).

5.3 WATERSHED CONTROL PROGRAM OF THE PORTLAND WATER DISTRICT

The recommendations from DHHS through the filtration waiver process and the SWAP have largely shaped the current Watershed Control Program (WCP). The major features of the program are detailed in the Watershed Control Program Annual Report for 2008–2009, which summarizes the watershed management activities and describes the threats to water quality and the measures taken to mitigate them. The major components of the program are (1) water quality monitoring, (2) security, (3) inspection and direct actions, (4) education; (5) source protection outreach, and (6) land acquisition and preservation. While the combination of all of these components results in an extensive watershed protection effort, the review provided here will focus on the components that relate to PWD as a major stakeholder with desires to incentivize private landowners in the Sebago Lake Watershed to manage land for surface water quality protection. The subsequent discussion here details land use issues that lead to particular tensions between PWD and landowners in the watershed.

Table 5.2 Sebago Lake Surface Water Assessment Findings Separated by Zone and Performance Measure

Zone	Measure	Findings	Risk Level
Watershed	Ambient Water Quality	Class GPA,[a] in full compliance for trophic status. Lake water quality is high. Certain tributaries have water with elevated phosphorus and coliform bacteria.	Low-Moderate
	Existing Conditions	Watershed is predominantly forested but includes high-density residential, commercial, and industrial uses. The intensity of development increases close to the lake.	Low-Moderate
	Future Development	Varying zoning control on future development. Approximately half of the towns in the watershed have no zoning, including two shoreline towns. Growth and development pressure is high.	Moderate
	Overall		Low-Moderate
Shoreland	Lake Classification	Oligotrophic	Low
	Soils	Erodible soils are present along certain shoreline segments. Camp roads also contribute to erosion.	Low-Moderate
	Activities Posing a Threat	The shoreline includes several areas with high-density residential development and many shoreline beaches, resorts, and marinas.	Moderate
	Potential for Future Threats	Future shoreland development controlled by zoning. Project review includes the Portland Water District. However, there are approximately 4,000 developed or developable shoreland lots around the lake.	Moderate
	Overall		Moderate
Intake	Raw Water Quality	Lake quality is good and stable. Filtration waiver granted.	Low
	Ownership/ Control	District owns shoreline at intake but there is little conservation ownership along other portions of the shore. District has comprehensive Watershed Control Program.	Low-Moderate
	Activities Posing a Threat	Recreation (boat launch, motorized vehicles, parking, ice fishing) adjacent to intake zone.	Low-Moderate
	Potential for Future Threats	Increased recreation in unrestricted areas, persistent contaminants (e.g. MTBE),[b] invasive plants, and increased shoreline development pose a risk that degraded water quality will encroach into the District's restricted intake zone.	Moderate
	Overall		Low-Moderate
Overall			**Moderate**

Source: SWAP, 2003.

[a] Class GPA is the sole classification of great ponds and natural lakes and ponds less than 10 acres in size. Maine Statute Title 38. See www.mainelegislature.org for more information.

[b] Methyl-t-butyl ether.

Table 5.3 Sebago Lake Watershed Town Land Use Zoning Regulation Summary Showing Lack of Zoning in Many Towns in the Sebago Lake Watershed

Zoning Summary for the Sebago Lake Watershed			
Town	**Status of General Zoning**	**Shoreline Zoning Overlay**	**PWD Review in Shoreland Zone**
Sebago	No specific zoning	Yes	Yes
Standish	Specific zoning adopted	Yes	Yes
Windham	Specific zoning adopted	Yes	Yes
Raymond	Specific zoning adopted	Yes	Yes
Frye Island	No specific zoning	Yes	Yes
Casco	Specific zoning adopted	Yes	Yes
Naples	Village district only	Yes	Yes
Harrison	No specific zoning	Yes	No
Otisfield	No specific zoning	Yes	No
Norway	No specific zoning	Yes	No
Waterford	No specific zoning	Yes	No
Albany Twp.	LURC zoning	Yes	No
Greenwood	No specific zoning	Yes	No
Bethel	No specific zoning	Yes	No

Source: SWAP, 2003.
Note: Although these are rural towns, this leaves watershed land susceptible to poorly planned high-density development, which could lead to impaired surface water quality.

5.3.1 Intake Zone Protection

5.3.1.1 Sharing a Multiuse Lake

In 2005, Paul Hunt, the environmental manager at PWD, wrote "Sebago Lake: Yours, Mine and Ours: On Sharing a Multi-Use Lake" (Hunt, 2005). In this paper, Hunt acknowledged the difficulty of advocating for increased protection when the water in the lake remains of such consistently high quality. Though the ideal situation would be to have one lake serve as the water supply and a second one for water recreation, Sebago Lake fills both roles for greater Portland. Remarkably, after nearly 150 years of serving multiple community uses, the quality of water drawn from the Lower Bay of Sebago Lake remains outstanding. Hunt notes the quality of the water is in part nature, as the size and depth of the lake enable it to absorb pollutants that would impact smaller lakes, and in part nurture, as the restrictions on access and undeveloped land around the Lower Bay of the lake have resulted in the best water quality where it is needed the most (Hunt, 2005).

Hunt also outlines PWD lessons learned in sharing a multiuse lake and offers recommendations to address this tension over the future of Sebago Lake water quality protection. These lessons include what Hunt (2005, p. 3) terms, (1) "separate incompatible uses," (2) "collect and share data," (3) "be an asset, not a monopoly," and (4) "have a vision." While the first three items emphasize reasons for water quality protection and the resulting efforts, the final lesson, "have a vision," states and defends PWD's position as a responsible party, not an authority, on access to the Lower Bay, especially with respect to ice fishing within the two-mile no-bodily-contact area. The vision states "the Portland Water District's Board of Trustees believe that 10 percent of Sebago Lake—the Lower Bay—should be set aside for the sole purpose of protecting the region's drinking water supply, leaving 90 percent for recreational access and responsible shorefront development" (Hunt, 2005, p. 7). If this vision were to become a reality for Sebago Lake, ice fishing would be prohibited in the Lower Bay and the Town of Standish boat launch would have to be relocated, prospects that, as discussed below, are met with resistance by many community members.

5.3.1.2 Standish Boat Ramp Relocation Efforts

Long-standing conflict over the PWD's efforts to relocate the boat launch in Standish were re-ignited when PWD received the watershed protection effort suggestions from DHHS in 1991. One of the suggestions was that the Standish boat launch should be relocated as part of an agreement between PWD and state agencies in order for PWD to obtain a waiver on a $20 million filtration plant (Williamson, 1994). In December of 1994 the Standish Town Council held a meeting originally intended to include a discussion with Standish residents about possible relocation sites for the boat launch located in the Sebago Lake Village section of Standish (Williamson, 1994). Residents and members of the Standish Tomorrow Association opposed the relocation of the boat launch altogether. Aside from holding on to the current boat ramp as a matter of pride, residents expressed the value of the boat ramp as a town asset. The relocation of the boat ramp was rejected in a town vote in the late 1990s and again in 2002—many residents viewed the PWD with suspicion, angry over its extensive land purchases in the area and its prohibition against entering the water at the beautiful sandy beach by the village (Connerty-Marin, 2002). In 2004, when PWD announced its position that all of Lower Bay should be closed to recreation to protect water quality, residents' negative sentiments again erupted (Balentine, 2004a). In response, Standish citizens formed a nonprofit citizen association to support legislation that protects boating and fishing access to the Lower Bay (Balentine, 2004b). Reacting to the town's resistance to the boat launch relocation, PWD, which owns the land directly surrounding the town's access to the launch, restricted parking and turnaround space by the boat ramp. Overall, the Standish boat launch discord represents the conflict of value in land management between PWD and lakeside residents that needs to be addressed if collaboration on responsible development and management of watershed land can occur.

5.3.2 Source Water Protection Efforts

In response to the DHHS suggestions that the PWD advocate for responsible land use practice in the upper watershed, PWD has an extensive source water protection education and outreach program, as well as significant land preservation efforts underway in the upper watershed, especially the Crooked River.

5.3.2.1 Source Water Protection Education and Outreach

PWD efforts addressing source water protection include the following (PWD, 2009):

1. *Watershed News*—a newsletter intended to educate watershed stakeholder on source protection issues of interest
2. The Rain Barrel Program—where PWD sells and educates community members on use of rain barrels for reuse of rainwater, reducing water demand in the district
3. Sebago Lake Ecology Center—a teacher resource and public education center in Standish that features visual demonstrations of watershed management methods to promote water quality protection
4. *State of the Lake Report*—a periodic publication highlighting the latest information on water quality, fisheries, and land development in the watershed
5. Education materials—including an Events Calendar, Sebago Lake Map, and other resources that are distributed around the watershed
6. Crooked River Reclassification Efforts—written and verbal testimony that PWD provided to update a small previously dammed portion of the Crooked River to class AA
7. Community Partnerships—supporting watershed and lake protection efforts such as Conservation Forestry Field Day, Forests to Faucets, Sebago to the Sea, and Partners in Protecting Sebago Lake.

These programs highlight watershed management education and support outreach activities throughout the community to inform wise land and water use practices to protect water quality.

5.3.2.2 Land Acquisition and Preservation

The PWD also continues to support land acquisition and preservation efforts in the watershed. Land acquisition is primarily funded by the Crooked River–Sebago Lake Fund, while conservation efforts in the upper watershed are spearheaded by land trusts under the auspices of the Upper Headwaters Alliance.

- The Crooked River–Sebago Lake Fund: The PWD's regulations restricts land acquisition efforts to properties within the two-mile boundary of Sebago Lake. The Crooked River–Sebago Lake Fund was established in 2009 to help pay for the protection of forests in the upper watershed (PWD, 2009). PWD designated $25,000 annually that can be available to help cover transaction costs for landowners who want to put a conservation easement on watershed land. Land within the Crooked River watershed will have highest priority for funding. PWD has also approached potential partners, (i.e. Casco Bay Estuary Partnership), for additional guaranteed funding (PWD, 2009).
- The Upper Headwaters Alliance: PWD has also joined the Upper Headwaters Alliance with local land trusts including the Western Foothills Land Trust, the Loon Echo Land Trust, the Greater Lovell Land Trust, and the Saco Valley Land Trust to protect the Crooked River watershed through outreach and preservation

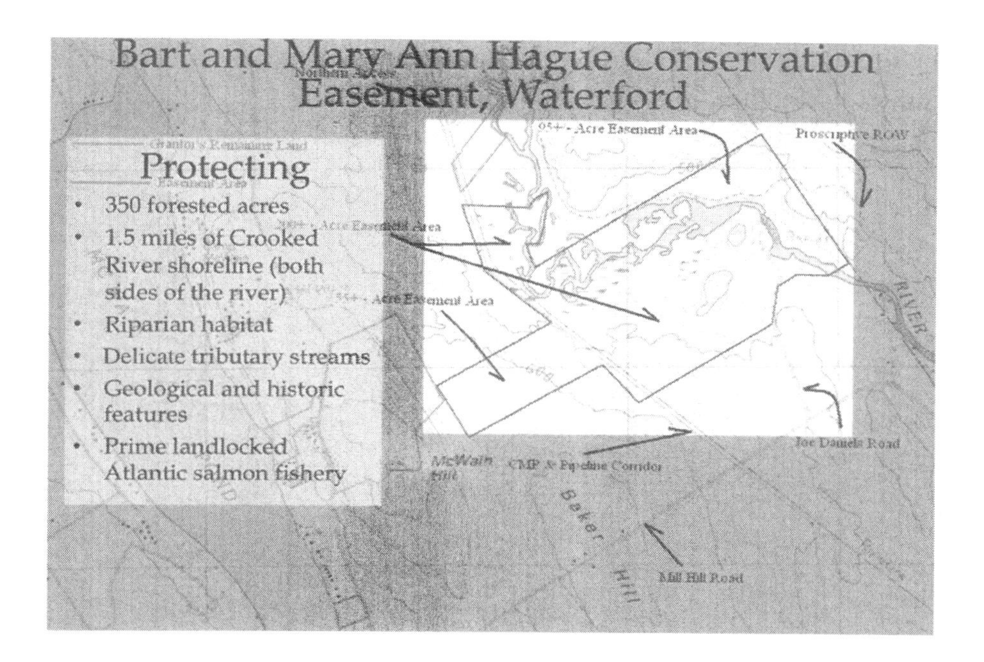

Figure 5.7 Property boundary of the Hague conservation easement (From Hunt, 2010).

(PWD, 2009). The Upper Headwaters Alliance developed GIS analyses to prioritize properties in the Crooked River watershed with high conservation value and to target potential conservation options toward those landowners. An example of this effort is the Bart and Mary Ann Hague conservation easement in Waterford, Maine, which secured 350 forested acres for water quality protection (Figure 5.7). As part of the education process, the Upper Headwaters Alliance highlights the benefits to landowners as well, including protection of valued property from future development, income tax benefits, and potential property tax savings.

5.4 WATER SUPPLY TRENDS AND THE VALUE OF WATERSHED PROTECTION

Despite PWD's best efforts as a responsible party in preservation and management of the Crooked River watershed for surface water protection, without the authority to enforce zoning requirements or unlimited funding to support landowner conservation efforts, PWD's abilities are limited. However, demands on water suppliers, both in Maine specifically and throughout New England generally, are changing. These trends include decreasing demand for drinking water, increasing stringency of regulations on drinking water quality, development pressures on watershed properties, the need throughout the water supply industry to quantify the benefits of watershed protection, and the facilitation of source water protection in the state of Maine.

5.4.1 Decreasing Demand for Drinking Water

One trend that has emerged from the last few years has been the decrease in demand over time for drinking water. In New York City, conservation efforts have led to decreased demand resulting in either increased rates for customers or decreased overall revenue for the water utility (see Freiberg et al., Chapter 4, this book). New York City depends on the rate payers to fund its programs; thus, as demand declines, utility funds decrease, potentially threatening the watershed protection budget and creating pressure to increase water rates to regain this loss (see Freiberg et. al., Chapter 4, this book). In Massachusetts, both the Massachusetts Water Resources Authority (the water supply utility for Boston) and the Worcester Water Department reference decreasing water demand as a future restriction on budget for watershed protection (see Alcott et. al., Chapter 2, this book). Likewise, the demand for drinking water supplied by PWD has been decreasing for several years, with a 19 percent decrease since 2001 (Figure 5.8) (Hunt, 2012, personal communication). This leaves PWD in a similar predicament as other Northeast water utilities, with further impetus to find alternative funding mechanisms for conservation and management for source water protection on the critical watershed properties, especially the Crooked River.

5.4.2 Increased Drinking Water Quality Regulation

A second theme that has emerged is the likelihood that water utilities will face regulation of a larger number of water contaminants in the future. The Connecticut

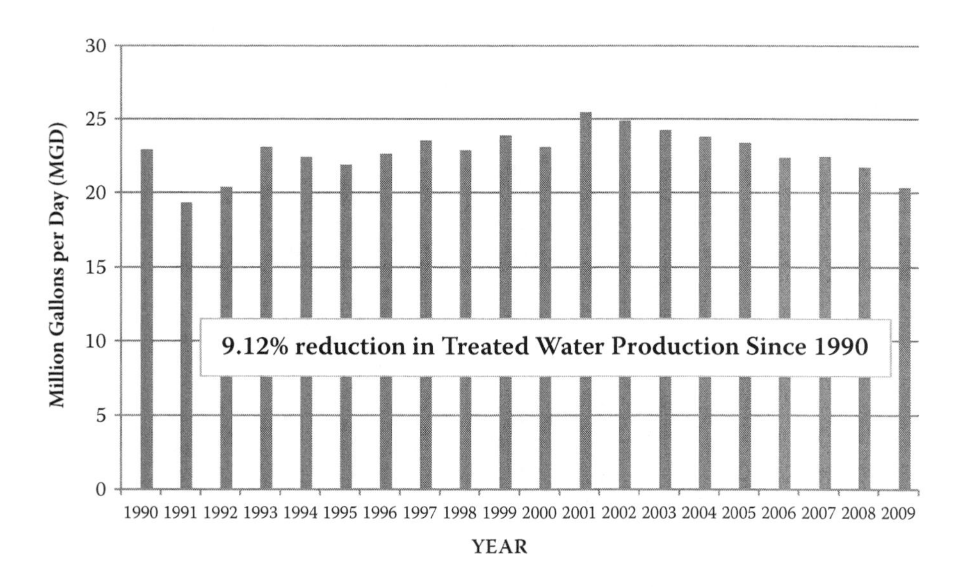

Figure 5.8 Declining treated water production due to decreasing customer demand at the Portland Water District (From Hunt, 2012, personal communication).

and Massachusetts case studies in this text (see Blazewicz et. al., and Alcott et al., Chapters 2 and 3 respectively, this book) highlighted both the potential for increased regulation and the difficulty in accounting for treatment of all of these contaminants with unfiltered supply systems reliant on watershed management for raw water quality.

Through the Unregulated Contaminant Monitoring (UCM) program, the EPA collects data for contaminants known or suspected to be present in drinking water but as of yet without health-based standards under the SDWA. These include pesticides, disinfection by-products, chemicals, waterborne pathogens, pharmaceuticals and personal care products (PPCPs), and biological toxins (EPA, 2010a). The UCM program covers no more than 30 contaminants and takes place on all large public water supply systems (EPA, 2010a). Every five years the EPA reviews monitoring information, updates the Contaminant Candidate List (CCL), and decides whether to regulate at least five contaminants, and prioritizes further research and data collection efforts for the next monitoring cycle (EPA, 2010c) (see also Blazewicz et. al., Chapter 2, this book). In addition, the most recent USGS national investigation of pharmaceuticals and other organic contaminants demonstrated the ubiquity of these contaminants in both ground and surface sources of drinking water (Focazio, 2008). Anthropogenic sources include "manufacturing releases, waste disposal, accidental releases, purposeful introduction (e.g., pesticides, sewage sludge application), and consumer activity (which includes both the excretion and purposeful disposal of a wide range of naturally occurring and anthropogenic chemicals such as PPCPs)" (Daughton, 2004; EPA, 2006; Alcott et al., Chapter 3, this book).

PWD does not currently monitor Sebago Lake or the Crooked River for PPCPs or other emerging contaminants identified by the EPA, so while the threat of contamination from PPCPs is unknown, PWD will most likely have to adjust water treatment strategies in the future to meet evolving treatment requirements. It is unclear whether this will provide more or less support for watershed management for surface water quality protection, but nevertheless this highlights the importance of enforcing limited and well-planned development on critical watershed lands to minimize concentrations of PPCPs in the watershed. Furthermore, given the trends of decreased funding and increased regulation, development of quantitative support for the value of watershed management can only help water supplier managers garner support for watershed protection efforts.

5.4.3 Efforts to Quantify the Value of Watershed Protection

While water utility managers throughout New England operate on the understanding that a multibarrier approach to security of drinking water quality, including a watershed management strategy, is critical for future water quality, there has been limited success in qualifying the incremental cost and value of watershed protection efforts. A series of studies led by the Trust for Public Lands (TPL) and partners since the late 1990s attempted to use watershed-monitoring data to understand the value of watershed protection in terms of increased water quality in runoff from well-managed lands. One study found that a 1 percent increase in turbidity can increase chemical treatment costs by 0.25 percent (Dearmont et. al., 1997). Another study conducted by TPL found that there were statistically significant relationships among percent land cover, source water quality, and drinking water treatment costs (Freeman et al., 2008). Decreased forest cover was significantly related to decreased water quality, while low water quality was related to higher treatment costs (Freeman et al., 2008).

The research effort by TPL highlighted two challenges to understanding the incremental cost of watershed protection. The first is that the analysis is inherently incremental; therefore it depends heavily on time span and how far ahead water suppliers are willing to look in their decisions about budgets and investments—watershed deterioration is incremental over time, while investment is a one-time big decision that leads to a conflict for water supply planners. Secondly, all federal programs focus on improving water quality—so programs have to show a measurable impact on water quality. If prevention of watershed deterioration is the goal, it can be difficult to show "measurable impacts" on water quality (Ernst, 2010). To optimize limited time and funds, management efforts must concentrate on a monitoring system that is efficient and most able to demonstrate measurable impacts from land use. Such a system has been developed to link water quality to engineered treatment costs (see Alcott et al., Chapter 2, this book).

Although the PWD has an extensive monitoring network in Sebago Lake and the supply tributaries including the Crooked River, increased strategic monitoring would be necessary should PWD be in a position where the incremental cost value of watershed protection must be defended to public constituents or customers (likely due to projections for future regulation and decreased revenue from declining demand as discussed above).

5.5 DEVELOPMENT PRESSURES ON WATERSHED LANDS

Although the Sebago Lake watershed is largely forested, it is at high risk for future development. A 1990 analysis of land uses in the Crooked River watershed showed that 89.2 percent of the watershed land water was forested (Figure 5.9), whereas an identical analysis in 2001 showed that 85.2 percent was forested (Figure 5.10), indicating that 4 percent of forest land in the Crooked River watershed was developed between 1990 and 2001 (Hunt, 2010).

According to the Forest-to-Faucet Partnership at the University of Massachusetts Amherst and USDA Forest Service, the larger Presumscot River basin surrounding the Crooked River watershed ranks very high among eastern New England watersheds in development pressure on private forests important for drinking water supply (Figure 5.11) (Gregory and Barten, 2008a). The trend of conversion of forests to development in the Crooked River watershed, combined with the lack of zoning regulations in towns within the Sebago Lake watershed and the risk for future development in the basin, applies increased pressure on PWD to create incentives for private landowners to manage their land for source water protection.

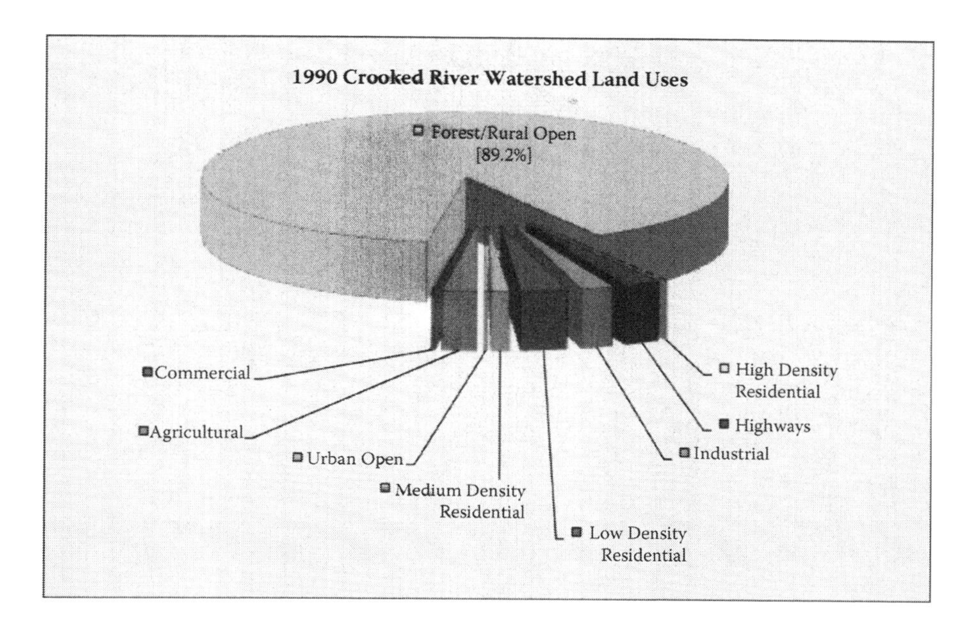

Figure 5.9 1990 Crooked River watershed land uses (From Hunt, 2010).

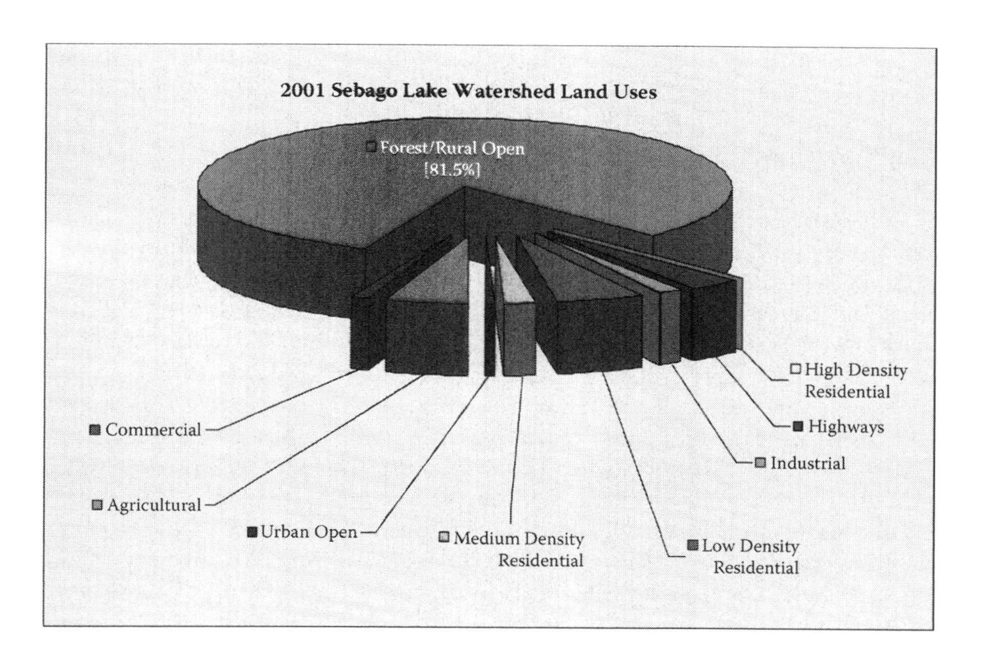

Figure 5.10 2001 Sebago Lake watershed land uses (From Hunt, 2010).

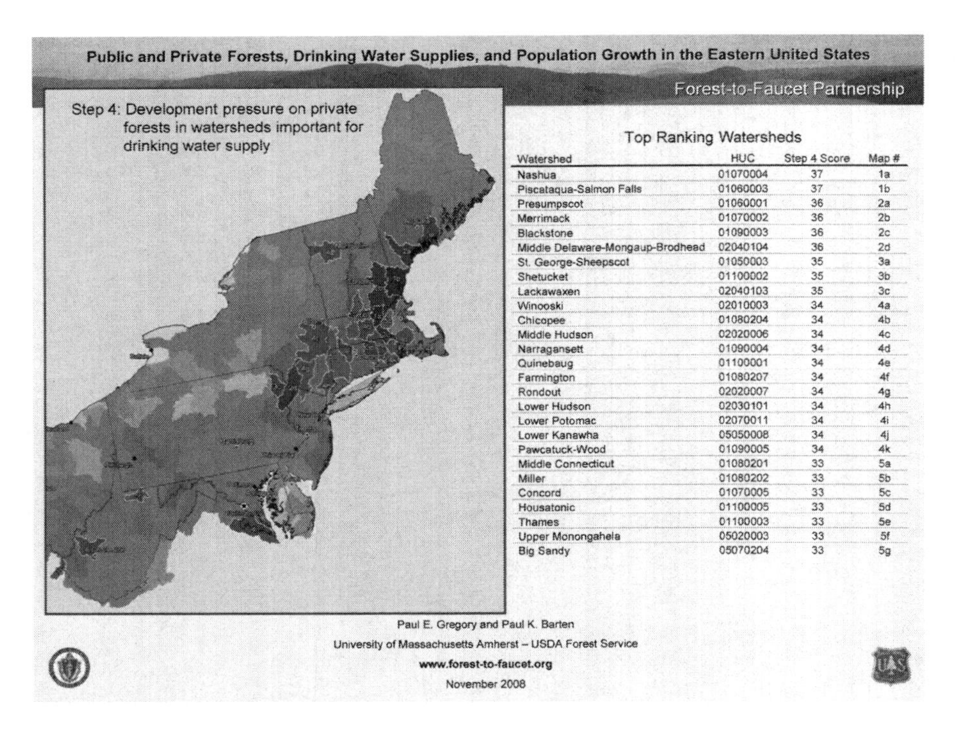

Figure 5.11 The Presumpscot River basin, which contains the Sebago Lake watershed and the Crooked River watershed, is ranked third among Eastern New England water supply watersheds at risk for development pressure on private forests (From Gregory and Barten, 2008a). (See color insert.)

5.5.1 Future of Source Water Protection throughout Maine

When the SWTR required filtration of surface water supplies throughout the country, 12 systems in Maine were granted waivers from filtration from the DHHS Drinking Water Program: Auburn, North Haven, Lewiston, Camden/Rockland, Great Salt Bay, Bethel, Portland, Bangor, Bar Harbor, Brewer, and Mount Desert North and South (Tolman, 2010). All of these systems still supply unfiltered drinking water except Camden/Rockland and Bethel (SWAP, 2003; Tolman, 2010). Camden/Rockland voluntarily built a filtration plant to address the Long-Term 2 Enhanced Surface Water Treatment Rule, the latest amendment to the SWTR addressing *Cryptosporidium* and other disease-causing microorganisms in drinking water (EPA, 2010b). Bethel experienced an extreme storm event in the watershed that led to unmanageable turbidity levels in the reservoir. The high number of unfiltered water supply systems in Maine has been attributed to regional differences in the population demographic (Tolman, 2010). High-quality raw water flows from sparsely developed forests in northwestern Maine to the population centers on the coast, and in general Maine has a lot of land with a small population. In addition there is a general willingness among the residents of Maine to value protection of water bodies for recreation (Tolman, 2010). While to date the DHHS Drinking Water Program has facilitated source water protection, throughout New England there is a need to coordinate drinking water regulations to encourage source water protection (Alcott et al., Chapter 2, this book).

Funded by a grant from the EPA, the Trust for Public Land (TPL), the Smart Growth Leadership Institute, the Association of State Drinking Water Administrators, and the River Network joined forces in a program called Protecting Drinking Water Sources, aimed at better integrating state land use policy, incentives, and drinking water programs (Protecting Drinking Water Sources, 2010). The program's mission is to "better align planning, economic development, regulation and conservation to protect drinking water sources at the local and watershed levels" (Protecting Drinking Water Sources, 2010, website). Protecting Drinking Water Sources selects pilot states and aims to improve collaboration among state agencies, local managers, and concerned stakeholders. This collaboration can maximize incentives and efforts aimed at protecting drinking water quality. Alumni states include Maine, Ohio, New Hampshire, Oregon, Utah, and North Carolina.

The result of the Protecting Drinking Water Sources effort in Maine was a prioritized action plan for enabling source water protection in Maine called An Action Plan to Protection Maine's Drinking Water Sources: Aligning Land Use and Source Water Protection (TPL, 2009). The program used a series of stakeholder meetings to identify opportunities to align state and local regulations toward source water protection, and then to prioritize them. The action plan process participants included representatives from most Maine regulatory agencies that deal with drinking water, as well as water supply utilities and environmental nonprofit groups (see Box 5.3).

The trends identified in this section will make source water protection more difficult for drinking water managers in general, and PWD in particular, include declining revenue from decreased demand for drinking water, increased development pressure of watershed land, and increased water quality regulation due to emerging drinking water contaminants (TPL, 2009). Combined, these trends will place more pressure on water suppliers to defend investment in source water protection through the comparison of incremental costs of water treatment via watershed protection and management versus engineered treatment solutions. However, such an analysis is not yet possible given the basic levels of watershed monitoring required in the current regulatory framework for drinking water quality. All of these pressures could be relieved by a well-conceived platform for payment for watershed service-based strategies to encourage private watershed land management for surface water quality protection in the Crooked River watershed.

The Portland Water District is the most prominent watershed stakeholder concerned with surface water quality in the Crooked River because the river supplies 40 percent of the surface water inflow to Sebago Lake, the water supply reservoir for PWD's customers in southern Maine. The trend of conversion of forests to development in the Crooked River watershed (4 percent of the land in the Crooked River watershed was developed between 1990 and 2001), combined with the lack of zoning regulations in towns within the Sebago Lake watershed and the risk for future

BOX 5.3 ACTION ITEMS DEVELOPED THROUGH THE PROTECTING DRINKING WATER SOURCES PROGRAM

The action items developed through the Protecting Drinking Water Sources program address incentivizing and facilitating landowners in drinking water supply watersheds to manage land for source water protection. Action items identified are listed and discussed briefly below by short- and long-term categories of priority.

SHORT-TERM

Action Item 1: Incorporate drinking water source protection among the assets considered for the quality of place investment. The Maine Quality of Place Initiative is an asset-based economic development strategy designed to grow Maine's economy while protecting natural assets. This action item outlines steps to include drinking water sources (watersheds, rivers, and forests) as assets.

Action Item 2: Streamline statewide GIS databases and develop protocols for collecting, analyzing, uploading, and managing data (a) to provide a one-stop center for the state, local, and regional governments (b) to reduce duplication efforts and funding. Stakeholders identified this as an opportunity so all participants in drinking water delivery and quality protection could be aware of trends in water quality.

Action Item 3: Develop overarching guidelines for compatible recreational opportunities on land and in waters critical to drinking water source protection. This action item recognizes the overlap between desirable recreational land and watershed and water bodies necessary for providing high-quality surface water and the need to determine and mandate guidelines for safe, low-impact recreation.

LONG-TERM

Action Item 4: Increase funding for drinking water source protection through the creation of a dedicated funding program. This action items addresses directly the need to alleviate financial pressure on water suppliers so as to enable source water protection.

Action Item 5: Enhance existing current use tax program to include landscapes important for drinking water source protection. Maine currently offers two programs that lower property taxes and encourage preservation of undeveloped land: the tree growth and farmland and open space tax laws. Both of these laws prevent development of watershed lands indirectly but could be modified or extended to explicitly enhance opportunities for source water protection.

Source: TPL, 2009.

development in the basin, apply increased pressure on PWD for creating incentives for private landowners to manage their land for source water protection.

Like other water utilities throughout New England, PWD is facing increasing regulation on water supply quality combined with decreasing revenue from declining demand for drinking water. The regulatory framework and guidance for watershed protection at PWD established by DHHS under the filtration wavier and the SWAP process has led to an exemplary watershed control program despite the fact that PWD owns only 1 percent of the land in the Sebago Lake watershed. However, given the increasingly limited resources available, PWD could benefit from the assistance of a separate payment for watershed services (PWS) program. Such a PWS program could augment PWD's efforts in the Crooked River watershed and better allocate and leverage PWD funds toward management of watershed ands for surface water quality protection.

5.6 POTENTIAL PAYMENTS FOR WATERSHED SERVICES PROGRAMS FOR THE PORTLAND WATER DISTRICT

5.6.1 The Basics

Well-functioning, intact ecosystems provide a number of services to society—from purification of air and water to aesthetic beauty and landscape enhancement—that are currently unvalued or undervalued. According to the Millennium Ecosystem Assessment, over 60 percent of the environmental services studied are being degraded faster than they can recover (Forest Trends, 2008). As a result, various kinds of markets are forming to attach a value to ecosystem services and allow investment in preservation and restoration of these environmental flows.

A payment for ecosystem service (PES) transaction is defined as a voluntary transaction in which a well-defined environmental service or form of land use likely to secure that service is bought by at least one buyer from a provider that continues to supply that service (Forest Trends, 2008). There are three major types of PES markets: (1) public payment systems for private landowners usually supported by the government, (2) formal markets with open trading voluntarily or under a regulatory cap, and (3) self-organized private deals (see Table 5.4 for examples).

All PES programs share the goal of providing high-quality water at reliable quantities through restoration, creation, or enhancement of wetlands, maintenance of forest cover, reforestation, and beneficial land use practices. These actions have the ecological goals of creating natural filters to reduce water pollution, maintain vegetation to regulate hydrological flows, flood control, and maintaining soil quality. The components necessary for an effective PES program can be categorized as (1) engaging and informing a variety of stakeholders, (2) defining ecosystem services to be considered, (3) valuing ecosystem services, (4) developing an agreement to guarantee delivery of services, and (5) establishing a payment mechanism (Forest Trends, 2008). These five program elements were chosen as a framework for providing recommendations for development of a PES program in the Crooked River watershed because they both encompass

Table 5.4 Examples of Water Market Payments

Water-Related Ecological Service Provided	Supplier	Buyer	Instruments	Intended Impacts on Forests	Payment
			Self-Organized Private Deals		
France: Perrier Vittel's Payments for Water Quality					
Quality drinking water	Upstream dairy farmers and forest landholders	A bottler of natural mineral water	Payments by bottler to upstream landowners for improved agricultural practices and reforestation of sensitive filtration zones	Reforestation but little impact because program focuses on agriculture	Vittel pays each farm about $230/ha/yr for 7 years. The company spent an average of $155,000 per farm, or a total of $3.8 million.
Costa Rica: National Fund for Forestry Financing (FONAFIFO) and Hydroelectric Utilities Payments for Watershed Services					
Regularity of water flow for hydroelectricity generation	Private upstream owners of forest land	Private hydroelectric utilities, Government of Costa Rica and local NGO	Payments made by utility company via a local NGO to landowners; payments supplemented by government funds	Increased forest cover on private land; expansion of forests through protection and regeneration	Landowners who protect their forests receive $ 45/ha/yr; those who sustainably manage their forests receive $70/ha/yr, and those who reforest their land receive $116/ha/yr.
Colombia: Associations of Irrigators' Payments (Cauca River)					
Improvements of base flows and reduction of sedimentation in irrigation canals	Upstream forest landowners	Associations of irrigators; government agencies	Voluntary payments by associations to government agencies to private upstream landowners; purchase by agency of lands	Reforestation, erosion control, springs and waterways protection, and development of watershed communities	Association members voluntarily pay a water use fee of $1.5–2/liter on top of an already existing water access fee of $0.5/liter.

(continued)

Table 5.4 Examples of Water Market Payments (continued)

Water-Related Ecological Service Provided	Supplier	Buyer	Instruments	Intended Impacts on Forests	Payment
			Trading Schemes		
United States: Nutrient Trading					
Improved water quality	Point source polluters discharging below allowable level; nonpoint source polluters reducing their pollution	Polluting sources with discharge above allowable level	Trading of marketable nutrient reduction credits among industrial and agricultural polluting sources	Limited impact on forests—mainly the establishment of trees in riparian areas	Incentive payments of $5 to $10 per acre.
Australia: Irrigators Financing of Upstream Reforestation					
Reduction of water salinity	State Forests of New South Wales (NSW)	An association of irrigation farmers	Water transpiration credits earned by state forests for reforestation and sold to irrigators	Large-scale reforestation, including planting of desalination plants, trees, and other deep-rooted perennial vegetation	Irrigators pay $40/ha/yr for 10 years to the State Forests of NSW, a government agency that uses the revenues to reforest on private and public lands, keeping the forest management rights.

Source: Excerpted from Scherr, Sara, Andy White, and Arvind Khare with contributions from Mira Inbar and Augusta Molar, 2004. For Services Rendered: The Current Status and Future Potential of Markets for the Ecosystem Services Provided by Tropical Forests. Yokohama, Japan: International Tropical Timber Organization, pp. 30–31.

factors that are currently missing from PWD's efforts, factors that have been identified as necessary for a successful PES program by industry analysts (Forest Trends, 2008).

5.6.2 Examples of Payment for Watershed Services Programs

Three payments for watershed services (PWS) programs considered innovative are (1) the Santa Fe, New Mexico, and U.S. Forest Service (USFS) partnership to protect the Santa Fe Municipal Watershed; (2) the Great Miami River Watershed Water Quality Trading Project designed for nutrient credit trading; and (3) the New York

City watershed management program. These programs, although at varying stages of development, are evaluated in terms of their attempts and/or plans to address each of the five components listed in Table 5.5.

5.6.2.1 Santa Fe, New Mexico, Municipal Water Supply Watershed

The Santa Fe, New Mexico, Municipal Water Supply's PWS program is a public payment program where the City of Santa Fe relies on water customers to directly support the benefits of maintaining a healthy, forested water supply watershed. The program is based on avoided costs of recovering the water supply quality after a major wildfire in the 17,000 acre USFS-maintained upper watershed (McCarthy, 2009) (see Table 5.5). The cost to retain the restored forest condition over 20 years is estimated at $4.3 million (an average of $200,000 per year with diminishing cost over time) (McCarthy, 2009). The avoided cost, the expenses from a 7,000-acre fire in the watershed, is $22 million. During the first five years of the program, the New Mexico Water Trust Board is providing a grant to pay for watershed vegetation management and will be listed as a credit on water customer bills (McCarthy, 2009). Once the grant funding is exhausted, the City of Santa Fe will assess a PES fee that will appear as a separate line item on water customer bills. This fee is projected to be $0.13 per 1,000 gallons per month. In addition, the PWS program has an outreach plan aimed at water customers, residents, and Santa Fe youth with a focus on general watershed ecology education and building support for the PWS approach (McCarthy, 2009).

While the Santa Fe PWS program is unique in that the threat of the wildfire in the watershed can lead to a strong argument for the avoided costs of siltation, it addresses a problem common to watershed protection efforts: the need for outreach focused on downstream water users, and use of water user funding to support upper watershed management. While PWD does not control their water supply watershed, the outreach to downstream water customers for separate PWS funding seen in the Santa Fe PWS example could present a transferrable strategy.

5.6.2.2 Great Miami River Watershed Water Quality Credit Trading Program

The Great Miami River Watershed Water Quality Credit Trading Program is a subset of the Ohio River Basin Trading Project. The program comprises a credit trading market focused on the filtration ecosystem service provided by watersheds. Landowners implementing management practices that reduce the discharge of nutrients (i.e., agricultural land uses, storm water management, or home sewage treatment systems) generate credits (Miami Conservancy District, 2009). Credits are purchased by National Pollutant Discharge Elimination System permit holders to offset discharge of nitrogen and phosphorus into watershed streams and rivers. The market determines the cost of a credit. In general the credit cost is the sum of the cost of the project (including administrative and transaction costs) divided by the number of credits. The trading ratios (total pollutant discharge physically allowed versus total

Table 5.5 Comparison of Three Types of Payment for Watershed Services Programs

	Components of Payments for Ecosystem Services (PES) Program		
	Santa Fe, New Mexico Municipal Water Supply Watershed[1]	Great Miami River Watershed Water Quality Credit Trading Program[2]	New York City Watershed Management Program[3]
Type of PWS Program	Public Payment Program	Credit Trading Market	Self-Organized Private Deal
Stakeholder Engagement	Protection of water quality is a shared goal of the City of Santa Fe and the Santa Fe National Forest (USFS), which manages the upper 17,000 acres of the water supply watershed. PES is based on avoided costs of high-intensity fire and overgrown (unmanaged) dense forest in the watershed. The PES program has a fully outlined outreach plan targeting residents of the City and County of Santa Fe, water customers of the City of Santa Fe Water Division, and Santa Fe youth with a focus on two areas: • Providing general watershed education, including forest and riparian ecology, natural and cultural history, and water issues • Building support for the PES model.	The Miami Conservancy District Water Conservation Subdistrict solicited input and formed partnerships throughout the development of the nutrient trading program. Major stakeholder groups included state and federal regulatory agencies, utilities holding National Pollutant Discharge Elimination System (NPDES) permits, community-based watershed organizations, and county soil and water conservation districts.	The Watershed Agricultural Council is a nonprofit funded in part by the NYC DEP with the mission of sustaining, maintaining, and protecting the New York City water supply through farm and nutrient management planning, conservation practice implementation, education, and economic development of the local agricultural community.

(continued)

Table 5.5 Comparison of Three Types of Payment for Watershed Services Programs (continued)

	Components of Payments for Ecosystem Services (PES) Program		
	Santa Fe, New Mexico Municipal Water Supply Watershed[1]	Great Miami River Watershed Water Quality Credit Trading Program[2]	New York City Watershed Management Program[3]
Type of PWS Program	**Public Payment Program**	**Credit Trading Market**	**Self-Organized Private Deal**
Definition of Ecosystem Services	Fresh water—supplies 30% of the water used in the City of Santa Fe Flood control and flow timing—forest cover maintains snow pack and combined with dams ensures year-round water Water purification—forest and woodland cover provide natural filtration Sediment regulation—plants and forest ground cover keep soil in place Fire protection—healthy lower-elevation forests will bum at low-intensity and reduce possibility of wildfire and sedimentation Invasives regulation—few sources for invasives species introduction Climate regulation—store carbon and low-intensity fires prevent a passive release of carbon	Prevention or discharge of pounds of phosphorus (TP) and pounds of nitrogen (TN) into watershed rivers and streams through— water quality credit trading program under OHIO DEP NPDES permit system Buyers: NPDES permit-holders Sellers: credit generated by landowners by implementing management practices that reduce the discharge of nutrients from agricultural land uses, storm water management, or home sewage treatment upgrades	Management of working landscapes to maintain or improve the quality of NYC's drinking water supply

Value of Ecosystem Services	Fresh water—$4.3 million over 20 years for management of 17,000 watershed acres Water purification—$1 million avoided cost of shutting down water treatment plant for 2 months after fire Sediment regulation—$10 million avoided cost of dredging 2 reservoirs to remove sediment Fire protection—$10 million avoided cost of 7,000 acre wildfire Invasives regulation—$500,000 avoided cost to control invasives spread after wildfire Total from regulating services—$22 million estimated total avoided costs from 7,000 acre fire	point to nonpoint source trading offers significant cost savings ($314M–$384.7M) during a 20-year period when compared to the expense of mandatory treatment plant upgrades (totaling $422.5M for traditional treatment-based regulatory approaches	NYC DEP has invested $1.49 billion in watershed management in the Catskill–Delaware watershed to avoid the $4 and $8 billion in upfront capital costs and up to an additional $350 million in annual O&M costs of construction of a treatment plant far this water supply
Service Agreement	Plan to draft a new Memorandum of Understanding between the City of Santa Fe and USFS for watershed management. Develop a collection agreement between Santa Fe and USFS every 5 years. Revise work plans, budget, and implementation annually.	A trade occurs when the Trading Program transfers water quality credits by agreement to an eligible buyer who uses them to comply with their NPDES permit. Buyer eligibility and trading ratios are predetermined based on condition of receiving waters.	NYC water users are taxed by NYC with supplemental funds provided by federal, state, and local governments to pay for land management improvements by upstream landowners.

(continued)

Table 5.5 Comparison of Three Types of Payment for Watershed Services Programs (continued)

	Components of Payments for Ecosystem Services (PES) Program		
	Santa Fe, New Mexico Municipal Water Supply Watershed[1]	Great Miami River Watershed Water Quality Credit Trading Program[2]	New York City Watershed Management Program[3]
Type of PWS Program	Public Payment Program	Credit Trading Market	Self-Organized Private Deal
Payment Mechanism	Phase 1: New Mexico Water Trust Board pays for ecosystem services for first 5 years via watershed restoration grant. During this time PES shows as credit on water bill for Santa Fe customers. Phase 2: Assess a fee to each water customer based on use, projected at $0.13 per 1,000 gallons per month. List the PES fee as a separate line item in water bill.	The cost of a water quality credit is determined by the market. In general the credit cost is the sum of the cost of the project (including administrative and transaction costs) divided by the number of credits.	Taxes on water users, NYC bonds, trust funds, subsidies, logging permits, differential land use taxation, development rights, conservation easements, development of markets of nontimber products, and certified wood.

[1] McCarthy, L., 2009. Santa Fe Municipal Watershed 20 Year Protection Plan: 2010–2029: Financial Management Plan. February 9.

[2] Water Conservation Subdistrict of the Miami Conservancy District, 2005. Great Miami Watershed Water Quality Credit Trading Program Operations Manual. February 8.

[3] Johnson, N., A. White, and D. Perrot-Maitre, 2000. Developing Market for Water Services from Forests: Issues and Lessons from Innovators. Forest Trends, World Resources Institute and the Katoomba Group, Washington, D.C., USA. Accessed April 26, 2010 at www.katoombagroup.org. Also, Freiberg, J., X. Hou, J. Nerenberg, and F. Samuel, 2010. NYC Watershed Management, Past, Present and Future, Yale School of Forestry and Environmental Studies Seminar: Emerging Markets for Ecosystem Services. April 23 (unpublished).

credited) are based on buyer eligibility and condition of discharge receiving waters. Overall the credit trading program is the least-cost solution for nutrient discharge control in the watershed. Point to nonpoint source trading offers significant cost savings ($314M–$384.7M) during a 20-year period when compared to the expense of mandatory treatment plant upgrades estimated to be $422.5M (Water Conservation Subdistrict of The Miami Conservancy District, 2005).

Similar to the Santa Fe PWS program, one key component of the Great Miami River Watershed program is outreach for landowner (or credit-seller) participation. The Miami Conservancy District, a political subdivision of the State of Ohio responsible traditionally for flood control in the watershed, acts as a clearinghouse for credit trading transactions. Landowners submit project proposals for funding for introduction of agricultural management practices that reduce nutrient discharge. The proposals are evaluated by the PWS programs' Project Advisory Group and ranked by the lowest cost per pound of nutrient reduction. The Project Advisory Group comprises representatives from a diverse group of stakeholders including wastewater treatment plants, agricultural landowners, Ohio Farm Bureau Federation, Ohio Water Environment Association, watershed organizations, county soil and water conservation districts, Ohio Department of Natural Resources (ODNR), and the USDA. Current landowner nutrient discharge-reduction projects in the watershed are illustrated in Figure 5.12 (Miami Conservancy District, 2009).

5.6.1.3 New York City Watershed Management Program

The New York City PWS program is a collection of self-organized private deals between the Watershed Agricultural Program, a nonprofit organization funded in part by the New York City Department of Environmental Protection (DEP) (drinking water supplier), and landowners in the Catskill–Delaware water supply watershed. The PWS is based on landowners receiving funding to improve land management practices for water supply protection. The origins of the program are summarized in the Long-Term Water Supply Protection Plan of the City of New York (DEP, 2006).

The Watershed Agricultural Program (WAP) provides funding and assistance to farmers willing to create pollution prevention plans known as Whole Farm Plans. The program has been operating since 1992 under the Watershed Agricultural Council (WAC), a nonprofit dedicated to the promotion of proper agriculture and forestry activities within the New York City water supply area. Whole Farm Plans have been completed for more than 390 farms in the Catskill–Delaware watershed (DEP, 2009a). This constitutes about 95 percent of the large farms in the region. Plans have been substantially implemented on 82 percent of these farms (Rush, 2009). In the Croton watershed 38 farms had approved Whole Farm Plans (DEP, 2009a). The program is designed to reduce or eliminate pollutant sources, prevent movement of runoff across landscapes, and promote the use of riparian buffers that serve to filter any polluted runoff that moves across the landscape. (See Freiberg et al., Chapter 4, this book)

The WAC also administers the Watershed Forestry Program in conjunction with the NYC DEP and the USFS. The goal of the program is to promote well-managed

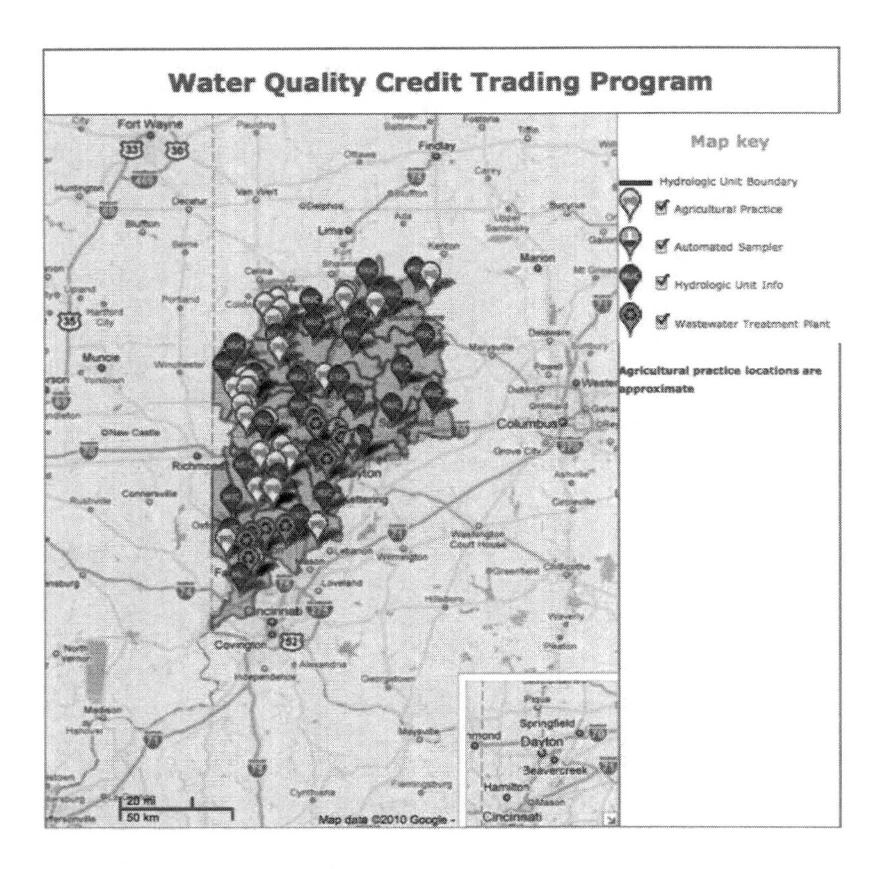

Figure 5.12 Map of current landowner nutrient discharge-reduction projects in the Great Miami River watershed (From Water Conservation Subdistrict of the Miami Conservancy District, 2005).

working forests as beneficial land cover for the watersheds. This is achieved through assisting landowners to create Forest Management Plans, conducting training for loggers and foresters, and providing direct technical assistance. The WAC Forest Management Plans cover more than 100,000 acres of private forestland. Approximately 75 percent of the watershed is forested, so the proper management of these forests is important in protecting water quality (Figures 5.13 and 5.14) (DEP, 2006; Freiberg et al., Chapter 4, this book).

The New York City PWS program, although paired with an aggressive land acquisition program, most closely resembles the scenario and conditions in the Crooked River watershed. Similar to the Santa Fe and Great Miami River examples, NYC DEP supports critical components of any successful PWS program: a full analysis of potential stakeholders, and outreach to both potential providers of ecosystems services and willing buyers.

(a)

(b)

Figure 5.13 (a) Bunk silo site before and (b) after implementation of nutrient management practices at a Delaware County dairy operation (From Watershed Agricultural Council, 2009).

5.7 SUMMARY RECOMMENDATIONS AND CONCLUSIONS

5.7.1 Develop a Payment for Watershed Services Program in the Watershed

5.7.1.1 Identify and Engage Stakeholders

As illustrated by all three PWS programs presented above, the process of identifying and engaging stakeholders is critical to achieving a functioning and effective PWS program in the Crooked River watershed. The multiuse values for the Crooked River watershed held by a diverse group of stakeholders throughout Maine make the due diligence of all possible stakeholders a significant challenge (see Box 5.4). To support water quality protection efforts for the area, use of PWS agreements similar to those used by the Watershed Agricultural Council in the New York City water supply watersheds, and incorporating the services already provided by PWD to the lakeshore residents in the Sebago Lake Watershed, would be desirable (Portland Water District, 2009).

Lessons learned from developing PWS programs worldwide suggest transfer of payments from downstream water user to upstream stakeholders for ecosystem conservation are the most common approach and account for the largest current source of financing by far (Johnson, 2000). This indicates that a good starting point for developing a PWS program in the Crooked River watershed would be a survey of PWD customers to determine both interest and willingness to pay for watershed protection. This may facilitate transition from informal alliances between downstream water users and upstream watershed land managers to formal, stable transactions.

While PWD customers would be a good starting point for PWS deals in the Crooked River watershed, eventually statewide commitment and support of economic incentives for watershed protection would provide a stronger regulatory framework for

BOX 5.4 POTENTIAL PAYMENTS FOR WATERSHED SERVICES STAKEHOLDERS

Below is a list of potential sources of funding for private deals in the Crooked River Watershed, grouped roughly by interest in watershed protection. These stakeholders could contribute financial resources or in-kind donations of time or resources to support land management projects in the Crooked River watershed.

A. PWD Customers: (1) Maine Medical Center and other public health services who rely on consistently high-quality drinking water for sanitation, (2) breweries and other food processing and restaurant industry members who rely on consistently high-quality water for food delivery, (3) local companies and corporations who could support watershed protection for improved public relations and corporate responsibility
B. Smart Development and Land Conservation Interests: (1) GrowSmart Maine, (2) Maine State Planning Office, (3) Land for Maine's Future, (4) Maine Association of Realtors, (5) The Trust for Public Land, (6) The Nature Conservancy, (7) Maine Coast Heritage Trust, (8) Smart Growth Leadership Institute, (9) Environmental Defense Fund, (10) City and Town Planning
C. Boards/Organizations: (1) City and Town Conservation Commissions
D. Government Agencies (Regulatory Support): (1) Department of Health Services Drinking Water Program, (2) Maine Department of Transportation, (3) Maine Department of Environmental Protection, (4) Maine Forest Service, (5) Maine Geological Survey, (6) Maine Department of Conservation, (7) Inland Fisheries and Wildlife
E. Watershed Associations and Similar Land Conservation Groups: (1) Maine Rural Water Association, (2) Saco River Corridor Commission, (3) Wells Natural Estuarine Research Reserve, (4) Cumberland Soil and Water Conservation District, (5) River Network, (6) Small Woodlot Owners Association

Drinking Water Utilities: (1) Maine Water Utilities Association, (2) Association of State Drinking Water Administrators, (3) Portland Water District, other unfiltered systems in Maine

the PWS program. The Maine Department of Health and Human Services (DHHS) Drinking Water Program has already recognized the importance of watershed protection in Maine via both support for the filtration waiver and the SWAP assessments. Early involvement and frequent contact with regulatory and legislative stakeholders could lead to support as the PWS framework develops in the Crooked River watershed.

5.7.1.2 Define and Value Watershed Services

There is a general understanding of certain relationships between land use and hyrdology, for example, that forests reduce soil erosion and sedimentation of waterways, that forests filter contaminants and influence water chemistry, and that forest loss shifts acquatic productivtiy (Johnson, 2000). As a result, assumed watershed services tend to focus on the provision of high-quality and reliable flow of water in a watershed. However, there is ongoing scientific controversy, as well as challenges to various elements of conventional wisdom related to water flow. For example, there is debate on the relationships between forests and flood control and between reforestation and water demand (Forest Trends, 2008). The take-home message from this debate is the requirement that a PWS in the Crooked River watershed include collection, compilation, and analysis of local watershed and site-specific data recording the impact of land use change on water quality. The PWS program should begin on deals that support known and proved relationships and build knowledge on enforceable PWS mechanisms before branching to other types of payments for ecosystem services.

All three PWS program examples discussed previously leverage an avoided cost or additional treatment or construction to encourage PWS transactions. In Santa Fe, the city and USFS were able to compare watershed management costs with avoided wildfire recovery costs. In the Great Miami River watershed, National Pollutant Discharge Elimination System permit holders are convinced to pay for mitigation to avoid costs associated with discharge water treatment upgrades to meet regulations. The New York City watershed taxes water users for contributions to land management projects based on the avoided cost of water treatment plant construction. To support this result in the Crooked River watershed, both the PWD and habitat conservation interests should invest in achieving a better understanding about Crooked River water quality and risks of additional cost.

To accomplish this, PWD should study and publish the costs of construction of a water filtration plant assuming the full extent of treatment (including clarification and sedimentation basins) and indicate the resulting increase in water rates to finance the construction, operation, and maintenance of the plant. In addition, an understanding of the tipping points of additional turbidity and bacterial loading on the current treatment plant, and related costs (chemicals, maintenance) and water rate increases would likely support the argument for watershed protection for downstream PWD customers. Similarly, the Maine Department of Inland Fisheries and Wildlife should use PWD baseline data on Sebago Lake trophic levels to gain a better understanding of the impacts of increased nutrient loading for populations in the lake and Crooked River watershed. The same is true for sediment levels and fish habitat in the Crooked River itself.

5.7.1.3 Develop a Nonprofit Organization to Facilitate a Landowner PWS Program

Identifying, supporting, and executing PWS transactions requires a significant amount of resources that are not necessarily available to either buyers or sellers of PWS. The most feasible approach may be for community-based and/or community-funded nonprofit organizations to play a role (Forest Trends, 2008). Formation of a nonprofit center focused on development, support, and eventual expansion of a

PWS program in the Crooked River watershed could play a role in (1) helping sellers assess an ecosystem service product and its value to prospective buyers; (2) assisting sellers with establishing relationships and rapport with potential buyers; (3) enabling sellers to get to know potential buyers by ensuring interactions detail prices paid for similar deals, buyer motivation for the deal, and challenges faced by potential buys that may inform their purchase decisions; (4) assisting with proposal development (pricing of services, managing transaction costs, assessing financing approaches, etc.); and (5) ensuring that the final agreement is in sellers' best interest and provision of risk management advice and services.

5.7.1.4 Establish a Payment Mechanism for Watershed Services

Establishment of an independent fund in the Crooked River watershed (potentially controlled and managed by the nonprofit discussed above) to serve as a collection point for contributions for all the potential buyers of PWS in the watershed might be the best way to most efficiently leverage the diverse potential funding streams for watershed protection in the watershed. This fund would be very similar to the Crooked River–Sebago Lake Fund included in PWD's Watershed Control Program discussed above, but could have a much wider array of contributors and potential project sites. Similar to the process in the Great Miami River watershed discussed previously, the nonprofit could use an independent board of stakeholders to review land management plan proposals from potential PWS sellers and distribute funds based on a predetermined ranking system. This ranking system could be based on the prioritization of properties in the Crooked River watershed and impact to water quality and could assess risk (resulting from the GIS watershed analysis). Payment mechanisms could include payment from water users, municipal bonds, trust funds (like the Crooked River–Sebago Lake Fund), subsidies, differential land use taxation, development rights, and conservation easements, and would be based on negotiated contracts between landowners in the Crooked River watershed and contributors to the fund or the fund controlling board itself.

5.7.2 Conclusions

The Portland Water District is responsible for delivery of safe drinking water to approximately 200,000 customers around Portland, Maine. Because provision of safe drinking water is inherently linked to the source watershed, PWD duties are linked to management of land in the water supply watershed for source water protection. Until this point, PWD has been able to use creative funding and extensive outreach and education, combined with forward-thinking leadership, to create a watershed control program to attempt to protect water quality in Sebago Lake. However, with increasing pressure on PWD from decreasing revenues due to a decline in demand for drinking water, and the threat of increased cost from more stringent water quality regulation, the future of watershed protection and management may necessarily reside outside of PWD. The emerging payments for watershed programs throughout the United States directly address this niche need in the public water utility sector. Themes emerging from several PWS programs functioning in the United States are the need for creative identification of stakeholders who value watershed land management at that specific watershed, creation of an arena to create, support, and negotiate deals between watershed landowners and potential funders, and exact definition of the ecosystem services traded. These components are nascent in the Crooked River watershed, and once fully developed, could lead to a much-needed PWS program in the region.

REFERENCES

Balentine, J., (2004a). PWD Wants Boat Ban in Lower Bay. *Scarborough Current*. February 5, 2004.

Balentine, J., (2004b). Standish Eyes Its Own 'Vision' for Lower Bay. *Lakes Region Suburban Weekly*. February 20, 2004.

Balentine, J., (2006c). Adding Access to Reduce Use—Brilliant! Editorial, *Lakes Region Suburban Weekly*. March 3, 2006.

Brakeley, S., and Z. Ezor, (2009). *The State of Rivers and Dams in Maine*. Environmental Policy Group in the Environmental Studies Program at Colby College in Waterville, Maine.

Caligiuri, P., J. Fitzpatrick, J., Hoyle, and K. Tracz, (2009). *Building an Integrated Ecosystem Service Marketplace in Oregon* (unpublished).

Colby, B., (2010). Water, Water, Everywhere: A Short History of the Portland Water District. *Memories of Maine, Southern Maine Edition*, Winter 2010: 9–13.

Connerty-Marin, D., (2002, November 6). Standish Voters Again Turn Down New Boat Launch. *Portland Press Herald*.

Dearmont, D., B. McCarl, and D. Toman, (1997, October). *Costs of Water Treatment Due to Diminished Water Quality: A Case Study in Texas*.

DEP (Department of Environmental Protection, New York City), (2006). *Long Term Watershed Protection Plan*. http://www.nyc.gov/html/dep/pdf/reports/2006_long_term_watershed_protection_program.pdf

Earnst, C., (2004). *Protecting the Source*. Trust for Public Land and American Water Works Association.

Earnst, C., (2010, February 24). Personal Communication.

Earnst, C., R. Gullick, and K. Nixon, (May, 2004). *Conserving Forests to Protect Water*. American Water Works Association: Opflow.

Electric Power Research Institute, (2009, April). *Regional Water Quality Trading in the Ohio River Basin: Program Summary*.

Electric Power Research Institute, (2009, November). *Ohio River Basin Trading Project: Frequently Asked Questions*.

EPA (U.S. Environmental Protection Agency), (2006). Comparison of Watershed and Water Quality for Systems Serving Over 100,000 and Avoiding Filtration.

EPA, (2010a). Unregulated Contaminant Monitoring Program. http://www.epa.gov/safewater/ucmr/index.html. Accessed April 20, 2010.

EPA, (2010b). Surface Water Treatment Rule. http://water.epa.gov/lawsregs/rulesregs/sdwa/swtr/index.cfm. Accessed October 20, 2010.

Focazio, M.J., D.W. Kolpinb, K.K. Barnesb, et al., (2008). A National Reconnaissance for Pharmaceuticals and Other Organic Wastewater Contaminants in the United States—(II) Untreated Drinking Water Sources. *Science of the Total Environment*, 402: 201–216.

Forest Trends, (2008). Payments for Ecosystem Services: Getting Started a Primer. Forest Trends, The Katoomba Group, and the United Nations Environment Programme, Washington, DC. www.katoombagroup.org. Accessed April 26, 2010.

Freeman, J., R, Madsen, and K. Hart, 2008. *Statistical Analysis of Drinking Water Plant Costs, Source Water Quality, and Land Cover Characteristics.* http://www.forestsforwatersheds.org/storage/Freeman%202008%20DW%20costs.pdf. Accessed April 21, 2010.

Gartner, T., (2009). *Northern Forest Watershed Services: Parallel Pilot Initiatives Providing Incentives for Forest Management and Conservation on Private Lands.* United States Department of Agriculture, Conservation Innovation Grant Application. American Forest Foundation.

Gregory, P., and P. Barten, (2008a). *Public and Private Forests, Drinking Water Supplies, and Population Growth in the Eastern United States, Northeastern Region (20-State) Overview.* University of Massachusetts Amherst, USDA Forest Service, Forest-to-Faucet Partnership.

Gregory, P.E., and P.K. Barten, (2008b). Forest-to-Faucet Partnership. University of Massachusetts, Amherst and the USDA Forest Service. www.forest-to-faucet.org

Hunt, P., (2005, September 20). *Sebago Lake: Yours, Mine and Ours: On Sharing a Multi-use Lake.*

Hunt, P.T., (2010). Presentation to Yale School of Forestry and Environmental Studies. Environmental Manager, Portland Water District, ME.

Hunt, P.T., (2012, August). Personal communication.

Johnson, N., A. White, and D. Perrot-Maitre, (2000). *Developing Markets for Water Services from Forests: Issues and Lessons from Innovators.* Forest Trends, World Resources Institute and the Katoomba Group, Washington, D.C., USA. www.katoombagroup.org. Accessed April 26, 2010.

Johnston, R., (2005). *Why Is Sebago Lake So Deep?* Maine Geological Survey; Geologic Site of the Month, February 1999. Updated October 6, 2005. http://www.maine.gov/doc/nrimc/mgs/explore/lakes/sites/feb99.htm. Accessed April 10, 2010.

Kenny, J.F., N.L. Barber, S.S. Hutson, K.S. Lindsey, J.K. Lovelace, and M.A. Maupin, (2009). *Estimated Use of Water in the United States in 2005.* U.S. Geological Survey Circular 1344.

Kim, A., (2008, February 6). Standish Loses Bid to Widen Road. *Portland Press Herald.*

Maine Department of Human Services Drinking Water Program, (1991, December 30). Letter to Portland Water District General Manager: Portland Water District, Waiver to Filtration under the Surface Water Treatment Rule.

Maine Department of Human Services Drinking Water Program, (1993, September 21). Compliance Agreement and Order Setting Forth a Schedule to Meet Avoidance Criteria.

Maine Department of Human Services Drinking Water Program, (January, 2000). *Maine Public Drinking Water Source Water Assessment Program.*

Maine Department of Human Services Drinking Water Program, SWAP (Source Water Assessment Program), (2003, March). *Maine Public Drinking Water Source Water Assessment Program Portland Water District Sebago Lake Watershed,* prepared by Drumlin Environmental.

Maine Rivers, (2009). *Crooked River Dam Appeal Denied.* Making Waves. http://www.mainerivers.org/newsletter.pdf. Accessed June 1, 2012.

McCarthy, L., (2009, February). *Santa Fe Municipal Watershed 20 Year Protection Plan: 2010–2029: Financial Management Plan.*

McCarthy, P., (2008, March 28). 100 Years of Quality Service. *Portland Press Herald.*

Miller, R., (2004, March 31). Water District Right to Protect Sebago. *Portland Press Herald.*

Ness E. 2009. Oregon Experiments with Mixed Credits. The Katoomba Group's *Ecosystem Marketplace*: News. http://www.ecosystemmarketplace.com/pages/dynamic/article.page.php?page_id=6611§ion=home&eod=1. Accessed June 01, 2012.

New Hampshire Department of Environmental Services, (2010). Environmental Fact Sheet: Pharmaceuticals and Personal Care Products in Drinking Water and Aquatic Environments—Answers to Frequently Asked Questions. http://www.mass.gov/dep/toxics/stypes/ppcpedc.htm. Accessed June 1, 2012.

Oregon Department of Environmental Quality, (2010). Water Quality Trading Program: Tools—Shade-a-lator. http://www.deq.state.or.us/wq/trading/trading.htm. Accessed April 30, 2012.

Protecting Drinking Water Sources, (2010). Enabling Source Water Protection. http://www.landuseandwater.org/. Accessed April 20, 2010.

PWD (Portland Water District), (1926). *Portland Water District Annual Report: Protection of the Water Supply-Filtration.*

PWD, (2009). *Portland Water District Watershed Control Program Annual Report for 2008–2009.*

Scherr, S., A. White, and A. Khare, (2004). *For Services Rendered: The Current Status and Future Potential of Markets for the Ecosystem Services Provided by Tropical Forests.* ITTO, Yokohama, Japan, pp. 30–31.

SDWA (Safe Drinking Water Act), (1974, December 16). 42 USC §§ 300(f), 88 Stat. 1660.

SDWA Amendments, (1986, June 19). Pub. L. No. 99-359.

SDWA Amendments, (1996, August 6). Pub. L. No. 104-182.

SDWAR (SDWA Regulations), (1989). Surface Water Treatment Rule of 1989, 40 CFR §§ 141.70 et seq. United States Environmental Protection Agency.

Sparten Environmental Technologies, LLC, U.S. Surface Water Treatment Rules for Drinking Water. Presentation. http://www.spartanwatertreatment.com/articles/US-Surface-Water-Treatment-Rules.pdf. Accessed June 1, 2012.

Stickney, F., (1942). *History of the Portland Water District.*

Tolman, A., (2010, February 12). Personal communication.

TPL (Trust for Public Land), (2009). *An Action Plan to Protect Maine's Drinking Water Sources: Aligning Land Use and Source Water Protection.* Protecting Drinking Water Sources Program Publication. http://www.smartgrowthamerica.org/documents/maines-drinking-water-sources.pdf. Accessed June 1, 2012.

Vickerman S. (2009). Building Oregon's Ecosystem Marketplace. The Katoomba Group's Ecosystem Marketplace: Opinion. http://www.ecosystemmarketplace.com/pages/dynamic/article.page.php?page_id=6924§ion=home&eod=1. Accessed June 1, 2012.

Water Conservation Subdistrict of the Miami Conservancy District, (2005, February 8). *Great Miami Watershed Water Quality Credit Trading Program Operations Manual.* http://www.miamiconservancy.org/water/documents/TradingProgramOperationManualFeb8b2005secondversion.pdf. Accessed April 29, 2010.

Williamson, J., (1994, December 21). District Trustees Take Some "Abuse." *American Journal.* Standish sec p. 14.

CHAPTER 6

Comparing Drinking Water Systems in the New England/New York Region

Lessons Learned and Recommendations for the Future

Caitlin Alcott, Emily Alcott, Mark S. Ashton, and Bradford S. Gentry

CONTENTS

6.0 EXECUTIVE SUMMARY

Drinking water suppliers in New England and New York City are diverse in terms of ownership type, populations served, and strategies used to maintain clean drinking water. This book explored six drinking water utilities in Massachusetts, New York, Connecticut, and Maine which primarily depend upon surface water sources. In synthesizing the findings from the examples of utilities in the New England/New York City region, we compare the current state of the drinking water providers researched. In particular, we compare the green infrastructure, the gray infrastructure, and the outreach efforts each utility employs. We then identify cross-cutting themes across the chapters, identify lessons learned, and offer some recommendations for future efforts to further improve the provision of clean and affordable drinking water. This information can be directly relevant and useful to much of the more populated parts of the region (EPA 2011). Such information also has a much broader geographic context when including other regions with similar water resource issues from around the world.

Ensuring resilience and clean drinking water delivery requires an integration of engineered technology and surface drinking water source watershed protection that is specific to the place and political context of the drinking water source. Commonly, support for engineered treatment operations can be gained using data gathered from filtration plants. However support for watershed protection can be more difficult to gain because benefits of watershed conservation on water quality can be diffuse and more difficult to quantify. Our general recommendations from this research include the following:

- Gain widespread support for clean water from the public by communicating the science and economic benefits of watershed protection
- Collect data that will help assess current conditions, where the benefits are, and where improvements or efficiencies could be created
- Support voluntary and regulated markets for water quality trading

- Strategically use regulation, landowner incentives, and land acquisition for watershed protection
- Restructure rates to finance management for source watershed protection

6.1 INTRODUCTION

Drinking water is delivered to most Americans relatively simply through surface watersheds. Rain or snow falls within a watershed. The water infiltrates into the soil and moves via streams into a reservoir where it can be stored and then treated and delivered by pipes to consumers. Environmental conditions and the history of policy development in an area shape the particular water delivery system. For example, parts of the New York City source water come from watersheds in the Catskill–Delaware region that are steep with erodible soils. These areas require forested protection or extra filtration through sediment catchments to maintain clean drinking water quality. Another New York City watershed, the Croton has a history of dense development. Here additional pollutants to the water supply must be removed through intensive engineered filtration by the water utility.

Drinking water providers must therefore manage threats and challenges along the delivery pathway. Major environmental threats to clean and abundant drinking water supplies come from pathogens, chemicals, excess sediment, nutrients, and unpredictable high or low flows. Each of these environmental threats can be addressed through watershed protection and/or treatment processes. Social or economic challenges to drinking water suppliers include (1) increasing water demand from population growth; (2) decreasing revenues as water users conserve resources; (3) lack of capital for infrastructure improvements; (4) increased occurrence of pollutants in drinking water watersheds (agricultural chemicals and fertilizers, runoff from urban areas, pharmaceutical use); and (5) increasingly stringent water quality regulations. Addressing social and economic challenges requires building political capital, forward-thinking land use planning, sustainable funding, and effective public outreach.

The multiple barrier approach to drinking water protection is described by the Environmental Protection Agency (EPA) as a "coordinated set of programs and requirements" and a "set of technical and managerial barriers" that help to maintain clean drinking water at every step of the process from source to delivery (EPA 2006). A multiple barrier approach was put into legislation with the 2006 Amendments of the Safe Drinking Water Act (SDWA) and the multiple barriers are described in an EPA (2006) document as the following:

1. *Risk Prevention.* Selecting and protecting the best source of water where possible or protecting a current source of water.
2. *Risk Management.* Using effective treatment technologies and properly designed and constructed facilities, and employing trained and certified operators to properly run system components.

3. *Monitoring and Compliance.* Detecting and fixing problems in the source and/or distribution system.
4. *Individual Action.* Providing customers with information on water quality and health effects so they are better informed about their water system.

Both green infrastructure and gray infrastructure are highlighted as important for the delivery of clean drinking water with various programs and requirements.

While individual water suppliers are required to use the multiple barrier approach, there are many different ways suppliers can allocate their resources and prioritize programs. In prior chapters we have described the various case studies of drinking water protection and delivery attempting to determine how managers of drinking water systems should optimize the use of concrete/steel treatment plants and forested/wetland areas to protect surface drinking water. According to EPA and drinking water managers, both types of assets protect the quality of drinking water; however, determining when and how ecosystem conservation or engineered solutions should be prioritized is complex. Comparing effectiveness of engineered treatment with a forested ecosystem can seem like comparing apples and oranges. The information collected by a filtration plant operator is simply much different from that collected by forest and watershed managers. The question of when to use watershed protection versus filtration plants to deliver clean drinking water is further complicated by social values, institutional inertia, and changing regulations as outlined in the case studies.

In synthesizing the findings from the examples of utilities in New York and New England, we compare the current state of the drinking water providers we researched. In particular we compare the green infrastructure, the gray infrastructure, and the outreach efforts each utility employs. We then identify cross-cutting themes across the chapters and offer some recommendations for future efforts to further improve the provision of clean and affordable drinking water.

6.2 SURFACE WATERSHEDS OF NEW ENGLAND AND NEW YORK: GREEN AND GRAY INFRASTRUCTURE

Drinking water suppliers in New England and New York are diverse in terms of ownership type, populations served, and strategies used to maintain clean drinking water. We explored six drinking water utilities in Massachusetts, New York, Connecticut, and Maine, which depend primarily upon surface water sources (Table 6.1). The drinking water utilities all serve urban areas, with the Massachusetts Water Resources Authority of Boston area and the New York City Department of Environmental Protection notably serving much larger populations than the other utilities. All of the drinking water systems analyzed for this report were publicly owned utilities except for Aquarian Water Company in Connecticut. In EPA region 1, which includes Massachusetts, Connecticut, and Maine, 71 percent of the population receives their drinking water from surface sources, particularly in more urbanized areas, and therefore the case studies can be directly relevant and useful to much

Table 6.1 Some Summary Statistics of Water Utilities Studied in Connecticut, Maine, Massachusetts, and New York

State	Regulatory Agency	Utility Name	Drinking Water Service Area	Population Served
Massachusetts	Massachusetts Department of Environmental Protection	Massachusetts Water Resources Authority	Boston metro area, Chicopee, South Hadley, and Wilbraham	2,200,000
Massachusetts	Massachusetts Department of Environmental Protection	Worcester Department of Public Works	City of Worcester	182,000
New York	New York State Department of Health	New York City Department of Environmental Protection	New York City	8,175,133
Connecticut	Connecticut Department of Public Health	South Central Connecticut Regional Water Authority	Twenty-three cities and towns in South Central Connecticut	400,000
Connecticut	Connecticut Department of Public Health	Aquarion Water Company	Fairfield, New Haven, Hartford, Litchfield, Middlesex and New London counties	580,000
Maine	Maine Department of Human Services	Portland Water District	Portland and surrounding communities	200,000

of the more populated parts of the greater Northeast and Atlantic seaboard, which are dependent upon surface drinking water supplies (EPA 2011). Such information also has a much broader geographic context when including other regions with similar water resource issues around the world (see Barrett et al., Chapter 7, this book).

6.2.1 Approaches to Green Infrastructure Assets

The evolution of each utility within their political and ecological contexts has helped define their commitment to watershed protection and resulting green infrastructure assets (Table 6.2). Three of the utilities have filtration waivers: the Massachusetts Water Resources Authority, the New York City Department of Environmental Protection, and the Portland (Maine) Water District, primarily because of undeveloped and/or protected watershed land. The remaining utilities in Connecticut must provide additional water filtration and treatment by state law. While the Catskill–Delaware system of New York has a filtration waiver, the Croton watershed system has required engineered filtration because of poor watershed protection and water quality.

Table 6.2 Comparisons Among Water Utilities of Watershed Protection Efforts

Utility Name	Filtration Waiver?	Watershed Names (if available)	Total Catchment Acres	Forested/ Undeveloped	Protected (i.e., owned or eased) by Water Utility
Massachusetts Water Resources Authority	Y	Ware, Quabbin, Wachusett	256,570	82%	57%
Worcester Department of Public Works	N	Ten reservoirs	26,000	no data	30%
New York City Dept. of Environmental Protection – Croton	N	Croton	240,000	15%	6%
New York City Department of Environmental Protection – Catskill-Delaware	Y	Catskill-Delaware	1,024,000	75%	36%
South Central Connecticut Regional Water Authority	N	Ten active reservoirs	77,000	no data	35%
Aquarion Water Company	N	Twenty-two active reservoirs in Conn	102,352	no data	19%
Portland Water District	Y	Sebago Lake Watershed	142,000	86%	1%

Note: Protected lands in the table refer only to those lands owned by the utility. This does not include state and other public lands or non-governmental lands (e.g. land trusts).

Patterns of land development in New England and around New York City have impacted formerly undeveloped watershed lands, increasing point and nonpoint source pollutants to waterways. New York City's drinking water system illustrates the impact of land development in the Croton watershed. The Croton has seen rapid growth in the last 50 years due to its proximity to New York City, while the Catskill–Delaware watershed has remained much less developed (see Freiberg et al., Chapter 4, this book). As a result, the Croton system could not maintain its filtration waiver, and a filtration plant is currently under construction for that part of the New York City water supply. Similarly, one primary concern of the Portland Water District is that their watershed lands are vulnerable to conversion. In the Sebago Lake watershed (the source watershed area for Portland), land cover is largely forested; but because this land is up to 99 percent privately owned, many of those forests could be developed and the Portland Water Department would lose its robust source water protection (see Hoyle, Chapter 5, this book).

But Massachusetts and the City of Boston took unusual steps using eminent domain and active land purchase to protect the Quabbin, the Ware River, and more recently the Wachusett watersheds. Finally, Connecticut (New Haven and Bridgeport) and Worcester, Massachusetts, represent the more standard approach of having both substantial forestlands that are owned by the utilities and filtration technologies. In addition, and as mentioned previously, it is state law in Connecticut to filter municipal water.

6.2.2 Approaches to Gray Infrastructure Assets

Land development, biophysical constraints, local regulations, and federal drinking water requirements all impact the kinds of water treatment infrastructure a utility employs. The engineered treatment facilities fall under the EPA's second barrier of *risk management*, or as we call it, *gray infrastructure* (Table 6.3).

Sedimentation and filtration processes are required for water systems that do not have a filtration avoidance determination. Varying levels of sedimentation and filtration are utilized, depending upon the quality of the incoming raw water and preference of the water utility at the time of construction.

Disinfection is required of all water systems, including those with filtration avoidance determinations, and can take several forms. The addition of chlorine is one of the most common disinfection methods but can lead to undesirable and hazardous disinfection byproducts. Ozone and ultraviolet (UV) light are more desirable disinfectants because they do not lead to disinfection by-products. Further ozone and UV disinfection help control the pathogen *Cryptosporidium,* which is the focus of EPA's recent Long Term 2 (LT2) Enhanced Surface Water Treatment Rule (EPA 2012). Many water utilities are currently building additional water treatments to follow the LT2 Rule including the Portland Water District (see Hoyle, Chapter 5, this book). A secondary disinfectant (e.g., chlorine) is typically used in conjunction with the ozone or UV treatment to maintain a disinfectant *residual* as water moves through the delivery and piping system.

6.2.3 Watershed Protection Partnerships

While water utilities are primarily tasked with delivering clean drinking water, their source water protection efforts can have ancillary benefits, and utilities can partner with organizations to protect source watersheds. Co-benefits could include recreation opportunities like hiking or birding on utility-owned land and fishing or boating in source water lakes and rivers. Benefits might also include environmental benefits like a diverse forest ecosystem, well-functioning wetlands and riparian buffers, a thriving fishery, or carbon sequestration. Finally, economic benefits can be realized when utilities harvest timber or other nontimber forest products from their lands or when utilities pay for land improvements for private landowners. Utilities differ in their efforts to market benefits additional to clean drinking water (see Table 6.4). Utilities have also cultivated a variety of different ways to develop partnerships that can create these benefits. Each of the utilities surveyed has some level of outreach with the public or with landowners in the source water watersheds, although the interactions can be quite different. In New York, for example, the majority of the interaction with stakeholders is in providing technical or financial assistance to private landowners to reduce their impacts on water quality. In both Connecticut examples, stakeholder

Table 6.3 Comparison of Treatment Processes and Facilities Among Water Utilities

Utility Name	Filtration Waiver?	Treatment Facility Name	Sedimentation	Filtration	Disinfectant(s)
Massachusetts Water Resources Authority – Boston	Y	John J. Carroll	NA	NA	Ozone, Chloramination
Massachusetts Water Resources Authority – Chicopee Valley Aqueduct	Y	Ware Water Treatment Facility	NA	NA	Chloramination
Worcester Department of Public Works	N	Worchester Water Filtration Plant	Rapid mixing and coagulation and flocculation	Filtration by anthracite coal and sand	Ozone, Chlorine
New York City Department of Environmental Protection – Croton	N	Croton Water Filtration Plant (2013 completion)	Mixing, coagulation and flocculation; stacked Dissolved Air Flotation (DAF) system	Filtered through 60 cm anthracite and 30 cm silica sand filter	UV, Chlorine
New York City Department of Environmental Protection – Catskill-Delaware	Y	The Catskill and Delaware Ultraviolet Disinfection Facility (2012 completion)	NA	NA	UV, Chlorine
South Central Connecticut Regional Water Authority	N	Lake Whitney treatment plant	Flash mixing, coagulation, flocculation, Dissolved air flotation (DAF)	Granulated Activated Carbon filtration	Ozone, Chlorine
South Central Connecticut Regional Water Authority	N	Lake Saltonstall treatment plant	Pre-chlorination, coagulation, flocculation	Sand/ Anthracite	Chlorine
South Central Connecticut Regional Water Authority	N	Lake Gaillard treatment plant	Coagulation, flocculation	Sand/ Anthracite	Counter-current Ozone
South Central Connecticut Regional Water Authority	N	West River treatment plant	Pre-oxidation, coagulation	Sand/ Anthracite	Chlorine
Aquarion Water Company	N	Nine surface water treatment plants throughout Connecticut	Conventional settling or dissolved air flotation	Filtration (not specified)	Chlorine
Portland Water District	Y	Water Treatment Plant near Standish	NA	NA	Ozone, UV (2013), Chlorine

Sources: McClellan et al., 1996.

Table 6.4 Comparison of Co-Benefits of Watershed Protection and Partners Among Utilities

Utility Name	Observed or Desired Co-Benefits	Primary Actions to Support Co-Benefits	Primary Partnerships
Massachusetts Water Resources Authority – Boston	Lake and land-based recreation, timber production	Land acquisition, conservation easements, forest management, species of concern management, recreation in less impactful areas, technical assistance, community outreach/ education	(not stated)
Worcester Department of Public Works	(not stated)	Land acquisition, forest management	Trust for Public Land
New York City (Croton and Catskil-Delaware systems)	Healthy forests, riparian and stream function, economic support for watershed residents, recreation	Land acquisition, conservation easements, forest management, wetlands protection, stream management, septic replacement, agricultural and forestry planning, technical assistance, public outreach and education,	Watershed Protection and Partnership Council (27 members including state and county agencies, and interests focused on economic development conservation, agriculture and forestry)
South Central Connecticut Regional Water Authority	Recreation, research, timber harvest, and non-timber products	Watershed and forest management, land acquisition, conservation easements, recreation, education	The Nature Conservancy, The Trust for Public Land, local land trusts, towns, other conservation agencies, Connecticut Department of Energy and Environmental Protection
Aquarion Water Company	Healthy forests, wildlife, fish, soils. Recreation, and aesthetics	Land acquisition, conservation easements	Connecticut Department of Energy and Environmental Protection, The Nature Conservancy
Portland Water District	A significant fishery (Crooked River), lake, river and land-based recreation	Land acquisition, outreach to partner organizations, landowner education and technical assistance	Crooked River-Sebago Lake Fund, Upper Headwaters Alliance, Casco Bay Estuary Partnership, local land trusts, Forests to Faucets, Sebago to the Sea, Partners in Protecting Sebago Lake

interaction is more geared toward local use of watershed lands for recreation, aesthetics, smallholder firewood, and maple syrup programs and general conservation and environmental protection goals. The Portland Water District (PWD) combines these two approaches by providing technical assistance to local landowners to reduce impact from land use or septic systems, and also manages recreation uses of the watershed and lake by local residents. The PWD is likely the one water supply we studied that is most vulnerable to contamination in the future because so much of its watershed is unprotected, with high recreation use and access that is relatively uncontrolled, and with a water supply that is unfiltered.

6.3 CROSS-CUTTING THEMES FOR THE DRINKING WATER UTILITIES

Six themes have emerged from our analysis of drinking water systems in the New England and New York City region. We identify them as (1) the need to adapt to ever-evolving challenges, (2) risk reduction through the multiple barrier approach, (3) raw water quality improvement to reduce treatment costs, (4) protection of source watersheds to reduce climate risk, (5) attachment of monetary values to source watershed protection to enable comparisons with treatment costs, and (6) protection of source watersheds to provide co-benefits.

6.3.1 Drinking Water Managers Face Ever-Evolving Challenges

Current trends and potential changes in drinking water supply and demands are providing additional challenges to managers. These include (1) regulations for drinking water treatment that are becoming stricter both in containing contaminant levels and in the growing list of contaminants to protect against (EPA, 2011); (2) revenue to water providers that is decreasing as demand for drinking water decreases with effective water conservation (Levin, et al. 2002); and (3) climate change that is adding to the uncertainty to surface water supplies (IPCC, 2007; Barnett et al., 2008).

EPA concern with emerging contaminants dates back to at least the 1990s when, in partnership with the U.S. Geological Survey (USGS), it began surveying water sources across the United States for unregulated contaminants of concern (Daughton, 2004). The most recent USGS national reconnaissance of pharmaceuticals and other organic contaminants, sometimes referred to as PPCPs (pharmaceuticals and personal care products), demonstrated the ubiquity of these contaminants in both ground and surface sources of drinking water (Focazio, 2008). An explanation for the apparent ubiquity of PPCPs is likely due to the large number of pathways for introduction. Anthropogenic sources include "manufacturing releases, waste disposal, accidental releases, purposeful introduction (e.g., pesticides, sewage sludge application), and consumer activity (which includes both the excretion and purposeful disposal of a wide range of naturally occurring and anthropogenic chemicals such as PPCPs)" (Daughton, 2004; EPA, 2006). The publications on the Connecticut and Massachusetts case studies highlighted both the potential for increased regulation, and the difficulty in accounting for treatment of all of these contaminants with unfiltered supply systems reliant on watershed management for raw water quality (see Alcott et al., Chapter 3, this book; Blazewicz et. al., Chapter 2, this book).

Decreasing demand for water is occurring across the Northeast as water conservation practices become more widely used. For example, in Massachusetts demand in Worcester has dropped 20 percent from 1986 to 2008 and water consumption in the Boston area has dropped about 33 percent in that same period (Alcott et al., see Chapter 3, this book).

A commonly anticipated result of climate change is more variable precipitation events, and greater uncertainty of timing and amount of precipitation can be difficult to plan for, especially in small systems (Kirshen et al., 2005; Frumhoff, 2007; Frumhoff et al., 2007). For example, in Massachusetts, the larger Massachusetts Water Resources Authority system for Boston is predicted to provide sufficient water through 2100, but the smaller Worcester system may be at greater risk to smaller climate changes like droughts or storms (see Alcott et al., Chapter 3, this book). In Connecticut the Governor's Steering Committee on Climate Change has investigated impacts of climate changes on drinking water infrastructure and has warned of both flooding and drought potential (Adaptation Subcommittee, 2010; Huntington, 2003).

6.3.2 Protecting Drinking Water Quality in the Face of Uncertainty Requires a Multiple Barrier Approach

The multiple barrier approach layers various water protection and treatment strategies to ensure that clean drinking water will be delivered and public health can be protected. EPA requires public drinking water suppliers to utilize a multiple barrier approach that includes risk prevention, risk management, monitoring and compliance, and individual action (EPA, 2006). Each utility prioritizes barriers differently, given other demands and opportunities. In New York City, for example, the drinking water utility is emphasizing different barriers in different parts of its system. The New York City Department of Environmental Protection (DEP) is building a filtration plant to process water from the more highly developed Croton watershed and more heavily pursuing upland watershed protections in the rural Catskill and Delaware system. However, in both cases DEP is also engaging other barriers with increased ownership or management of lands in the Croton catchment and a UV disinfection facility under construction for the Catskill and Delaware water sources (see Chapter 4, this book).

For Connecticut, the multiple barrier approach has been reinforced by state law whereby all surface waters need to be filtered. However, land protection is a priority, with state law again preventing sale of watershed lands for development. Both Bridgeport (Aquarion) and New Haven (South Central Connecticut Regional Water Authority) use these laws to build more cost-effective and cheaper filtration plants on watersheds that are protected from development (see Blazewicz et al., Chapter 2, this book).

By contrast, the Portland (Maine) Water District relies almost entirely upon a largely undeveloped and forested watershed and a deep natural lake for its clean water source—neither of which are protected. Without additional and more robust barriers this water source may be at risk (see Hoyle, Chapter 5, this book).

6.3.3 Improving Protection of Raw Water Quality to Reduce Treatment Costs

Raw—or untreated—surface water enters a treatment plant after falling as rain or snow and passing through a watershed. The land uses of a particular source watershed will influence the quality of raw water when it enters a treatment plant. If that watershed is mostly undeveloped, raw water will not pick up many of the contaminants associated with urban or agricultural areas. Further, if soils in that watershed are stable, raw water will have less suspended sediment and be lower in turbidity (Ernst et al., 2004; EPA, 2006).

Maintaining undeveloped, high-functioning land in surface drinking water supply watersheds is one way to protect raw water quality (Dudley and Stolton, 2003; Ernst et al., 2004). Forests and soils function to filter particulates from air and water including nitrogen and mercury deposition (Satterlund and Adams, 1992; Likens and Bormann, 1995), slow or eliminate overland flow, and provide high-quality water relative to more urbanized areas (de la Cretaz and Barten, 2007). Additional ways to maintain clean raw water include maintaining healthy riparian and wetland vegetation to filter water, mitigating contaminated stormwater runoff in urban areas, fixing leaky septic systems, and regulating chemicals that are applied to lawns or agricultural areas (Johnson et al., 2001). All drinking water providers studied for this book were aware of the economic benefits of keeping raw water uncontaminated and each attempted to maintain clean raw water given the different circumstances and histories of each place. For example Boston's Massachusetts Water Resources Authority (MWRA) owns or has protected the majority of its watersheds where as New York City utilizes direct land acquisition and private land protection incentive programs. Portland currently benefits from a large undeveloped forested watershed and maintains some landowner incentives, although land acquisition and land owner incentive programs are less developed than Boston or New York City.

6.3.4 Protecting Source Watersheds to Increase Resilience to Climate Change

As climate adaptation becomes a priority focus, water managers may look to source watershed protection to reduce impacts of severe weather events or changing climate patterns on raw water quality, quantity, or seasonal timing. A common prediction of climate models is that increased precipitation will fall as rain rather than snow (Barnett et al., 2008). In this case, the water that traditionally was "stored" in the snowpack over the winter can run off and may be lost for municipal use during dry months. However, a well-functioning ecosystem that allows winter precipitation to infiltrate into the soil profile may not see such dramatic losses of water supplied during the winter (Mote et al., 2005).

Further, although changes in climate are predicted to bring more amounts of annual precipitation, it is thought that annual and interannual precipitation patterns will be more erratic with extended periods of drought followed by large amounts of rain and with storms that will be unpredictable. This possibility makes planning for a resilient forest watershed, that can continue to function as filters under various weather extremes, important (Adaptation Subcommittee, 2010). For example, the MWRA currently utilizes active management regimes for developing resilient forested landscapes in the Quabbin and Ware watersheds with the aim of protecting against catastrophic events including hurricanes, pest outbreaks, and fire (see Alcott et al., Chapter 3, this book). These principles of managing for unexpected natural events can be incorporated into a climate change adaptation strategy.

Supporters of watershed protection often use the precautionary principle as an important guiding ideal. That is, if there is a suspected risk that water supplies may be in danger from climate shifts or severe weather, it is better to be conservative and enact more barriers to public health risk from a contaminated water supply.

6.3.5 Assigning Monetary Values to Source Watershed Protection

Putting a precise value on raw water quality protection or the prevention of contamination is extremely difficult. Ecosystems are complex, and tracking specific water filtering effects, particularly belowground, is not currently feasible. One way values can be assigned is to compare cost offset by protecting a watershed with costs to build an additional filtration plant. This technique was used successfully in New York City source watersheds where infrastructure upgrades would have cost approximately $9 billion (EPA, 1996) (see Freiberg et al., Chapter 4, this book).

The operations and maintenance costs of filtration plants are substantially increased by very turbid, or sediment-rich, raw water, which can often result from developed source watersheds. One study found that a 1 percent increase in turbidity created a 0.25 percent increase in chemical treatment costs (Dearmont et al. 1998). The Connecticut case study also found indications of this trend (see Blazewicz et al., Chapter 2, this book). The operation of a basic filtration plant for the South Central Connecticut Regional Water Authority cost $220 per million gallons in a watershed that was 48 percent protected, whereas treatment cost $600 per million gallons in a watershed where only 6 percent of the land was protected.

Attempts to determine the overall “ecosystem service value” of watersheds and watershed management have been made with some success. For example in the Great Miami River watershed in Ohio, management practices that reduce harmful nutrients from entering the stream generate “credits.” The credits are purchased by National Pollutant Discharge Elimination System (NPDES) permit holders to offset discharge of nitrogen and phosphorus into watershed streams and rivers. The price of the credits is determined by the market and thus reflects the value of conservation management practices (WCSMC, 2005). In New England, the Portland Water District has been exploring ways they may be able to use ecosystem service valuations to identify the value the forested Sebego Lake watershed adds to drinking water protection (see Hoyle, Chapter 5, this book).

6.3.6 Protecting Undeveloped Source Watersheds Can Provide Other Valuable Co-Benefits

Protecting source watersheds from development can create additional benefits for the surrounding communities as well as for the protected water source. Some benefits include recreational opportunities, air filtration, aesthetic beauty, cleaner well water, and increased property values. These additional co-benefits can often bring support from community members who desire open land protection for reasons other than drinking water (Dudley and Stolton, 2003; Ernst et al., 2004). However, source watershed protection may also bring restrictions on the activities to a protected area. These restrictions may include no development, no motorized vehicles, or no public access at all. Community members may be frustrated with restrictions on what they may see as public lands and the burden of water protection for downstream users (Dudley and Stolton, 2003).

Nonetheless, promoting watershed co-benefits often is deemed a net benefit for utilities. For example, the South Central Connecticut Regional Water Authority (SCCRWA) manages its land holdings for recreational benefits when possible and partners with local groups that are concerned with values other than drinking water quality. This includes a smallholder firewood program for neighbors and sugar bush management for maple syrup taps by enthusiasts. The utility states that although "these users are a low priority as a whole, these people are a high priority target audience for natural resource issues. Their activity provides them an inside view of forestry operations and they are likely, due to their first-hand experience, to share their understanding with others" (SCCRWA 1989).

Other utilities have found ways to connect with locals and rate payers as well. For example, Aquarion in Connecticut jointly manages a tract of forestland with the Nature Conservancy and the Connecticut Department of Energy and Environmental Protection that is used for recreation. Similarly, the areas around Quabbin Reservoir, Massachusetts, and Sebago Lake, Maine, are used for recreation as well, though the Quabbin is quite restricted and Sebago Lake has almost free access.

6.4 RECOMMENDATIONS

Forward-thinking source water management will be necessary in the face of continued challenges for the drinking water industry and more demand from urban users. Recommendations will vary by place; however, below we identify general findings that may guide managers and stakeholders in providing clean drinking water at a low cost with an eye to the precautionary principle and a whole watershed planning approach. Ensuring resilience and clean drinking water delivery requires an integration of engineered technology and surface drinking water source watershed protection. The following is a set of synthesized recommendations that result from the case studies.

6.4.1 Gain Widespread Support for Clean Water and for Watershed Protection

Gaining support for strategic source watershed protection will require many partners to work together. To achieve cooperation, it is imperative to communicate the importance of the science of watershed protection and hydrology and the concept of ecological and social resilience to changing demands and changing climate. Such communication needs to involve all stakeholders including state, local, and federal agencies; consumers; and water providers (EPA 1999).

Regulations, incentives, or organizational values may drive a drinking water provider to support social well-being and environmental stewardship. Engaging with many different types of interest groups has been an effective way to further gain support for watershed protection. For example, marketing watershed protection efforts to provide clean water to kids or protecting general public health or the beauty of a community can bring support from different groups. Regional resources or icons have also been used to gain attention such as Seattle's salmon protection efforts. Additionally, in most areas recreational groups benefit from watershed protection. For example, hikers or fishermen can be engaged to support watershed protection efforts. However, when many interest groups become involved, the primary goal of delivering affordable, clean drinking water can become diluted.

Watershed-based groups or councils have been used to organize and communicate local community support. Watershed councils can help coordinate management of watershed lands, take an active role in restoration and fundraising, and participate in data collection and monitoring of the ecosystem (EPA 1999).

6.4.2 Collect Data That Will Help Assess Current Conditions, Where the Benefits Are, and Where Improvements or Efficiencies Could Be Created

Although precisely calculating the optimal investment in green versus gray infrastructure may not be possible at this time, decision support tools can be developed to aid in management of these complex systems. Additional targeted data collection and analysis for engineered and natural systems will improve understanding and optimization efforts.

Tracking engineered assets and costs of construction, operations, and maintenance helps managers maintain functioning filtration plants and find efficiencies. Different water utilities collect and analyze data differently. Computer-based frameworks have emerged to help track and visualize built assets in real time. The Portland Water District uses this type of tool (see Hoyle, Chapter 5, this book).

Watershed uplands can also be monitored like an asset, and the results of watershed quality indicators can be correlated to the impacts on water quality. However, because ecosystems are inherently more complex than an engineered filtration plant, upland monitoring can be more difficult to design. Site-specific qualities like soil type, geology, geomorphology, and climate will drive the water filtration qualities of a particular watershed. In general however, important natural assets to monitor might include forest cover, riparian vegetation condition, wetlands, and stream water quality. Human land use change can also be tracked including zoning and development, conservation easements, protected area designation. Finally, monitoring at high-sensitivity areas such as the intake into the filtration plant or near suspected polluters can be used to inform management decisions. Further, mapping source watershed health indicators and comparing those indicators to the raw water quality could help managers identify downstream consequences of upstream changes. Philadelphia's water district publishes a forecast of river water quality based on recent weather events (Philadelphia Water Department 2011).

Ultimately, integrating the tracking systems for engineered and watershed assets will allow managers to quickly react to watershed changes with the built infrastructure and develop more effective long-term planning for natural areas.

6.4.3 Support Voluntary and Regulated Markets for Water Quality Trading

Establishing payment structures for ecosystem service stewardship has been a regular policy discussion, although there is little regulatory framework for these markets. If a policy framework or even a volunteer trading system for source watershed stewardship emerges, it could be a very useful mechanism for protecting raw water quality. In a credit trading system, costs of watershed stewardship would be offset either by a municipality, a water provider, or rate payers. Denver, Colorado, and Santa Fe, New Mexico, have experimented with rate increases to pay for upland source watershed protection (Stanton and Zwick 2010).

6.4.4 Strategically Use Regulation, Landowner Incentives, and Land Acquisition for Watershed Protection

Government incentives and land management subsidies can help support the costs of source watershed management. The U.S. Department of Agriculture's Environmental Quality Incentives Program (EQIP), Conservation Reserve Program (CRP), and Conservation Reserve Enhancement Program (CREP) are examples of federal programs that can be expanded to explicitly target source watersheds. Additionally, like the New York City example, state funding can be used for technical assistance for landowners (see Freiberg et al., Chapter 4, this book).

Some lands may be important enough for maintaining clean water supplies that municipalities choose to own and protect them outright, rather than allowing for other uses. Purchasing lands is expensive, and identification of these areas must be strategic to get the most benefit for the money. Areas to focus on would include land adjacent to the intake, riparian corridors, and unstable slopes. Connecticut, for example, prioritizes land for acquisition or stewardship incentives with the ratings Class I, II, or III (see Blazewicz et al., Chapter 2, this book).

6.4.5 Restructure Rates to Finance Management for Source Watershed Protection

A major obstacle to water managers across the country is the relatively low rates consumers pay for water, and the dropping consumption of water with improving conservation measures (Johnson et al., 2001; Levin et al., 2002). Overcoming this hurdle is difficult, but a few municipalities have been successful. Three options for restructuring include block-rate pricing, a portfolio approach, and separating types of costs. Alternatively, rates could simply be increased to reflect greater costs (Levin et al., 2002)

In block-rate pricing structures, the largest consumers are charged higher per-unit rates than consumers who use less. This structure could further charge higher rates to users during peak periods and seasonal or unexpected periods of drought. In the 1990s, the number of water utilities using block rates increased from 17 percent to 37 percent. However, this strategy can drive conservation and reduce revenues, leading some utilities to avoid it (Levin et al., 2002).

A portfolio approach for funding would look to sources of income other than rates charged to consumers. System development charges can be used to help offset costs of infrastructure or watershed protection. Separating capital costs from operational costs would help to reduce dramatic rate increases. If the capital costs are paid by property tax or other state funding, the rate payers will be responsible only for operational costs, which are more aligned with water usage.

Finally, specifically identifying watershed management as a line item on bills can be a useful communication tool. Success of this strategy is seen in areas like Denver, Colorado, and Santa Fe, New Mexico, where risk of fire is well known to water consumers and threat of water contamination after catastrophic wildfires may be required for acceptance of watershed management with ratepayer dollars (Stanton and Zwick 2010).

6.5 FUTURE RESEARCH AND CONCLUSIONS

Based on our observations and analyses from preceding chapters, we suggest that there are some common unifying questions and research themes that need to be further investigated. They span a set of questions that require more research and further technical refinements given today's development and drinking water issues. The questions and associated remarks are listed below.

1. What are the best ways to manage storm water runoff in urban or suburban areas? Runoff from developed areas contains complex contaminants from petroleum products, pharmaceuticals, and intensive fertilizers, which can be delivered to streams in a quick pulse during a storm.
2. What is the state of our water delivery infrastructure? Where and how should we upgrade aging systems? Clean water delivery depends on an expensive network of

underground pipes. As these systems age it is difficult to find resources to replace them.

3. What monitoring and reporting tools can be developed to help decision makers? Monitoring and reporting are often seen as prohibitively expensive; however, without information about how a system is functioning, managers are operating blindly. New computer technologies developed for other fields can be adapted to the water management field.
4. What are the emerging contaminants of concern? How should managers approach this looming problem? Many chemicals we use every day, including cleaners, pharmaceuticals, and fertilizers as well as natural contaminants like cryptosporidium, are becoming easier to monitor in drinking water. Because of improved detection, it is likely that water managers will be required to prevent these contaminants. Protecting source watersheds from development is one technique for preventing contamination. Are there other options for keeping water clean?
5. What are suitable economic models and techniques to allow comparison between green and gray infrastructure? Today, techniques in financial and economic comparisons between watershed protection and conservation to preserve or enhance drinking water yields and qualities are difficult to compare with engineered techniques in filtering water. Clearly, to better inform decision making and policies for future investments in drinking water supplies, such comparisons need to be improved.

Drinking water providers in the Northeastern United States are faced with the daily challenge of providing clean, abundant, and affordable water to consumers in the face of increasing challenges like decreasing revenues, increasing regulations, greater uncertainty of water supplies, and tightening government resources. Using a multiple barrier approach, drinking water providers engage watershed protection, engineered treatment technology, monitoring and data management, and consumer outreach to help deliver clean water. Determining which barriers to emphasize with limited budgets is a challenge for all drinking water providers. The case studies in this text outline similar and diverse ways of using those programs to protect drinking water. We found that quantifying and justifying the benefits of source water protection and consumer engagement is difficult, and we strongly advocate for ever-increasing emphasis on data collection, analysis, and communication to consumers.

REFERENCES

Adaptation Subcommittee (2010). The Impacts of Climate Change on Connecticut Agriculture, Infrastructure, Natural Resources and Public Health. Connecticut Climate Change Adaptation Subcommittee Report to the Governor's Steering Committee on Climate Change.

Barnett, T., D. Pierce, H.G. Hidalgo, C. Bonfils, B.D. Santer, T. Das, G. Bala, A.W. Wood, T. Nozawa, A.A. Mirin, D.R. Cayan, and M.D. Dettinger (2008). Human-Induced Changes in the Hydrology of the Western United States. *Science* 319, 1080–1083.

Daughton, C.G. (2004). Non-Regulated Water Contaminants: Emerging Research. *Environ. Impact Assess. Rev.* 24(7-8):711-732; doi: 10.1016/j.eiar.2004.06.003

de la Cretaz, A. L., and P. K. Barten (2007). *Land Use Effects on Stream Flow and Water Quality n the Northeastern United States*. Boca Raton, FL: CRC Press.

Dearmont, D., B.A. McCarl, and D.A. Tolman (1998). Costs of Water Treatment Due to Diminished Water Quality: A Case Study in Texas. *Water Resources Research* 34(4): 849–853.

Dudley, N., and S. Stolton (2003). Running Pure: The Importance of Forest Protected Areas to Drinking Water. World Bank/WWF Alliance for Forest Conservation and Sustainable Use.

EPA (U.S. Environmental Protection Agency) (1996). Watershed Progress: New York City Watershed Agreement.

EPA (1999). Protecting Sources of Drinking Water Selected Case Studies in Watershed Management. EPA 816-R-98-019, April 1999.

EPA (2006). The Multiple Barrier Approach to Public Health Protection. Water. EPA 816-K-06-005.

EPA (2011). Drinking Water Regulations under Development. Retrieved August 12, 2011, from http://water.epa.gov/lawsregs/rulesregs/sdwa/regulationsunderdevelopment.cfm.

EPA (2011). New England's Drinking Water. Retrieved March 9, 2012, from http://www.epa.gov/region1/eco/drinkwater/ne_drinkwater.html.

EPA (2012). Long Term 2 Enhanced Surface Water Treatment Rule (LT2). Retrieved June 21, 2012, from http://water.epa.gov/lawsregs/rulesregs/sdwa/lt2/regulations.cfm.

Ernst, C., R. Gullick, and K. Nixon (2004). Protecting the Source: Conserving Forests to Protect Water. *Opflow* 30:1–7.

Focazio, M.J., Kolpin, D.W., Barnes, K.K., Furlong, E.T., Meyer, M.T., Zaugg, S.D., Barber, L.B., and Thurman, E.M. (2008). A national reconnaissance of pharmaceuticals and other organic wastewater contaminants in the United States--II. Untreated drinking water sources. *Science of the Total Environment*, 402(2-3), p. 201-216, doi:10.1016/j.scitotenv.2008.02.021.

Frumhoff, P.C. (2007). Confronting Climate Change in the U.S., Northeast: Science, Impacts, and Solutions. Synthesis Reports of Northeast Climate Impacts Assessment. Union of Concerned Scientists. Cambridge, MA.

Frumhoff, P.C., J.J. McCarthy, J.M. Melillo, S.C. Moser, D.J. Wuebbles, C. Wake, and E. Spanger-Siegfried (2007). Confronting Climate Change in the U.S., Northeast: Science, Impacts, and Solutions. Synthesis Reports of Northeast Climate Impacts Assessment. Union. of Concerned Scientists, Cambridge MA.

Huntington, T. (2003). Climate Warming Could Reduce Runoff Significantly in New England, USA. *Agricultural and Forest Meteorology*, 117: 193–201.

IPCC (International Panel on Climate Change) (2007). Climate Change 2007: Synthesis Report. Contribution of Working Groups I, II, and III to the Fourth Assessment Report of the International Panel on Climate Change. Core Writing Team, R.K. Pachauri, and A. Reisinger (Eds.). Geneva, Switzerland, IPCC: 104.

Johnson, N., C. Revenga, and J. Echeverria (2001). Managing Water for People and Nature. *Science* 292: 1071–1072.

Kirshen, P., M. Ruth, and D. Coelhol (2005). Climate Change in Metropolitan Boston. *New England Journal of Public Policy* 20: 89–103.

Levin, R.B., P.R. Epstein, T.E Ford, W. Harrington, E. Olson, and E.G. Reichard (2002). U.S. Drinking Water Challenges in the Twenty-First Century. *Environmental Health Perspectives* 110: 43–52.

Likens, G.E., and F.H. Bormann (1995). *Biogeochemistry of a Forested Ecosystem*. New York, Springer-Verlag.

Mote, P., A. Hamlet, M.P. Clark, and D.P. Lettenmaier (2005). Declining Mountain Snowpack in Western North America. *American Meteorological Society* 86: 39–49.

Philadelphia Water Department (2011). Philly RiverCast. A daily forecast of Schuylkill River water quality in Philadelphia. Retrieved August 8, 2011, from http://www.phillyrivercast.org/.

Satterlund, D.R., and P.W. Adams (1992). *Wildland Watershed Management*. New York, Wiley.

South Central Connecticut Regional Water Authority (SCCRWA), (1989). Forest Management Plan for the South Central Regional Water Authority.

Stanton, T., and S. Zwick (2010). Why Denver Spends Water Fees on Trees. *Ecosystem Marketplace*. http://www.ecosystemmarketplace.com/pages/dynamic/article.page.php?page_id=7706§ion=home. Accessed December 5, 2012.

WCSMC (Water Conservation Subdistrict of the Miami Conservancy) (2005). Great Miami Watershed Water Quality Credit Trading Program Operations Manual.

CHAPTER 7

Global Relevance of Lessons Learned in Watershed Management and Drinking Water Treatment from the Northeastern United States

Alex Barrett and Mark S. Ashton

CONTENTS

7.0 EXECUTIVE SUMMARY

We use the synthesis and main conclusions of the case studies discussed in this book to highlight their relevance to drinking water supply issues from forested surface watersheds of the New England and New York City area to other regions of the world. We lay out a framework under which lessons learned might help to inform management decisions elsewhere. The northeastern United States is an appropriate region upon which to base our analysis because of its long history of industrialization and urbanization. This region has been witness to countless development pressures that have made it a center for novel thinking and management around surface drinking water. It remains a core region in the United States that still maintains and supplies some of the cleanest surface drinking water within the country, even given its long history of development. In this chapter we first use three lenses of analysis, (1) biophysical factors; (2) forest history, land use, and infrastructure development; and (3) human institutions, that is, legal, social, governmental, and regulatory environments, to discuss the watershed issues as applied to the northeastern United States. We then demonstrate how watershed managers elsewhere might structure an analysis of their own system. No region will be the same as another, but based on a comparative analysis, the solutions highlighted in the case studies can be used to guide improved watershed systems management in other regions of the world. With the three lenses as an analytical framework, we group regions of the world and discuss how to structure a comparison with New England and New York City that will highlight the most effective areas for immediate action. We use these lenses to articulate what needs to be known about a watershed before a manager begins to balance the green and gray components of infrastructure to deliver the highest quality drinking water possible.

7.1 INTRODUCTION

We here use the synthesis and main conclusions of the case studies discussed in this book to highlight their relevance to drinking water supply issues from forested surface watersheds in other regions of the world. We attempt to draw out the similarities and differences in social and biophysical constructs between the northeastern United States as compared to other regions of the world, in particular to temperate forested biomes that are highly urbanized. The Northeast is a region in the country with the longest history of land conversion, industrialization, and urbanization. Yet its surface drinking water supply has still some of the best water quality (unfiltered) in the country. Given this history, the New England / New York region is a suitable template upon which to base our analysis and to transfer watershed management knowledge to countries elsewhere with similar climates but emerging economies and more recent water resource development histories.

In preceding chapters, three important contextual considerations emerge to guide this discussion:

1. Biophysical factors
2. Forest history, land use, and infrastructure development
3. Human institutions: legal, social, governmental, and regulatory environments

For managers of forested watershed systems in other regions, these three considerations can serve as lenses to focus, facilitate, and strengthen comparisons and analyses. Looking through each lens in turn, managers can more readily make comparisons across seemingly dissimilar watershed issues and gain important insight aimed at improving watershed management outcomes in their respective region.

7.1.1 Rationale: The Global Relevance of Surface Drinking Water Issues from the New England/New York City Region

The case studies presented in this book focus on a geographically small and biophysically unique region in the northeastern United States. In general, the New England/New York City region comprises coastal cities sourcing their drinking water from upland, largely forested areas where precipitation is relatively consistent throughout the year and where rural populations of people are widely distributed to varying degrees from dense suburb to sparsely rural. While there are different approaches to watershed management in the case studies, they all share a common goal of efficiently and effectively supplying high-quality drinking water while providing local, regional, and national stakeholders the co-benefits that accrue from well-managed watersheds.

Even in as geographically small a region as the Northeast, there is a wide array of successful approaches to forested watershed management. The strength of presenting these case studies in one book lies in the fact that together, they help to identify and describe a set of guiding principles for watershed management in the region (see Alcott et al., Chapter 6, this book). Each approach is different, and the results vary across multiple scales. Each system faces constantly changing constraints on management that can be unpredictable and that can be organized into the following categories (see Alcott et al., Chapter 6, this book):

1. Changes of ownership between the public and private spheres
2. Changing regulatory structures
3. New forest conservation tools, mechanisms, and techniques
4. A long-term interchange between researchers and managers on the ground
5. A changing climate

These changing factors create a common set of experiences and challenges that have required management systems to develop in resilient and sometimes dramatic ways. The synthesis chapter (Alcott et al., Chapter 6, this book) highlights the commonalities; we here offer a framework for using these commonalities as a way to seek greater global applicability for the lessons learned in these case studies.

Land management and source water quality are central themes throughout the case studies of this book. They are important factors to consider when we apply our understanding of the commonalities within these themes to help guide management decision making in other regions of the world. To reasonably apply any of the lessons related to source protection, watershed managers must first understand the biophysical, social, political, and historical context in which they are operating. Hence the importance of understanding your watershed (Ernst et al., 2004b); and therefore the importance of our analysis for practitioners looking to gain insight into their specific regions based on the combined experiences of watershed managers in the Northeast.

7.2 METHODS

The preceding chapters of this book form the basis for our framework. They provide the analyses, discuss the implications, and look across systems to synthesize and tease out the lessons learned. Looking beyond the Northeast, we have relied on technical reports, peer-reviewed literature, books, and geographic information system (GIS) mapping to guide our thinking on how to best compare watershed management constraints and future possibilities for improvement.

7.2.1 Analytical Relevance and Applicability

While the New England / New York City region in the United States is indeed a unique region, the series of lessons that the authors highlight in their case studies appears to have wider, more global applicability. Natural systems are inherently complex and are made more so by the demands that drinking water supply managers place upon them. As we look to other regions with forested watersheds that feed into surface drinking water systems, we can reference the guidelines and key findings presented by the case studies to ground our analysis, but we keep in mind the variability and the uniqueness of place. To do so requires the artful application of general principles to specific situations on the ground.

For these recommendations to have analytic relevance beyond the small geographic scope of this book, practitioners need to strike a balance between broad, easy-to-make statements about similarities between regions, and specific, on-the-ground applications of management approaches that work well in one area and might help in another. For example, while the intricacies of local politics and economics in northeastern China are beyond the scope of this book, it is important to assess the value that more protected forests in a temperate continental region might mean for water filtration, purification, and delivery requirements. It would become especially more important if factors like land use history, forest type, and regulatory structures were similar.

We therefore encourage a focused and specific understanding of local intricacies that would allow watershed managers to most effectively manage forested watersheds and to make sensible decisions that will yield the best results possible. But, it is a broad viewpoint, and therefore a necessary willingness to experiment is required to allow managers to bring creative new insights to bear on how to improve these decisions and the mechanisms and institutions that make them possible. Our analysis and framework serve to orient managers' and policymakers' thinking about how this book's themes and recommendations are applicable outside the New England/New York City areas.

7.2.2 Context for Comparison

The multiple barrier approach to water quality provision focuses on the combination of protective barriers against contamination based both on green infrastructure such as forests and on gray infrastructure such as water treatment facilities (EPA, 2006). Each case study presents a situation where these two types of infrastructure are managed in tandem to deliver high-quality drinking water to end users. The most fundamental factors that determine the needed mix of green versus gray infrastructure are based on physical, biological, and climatic factors of a region. Political and regulatory climates, as well as creative source protection mechanisms such as payments for watershed services (PWS), develop in and are constrained by the biophysical factors of a region. Over time, these same factors have largely driven the land use history of an area. The relevance and hence applicability of the recommendations for future management presented in this book depend in large part upon the biophysical similarities between regions where we seek to make comparisons. The structure for our comparison hence begins with this biophysical starting point. It is the most fundamental of the three lenses.

7.3 BIOPHYSICAL CONSIDERATIONS AND CONSTRAINTS

It is beyond the scope of this book to analyze in depth the intricacies of forested watersheds and how conditions on the ground might approximate those of the case studies presented. Instead, by offering a series of structuring questions and answering them from the standpoint of the New England/New York City region of the United States, we seek to model a comparative approach to analyzing the applicability of the lessons learned to other watersheds. Relevant characteristics of the biophysical attributes of a watershed can be characterized broadly into eight categories. First is the classification of the biome type by large-scale geographic region. Second, third and fourth the classification can be defined by the physical nature of the topography (landform), the geology and related soils respectively. The fifth and sixth consideration is climate, which should especially include the amount and seasonality of precipitation and the nature of the diurnal and seasonal flux in temperature. Seventh is the organismal product of climate and physical aspects of topography, soils and geology, namely the structure and composition of the vegetation. Lastly, the product of all prior watershed categories is the hydrology, which is shaped by both the nature of the climate, the physical attributes of landform, geology and soils of the watershed, and also the influence of the ameliorating effects of the vegetation (see Table 7.1).

Taking the Northeast region of the United States more broadly (beyond New England), as an example it would be considered to be part of the temperate broadleaf and mixed forest biome. Many physical and climatic similarities therefore exist to other regions of the world that comprise the temperate broadleaf and mixed biome. For instance, given this biome's moist climate and relatively fertile soil, it is no coincidence that that bulk of humanity lives within it. And like the Northeast, the

Table 7.1 An Analysis of a Watershed and the Amounts of Green and Gray Infrastructure Needed in Order to Deliver High-Quality Surface Drinking Water Can Begin with an Understanding of the Biophysical Considerations of the Watershed. Eight Categories are Identified.

1. Biome	• Type, frequency of type globally • Defining characteristics • Type of vegetation • Proximity to ocean
2. Landform	• Topography • Aspect, slope • Facility of movement for people and machinery • Amount of land physically altered and shaped by human activity
3. Geology	• Surficial and bedrock • Relationship to soil • Relationship to climate and hydrology • Current or future resource extraction possibilities
4. Soils	• Physical and chemical properties • Profile development • Soil development (*in situ, ex situ*)
5. Precipitation	• Amount, temporal distribution • Form (rain, snow) • Relationship to landform • Climate change modeling predictions
6. Seasonality	• Number and length of seasons • Characteristics of each season • Relationship to landform and small-scale, local variations
7. Vegetation	• Structure and composition • Stratification, Leaf Area Index • Successional floristics • Species demographics
8. Hydrology	• Annual flow distribution • Residence time and location of water on the landscape • Stream channel morphology • Pond, lake, and reservoir morphology

majority of the population and the big cities are close to the coast with a dependency on drinking water from surface watersheds in the adjacent uplands (Ellis and Ramankutty, 2008). Applying these categories to form a baseline for a watershed analysis can be useful (see Box 7.1).

7.4 FOREST HISTORY, LAND USE, AND INFRASTRUCTURE DEVELOPMENT

There are many common trends in forest conversion, land use, and development across moist temperate and tropical systems worldwide (Tyrrell et al., 2012). These systems tend to see human impact because they are pleasant, convenient places to live, with high rainfall and relatively fertile and/or irrigable soils that can sustain intensive agriculture. Aside from human effects on the forests, human infrastructure developments constrain watershed managers' range of options. Impervious surfaces,

BOX 7.1 AN ANALYSIS OF THE NORTHEAST WOULD BEGIN WITH THE FOLLOWING OUTLINE BASED ON THE FRAMEWORK FROM TABLE 7.1

1. Temperate continental climate, weather moves west to east, weather patterns affected by polar front, situated on eastern edge of a continent near a large, open ocean (Whitney, 1994; Keim, 1998; Barry and Chorley, 2003).
2. Precipitation is spread evenly throughout the year (1090 mm per year, on average), water is abundantly available and precipitation is shifting toward more rain and less snow (Barry and Chorley, 2003; de la Cretaz and Barten, 2007; Huntington et al., 2010).
3. Young, forest soils, little profile development, formation influenced by vegetation postglaciation, soil depth shallow, permeability constrained in many cases by dense substratum of basal till and/or bedrock (Ciolkosz et al., 1989; MDC/DWM, 2001; Brady and Weil, 2008).
4. Peak streamflows in spring and fall, average annual streamflows relatively homogenous, trellis and dendritic stream networks, major rivers, lakes and ponds, incised stream channels, average runoff approximately half of total precipitation (Bailey et al., 1994; Kirshen and Fennessey, 1995; Vogel et al., 1995; Hodgkins et al., 2003).
5. Region characterized by marked seasonality, increasing variation at smaller local scales, hurricanes are a common disturbance, periodic droughts occur but water is generally not limiting, ocean has limited effects on seasonality along coast (Whitney, 1994; Zielinski and Keim, 2003; de la Cretaz and Barten, 2007).

intense agriculture, and wastewater management techniques ranging from municipal systems down to individual septic tanks all impact water quality and should be assessed by managers seeking to balance gray and green infrastructure improvement (Burns et al., 2005; Arscott et al., 2006).

The forests of the Northeast, and more specifically the New England/New York City region today, represent the legacy of human activities on the landscape (Irland, 1999). From Native Americans' use of prescribed fire, through the waves of land clearing for agriculture in the eighteenth and nineteenth centuries, and on to forest recovery and regrowth of the twentieth century, the forested watersheds examined in these case studies continue to develop based largely on a legacy of human disturbance (Foster, 1992; Whitney, 1994; Irland, 1999). This intensely resilient regrowth following human disturbance is a fundamental characteristic of northeastern forests and to a large degree enables us to focus on the green infrastructure component of watershed management. In the northeastern United States we can rely on the forests to bounce back following disturbance, but in many biomes and regions such conditions may not be conducive to forest recovery.

The cycle of forest clearing for human settlement and agriculture, followed by population growth, industrialization, and the depopulation of rural lands, followed by the resurgence of young second-growth forest is one that can be seen elsewhere

Table 7.2 With Biophysical Considerations Identified, Forest Watershed Managers Can Then Characterize the Specific Type of Forest, Its History, and How Human Development within the Watershed Impacts the Forest's Ability to Be a Resilient and Reliable Component of the Green Infrastructure; Seven Descriptors for Characterization Are Given

Forest Type	Tree and understory species and their autecology Rainfall on vegetation (de la Cretaz and Barten, 2007) Primary or secondary forests Forest composition changes over time
Disturbance Regime	Type, frequency, scale Human impacts on disturbance regime Level and speed of forest resilience Climate change impacts New disturbance elements: exotic invasives
Forest Development	Stage of forest development (Oliver and Larson, 1996) Historical vs. present forest cover Level of scientific understanding of place-specific forest development patterns
Resource Extraction (Past and Present)	Type Extractors and their relationship to the land Ecological implications Physical implications Relationship to natural disturbance regime Economics and global importance of resource
Impervious Surfaces	Amount and proportion to total area Effects on runoff, groundwater recharge, and mean water residency times (Burns et al., 2005) Human development build out and local land use law
Wastewater Management	Residential, commercial, and industrial Treatment facilities: location, age, connectivity Residential septic: quantity, condition, distribution
The Built Environment	Transportation infrastructure Energy infrastructure (old and newly developing) Communications infrastructure Solid waste disposal

across temperate mixed and tropical moist forest biomes that went through similar histories as the Northeast did. Good examples for temperate regions would be Northeast Asia (e.g., Japan, South Korea), and for tropical regions such as Central America, the Pacific coastal region of the north central Andes and the Atlantic forest region of Brazil. All such regions have large cities juxtaposed with second-growth forests that are the drinking water supplies for people (e.g., São Paulo, Rio de Janeiro, Panama City, Seoul, Tokyo, Yokohama).

Knowing that the biofilter provided by forests has a regenerative capacity, a series of guiding questions for watershed managers can be used to characterize the development history and current integrity of today's drinking watershed supply. Seven categories can be used to help characterize watershed integrity: (1) forest type, (2) disturbance regime, (3) forest development, (4) resource extraction, (5) impervious surface, (6) wastewater management, and (7) the built environment (Table 7.2). Such categories can be used to evaluate the integrity of the watershed based on a knowledge of forest composition; age class; and its ability to

BOX 7.2 USING THE CASE STUDY WATERSHEDS AS A COMPARISON AND THE FRAMEWORK FROM TABLE 7.2 AS A GUIDE, THE FOLLOWING BASELINE DESCRIPTION OF THE FOREST TYPE, ITS HISTORY, AND THE IMPACTS OF HUMAN SETTLEMENT ON THE WATERSHED IS PROVIDED BELOW

- The Catskill–Delaware system features predominantly northern hardwoods and is 80% forested (Bailey et al., 1994).
- The other systems feature mainly central hardwoods with oak, hickory, and white pine as the main species (McWilliams, 2005; Kyker-Snowman et al., 2007; Griffith et al., 2009).
- Main disturbance comes from wind, hurricanes, ice storms, and anthropogenic influences (Lorimer and White, 2003), climate change impacts have been modeled and potential impacts and solutions reviewed (Evans and Perschel, 2009).
- Forest development patterns are well-understood.
- Current natural resource extraction pressures (including timber) are minimal, shale gas development could complicate the situation for New York's watershed (Kargbo et al., 2010).
- History of land clearing for agriculture followed by abandonment and then extensive timber extraction followed by natural regeneration and regrowth (Foster, 1992; Irland, 1999).
- Current forest threats are driven mainly by human settlement development pressures (Aber et al., 2010).
- Relationships between impervious surface area and water quality are well studied, analytical frameworks are replicable, data for specific watersheds are readily available (McMahon and Cuffney, 2000; Coles et al., 2004).

regenerate in response to unpredictable episodic and often catastrophic disturbances (drought, fire, hurricanes) and to chronic effects of insects and diseases (often introduced). In addition forests have been exploited directly for a variety of extractive resources over time and face an ever-increasing trend of permanent land conversion to impervious surfaces due to infrastructure development (roads, energy, housing, industry). All of these impact watersheds through wastes that come directly from such development (septic systems, petroleum pollutants, agricultural wastes, and leachate) (see Box 7.2 for an example of an initial analysis of these considerations for the case study watersheds in the New England/New York City region).

7.5 HUMAN INSTITUTIONS: LEGAL, SOCIAL, GOVERNMENTAL AND REGULATORY

The human institutions that have developed in and around land control and land management further determine the level of similarity between watershed systems.

To make helpful comparisons, watershed managers must analyze the legal, social, governmental, and regulatory environments in which they operate. When a water authority owns its entire watershed, it has a high level of control and can more easily balance green and gray infrastructure requirements. The city of Seattle, Washington, provides a case in point (Ernst et al., 2004b).

In the case studies presented in this book, and in most places worldwide, watershed managers do not have the luxury of direct control over human activities at the watershed scale. Instead, they must deal with the complexities of governance structures, regulations, local land use law and custom, water rights, and economic drivers as well as other complicating factors such as familial ties, sociopolitical history, and national natural resource management strategy. Central governments often retain ultimate power over land use decisions, but the on-the-ground consequences of exercising this power can fester for years and complicate watershed management. Eminent domain takings in the United States gave rise to the Quabbin Reservoir, and the social implications of that action are still felt more than 80 years later. Key factors related to the human institutions that facilitate and constrain watershed management need to be considered (Table 7.3). Factors can be broadly categorized as governance structures (national, state and local); rights and tenures to land, water, and other resources; the engineering of the water storage and delivery system; and lastly economic systems and markets. Layered on top of biophysical, forest, and land use history considerations, these topics attempt to guide an analysis and a comparison of how the current status of forested watersheds in the New England/New York City area might help guide management decisions in other regions of the world (see example of the Northeast in Box 7.3).

Compared with the similarities in land use history and economic development, followed by forest regrowth of populated regions of the moist temperate and tropical biomes and the development of surface drinking water supplies, similarities in legal structure and land tenure among these regions can be very different. For example, common law is the mainstay of the legal system throughout North America and the United Kingdom and its former colonies that make up the Commonwealth. This legal system is based on precedent and judge-made decisions. This is very different from all of Latin America and Northeast Asia, which have a more codified system that can be classified as civil law in which its core principles are codified. Such frameworks have important repercussion in water rights, property and resource rights, and water and pollution laws. Differences in civil law clearly exist between Latin America and Asia based on power and governance structure. Countries like China have very strong central powers in decision-making compared with most of Latin America.

7.5.1 Land Conservation and Payments for Watershed Services

The complexity of human systems can be overwhelming. For watershed managers, an effective technique that has proven helpful across all case study watersheds in this book for protecting water quality is to restrict or incentivize certain uses of land in a watershed. The main tools that have emerged to facilitate this type of

Table 7.3 The Legal, Social, Governmental, and Regulatory Framework. Ten Social and Legal Considerations Must Be Understood to Make More Effective Institutional and Watershed Management Decisions.

National Government	Form, structure, historical development Legitimacy and authority Responsibility for drinking water provision Level of authority in local land use decisions
State/Provincial Government	Form, structure, historical development Relationship to national and to local governing bodies Level of authority in local land use decisions
Local Government	Form, structure, historical development Relationship to state and national governing bodies Level of authority in local land use decisions
Land Tenure	Type of person/organization that can own land Scope of eminent domain powers Level of certainty in title to land Record keeping Land title as basis for contractual agreements Legal constraints on uses
Legal Structures around Water	Ownership of water (present and historic) Rights of ownership around water Legal connections between land and water
Economic System	Type of system (theoretical and actual) Wealth distribution Transaction mechanisms Roles for water utilities to play
Payments for Watershed Services	Voluntary market-based transactions aimed at improving water quality (Hanson et al., 2011) Level of importance for local economies Transactional infrastructure and know-how
Water Regulations	Level of water quality regulation Regulator, level of authority, legal framework Regulation of activities in and around water Monitoring and control of extraction
Water Extraction, Storage, and Use	Source of drinking water Storage and transportation Infrastructure ownership (present and historic) Competing uses for potable water
Upstream–Downstream Social, Political and Economic Dynamics	Socioeconomic history and current dynamic Demographic shifts Political and jurisdictional boundaries

management and watershed improvement in areas with high levels of private land ownership are

1. Land conservation via legal mechanisms
2. Payments for watershed services

These two tools blend together many of the complex facets of human institutions and allow us to efficiently discuss similarities and differences between systems.

BOX 7.3 AN ANALYSIS OF THE LEGAL, SOCIAL, GOVERNMENTAL AND REGULATORY FACTORS IN THE NORTHEAST BEGINS WITH THE FOLLOWING BASELINE ASSESSMENT; USING THE FRAMEWORK IN TABLE 7.3 THIS ANALYTICAL LENS IS BROAD

- Nationwide, federal system of representative democracy, states retain many powers, local governments exercise high levels of control over land use decisions (Meck, 1996; Nolon et al., 2008).
- Strong land tenure, rights to property are divisible, which enables conservation easements as a watershed management tool, forestland predominately family owned, history of governmental takings, governmental takings of still land occur (but as the exception), PWS can operate on sound legal and financial footings (Ernst et al., 2004b; Butler, 2008; Porras et al., 2009).
- Water historically owned under a system of riparian rights where adjacent landowners have access to water, extraction is based on the vague term *reasonable use*, no state has a comprehensive water code, but many have moved to a hybrid system of legal rights to water where users must apply for and obtain an extraction permit after a certain date (Ausness, 1982; Sherk, 1989).
- Drinking water quality governed by federal legislation that is becoming increasingly stringent, pollution of waterways also governed by federal law, most enforcement activities delegated to each state but federal agencies maintain right to intervene (Paddock, 1990; Wickham et al., 2011).
- Regulated, capitalist market, government involved in ensuring that drinking water standards are met, legal maneuvers and market-based transactions involving landownership and control are common and accepted, some PWS are already operating (Ernst et al., 2004a; Hanson et al., 2011).
- Reservoir-based systems effectively regulating and storing flow.

As in the rest of the United States, landownership in the Northeast can be envisioned as a "bundle of sticks" where various ownership rights to use land are conceptualized as individual "sticks" that can be removed, individually, from the bundle without eliminating the basic, underlying ownership of the property. *Fee-simple ownership* is the standard and most widely understood type of private landownership in the United States. This is undivided ownership, where the landowner owns all the rights (or "sticks") pertaining to the property. The outright purchase of fee-simple title to land is an effective, yet expensive, way to control activities on lands within a watershed (Ernst et al., 2004b).

Increasingly, forested watershed conservation efforts are relying on the purchase of development rights to protect forested watersheds (Barten and Ernst, 2004). These transactions are voluntary and have helped to ameliorate upstream–downstream social conflicts around source protection efforts because upstream populations can retain title to their land. Gone are the days when governments in the Northeast could relocate entire villages in the name of watershed improvement. With the sale or donation of development rights, the landowner forever encumbers the title to the property and eliminates the possibility of future owners developing the land. By purchasing the "right to develop" and removing that stick from the bundle, watershed managers seek to efficiently and effectively protect forests and the filtration and buffering that they provide.

Payments for watershed services (PWS) are market-based incentives that seek to incentivize land management practices that improve the quality of water flowing off of a watershed (Porras et al., 2009). PWS schemes involve legal, economic, regulatory, and social dimensions. They therefore provide a relatively comprehensive way to compare watersheds. There have been a number of worldwide surveys cataloging and analyzing the wide range of PWS programs operating today (Porras et al., 2009; Stanton et al., 2010).

Latin America, China, and the United States are at the forefront of PWS: Latin American for the number and duration of its programs; China for the scale and financial size of its transactions; and the United States for the amount of land it protects through these transactions (Stanton et al., 2010). Given the diversity of political, legal, economic, and social structures governing each of these regions, it would seem that PWS schemes are a transferrable and widely applicable mechanism for protecting watersheds. Indeed they are, but they are so diverse in their implementation and in their results that they alone do not provide a simple template for developing successful resource protection strategies across watersheds. Instead, as watershed managers evaluate the applicability of PWS to their system, they should step back a level and rely on the three lenses for comparison that we have presented here.

7.6 GLOBAL APPLICATIONS OF THE LESSONS LEARNED

The growing understanding around the world that land conservation as well as market-based incentives like PWS can help to protect the source of drinking water has led to a boom in watershed analyses, comparisons, and policy as well as land management recommendations. Think tanks, international development organizations, and academic institutions have led the charge and have built up a wide and in some cases quite comprehensive body of research and literature (Dudley and Stolton,, 2003; Ernst et al., 2004b; Porras et al., 2009; Stanton et al., 2010).

7.6.1 Understand Your Watershed

In the case studies presented in this book, we have seen how densely populated urban areas and cities rely on forested watersheds to provide them with high-quality drinking water in an affordable and resilient way. Forest cover makes the most sense for drinking water watersheds (Hamilton, 2008). The temperate, second-growth forests of the Northeast provide this cover and are the fundamental component of the "green infrastructure" that these case studies have highlighted. The unique biophysical and land use history and current situation of the northeastern United States, paired with a legal system that allows a high level of local and state land use and water regulation control, have led to each of these systems pursuing the protection of these forests as a central strategy for high-quality water production.

As the report *Protecting the Source* notes, understanding all aspects of a watershed is the first and most fundamental step in determining the suite of solutions to apply to secure a safe drinking water at acceptable social, economic, and environmental costs (Ernst et al., 2004b). Over the course of this book, we have come to intimately understand several watersheds that are all within 400 miles of each other. In this chapter, we have expanded on this understanding and placed these watersheds into the biophysical, legal, and political contexts in which they have developed and in which they currently operate.

Out of this understanding and based on the lessons learned from the case studies, Alcott et al., (Chapter 6, this book) draw five central recommendations for watershed managers in other regions to consider:

1. Gain widespread support for clean water and watershed protection
2. Collect data that will help assess current conditions
3. Support voluntary and regulated markets for water quality trading
4. Strategically use regulation, landowner incentives, and land acquisition for watershed protection
5. Restructure rates to finance management for source watershed protection

These recommendations are based on the seven emerging themes that Alcott et al., (Chapter 6, this book) identify across our watershed case studies. The question now becomes: How and where might these recommendations help to improve drinking water outcomes for forested watershed managers in other parts of the world? Where can we find similarities, and how can these similarities help us to analyze and improve watershed management?

7.6.2 BIOPHYSICAL STARTING POINT

The multibarrier approach is based fundamentally on first assessing condition and then protecting the surface drinking water supply. This has emerged as a pivotal factor in all the case studies. It is therefore reasonable to begin to answer this question of wider applicability by examining other regions of the world that share similar biophysical conditions with the Northeast. Our main focus is on temperate, broad-leaved forested biomes near coasts (see Figure 7.1). This particular biophysical lens allows water systems managers to better understand the constraints of the system in which they operate. It also allows for a more meaningful comparison with the Northeast.

Remote sensing and mapping technologies have revolutionized our ability to analyze biophysical data in relation to human population distributions. Organizations like the United Nations, the World Bank, the World Resources Institute, the World Wildlife Fund, and the U.S. Natural Resources Conservation Service provide a wealth of publicly available maps and data sets that can help to guide watershed analysis.

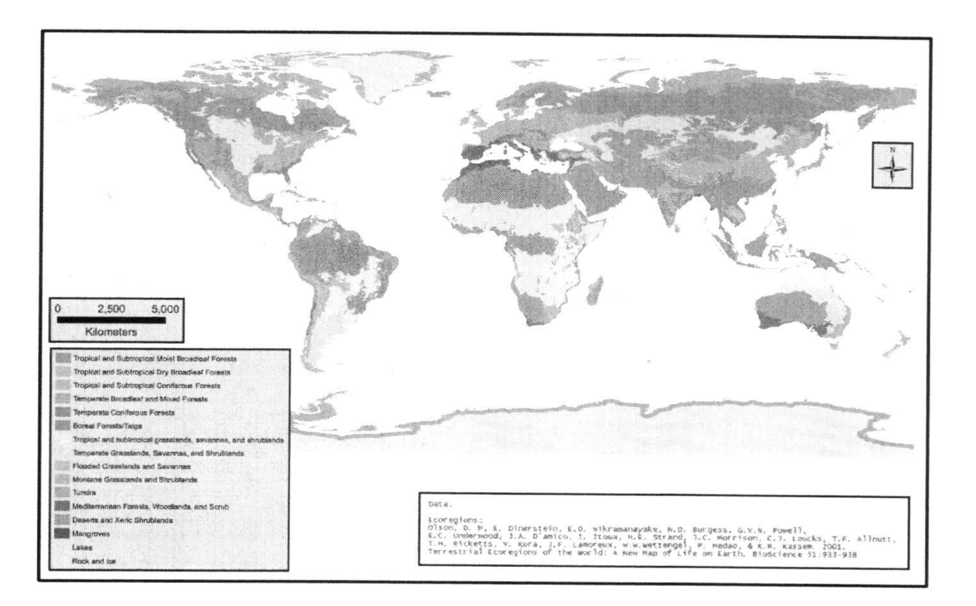

Figure 7.1 A worldwide map of biomes depicting the major vegetation types. The temperate broadleaf and mixed biome and the tropical and subtropical moist broadleaf biome represent only a small part of the earth's terrestrial systems but account for a disproportionate amount of the world's population and rainfall (Modified after Olson et al., 2001). (See color insert.)

For example, the Food and Agriculture Organization (FAO) has performed a sophisticated mapping analysis of the biophysical factors that influence what plants grow where (Figure 7.2). Their focus is on agriculture, but the results are applicable to temperate forested biomes as well. The authors of the FAO report map many factors including thermal climate zones, length of growing period, soils, and terrain (van Velthuizen et al., 2008). Information of this kind is crucial because it determines the feasibility and the constraints on developing the first barrier in the multibarrier approach: the forest filter.

Human population centers are predominantly located in coastal areas, often where large river systems empty into the sea (Small and Nicholls, 2003; McGranahan et al., 2007). New York, Boston, Portland (Maine), and much of Connecticut are all coastal areas that fit this description. As we have seen in the case studies and elsewhere, these coastal areas rely on upland forested systems to provide them with drinking water. With these coastal cities worldwide expected to continue their growth, an understanding of how to manage their water supply systems becomes even more important (Culliton, 1998; Tibbetts, 2002) As expanding coastal centers come into increasing contact with the less intensely settled forested areas currently protecting their drinking water sources, watershed protection efforts will need to strengthen and managers will need to work in increasingly creative and novel ways (Burns et al., 2005). More than one-third (37 percent) of the world's population lives within

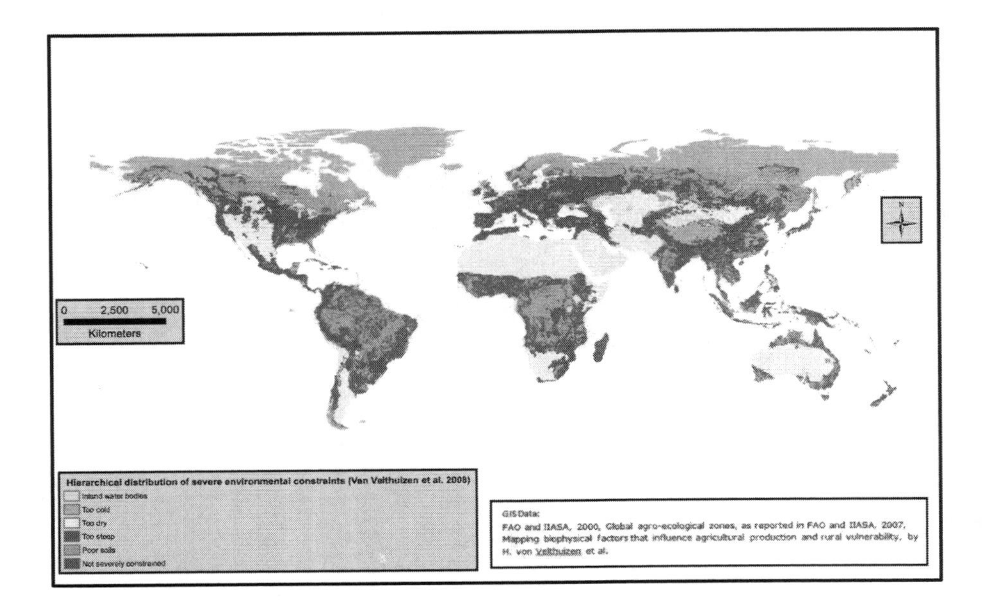

Figure 7.2 A worldwide map of climate and environmental constraints. The map depicts the results of a GIS analysis identifying areas of the world that are most favorable for growing plants. These areas would also be most favorable for developing resilient forested biofilters around drinking water systems (Modified after van Velthuizen et al., 2008, with permission). (See color insert.)

100 km of the coast (Small and Cohen, 2004) and two-thirds of the world's population lives in a (formerly) forest(ed) biome. The overwhelmingly dominant source of drinking water for such regions is the upland surface watershed, which can deliver large volumes of clean water efficiently downhill to population centers on the coast.

7.6.3 The Northeast in the Global, Biophysical Context

The Northeast falls in an area of low environmental constraint (van Velthuizen et al., 2008) (see Figure 7.2). On a basic, biophysical level, the northeastern United States is a relatively unique environment on a global scale. Mid- and northeastern China, as well sections of Korea, Japan, and much of Europe may most closely approximate the Northeast's temperate continental biome (see Figure 7.2). But other factors such as land tenure patterns, soils, land use history, human population distribution, and current government regulation and enforcement patterns complicate the comparison. Overall, more than 350 major cities worldwide are in temperate broadleaf and mixed biomes (see Figure 7.3).

Temperate forests currently cover approximately 7 percent of the earth's land mass (Chapin et al., 2011). These forests comprise approximately 25 percent of forested area worldwide (Tyrrell et al., 2012). In areas like North American and Europe, only 1 percent of temperate forests that remain have not been dramatically disturbed

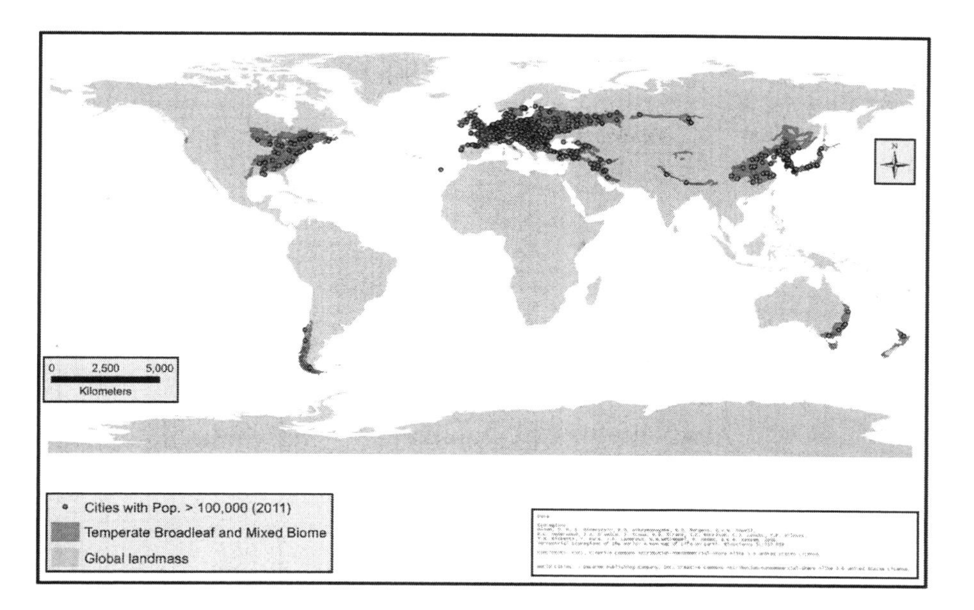

Figure 7.3 Major cities in temperate broadleaf and mixed biomes. This map shows cities with 2011 populations >100,000 that are located in temperate broadleaf and mixed biomes. These cities would be the most comparable to the case study surface drinking water systems presented in this book (Modified after Dudley and Stolton, 2003).

and altered by human activities (Reich and Frelich, 2002). Most temperate forest regions are either stable or increasing their forest cover as they recover from a similar history of land clearing and subsequent forest regrowth via either natural regeneration or afforestation programs (Tyrrell et al., 2012).

In a similar theme, tropical moist forests of Latin America and Southeast Asia have undergone a more recent period of deforestation (within the last 50 years) after which new secondary vegetation has regrown, much of this being very different in species composition and structure to the original forest (Wright, 2005; Dent and Wright, 2009; Hodgman et al., 2012) However, some of the patterns in land use history and development are similar to that in temperate moist biomes. This is especially the case for Central America (Chazdon, 2003; Hodgman et al., 2012), the Pacific coastal regions of the Northern Andes, and the Atlantic forest region of Brazil, where large cities and population centers (often coastal) are adjacent to uplands that supply surface drinking water. This is also the case for parts of Southeast Asia, especially Java, peninsula Malaysia, and the Philippines. Examples of cities and regions can be listed as follows: (1) Central America (Panama City, San José); 2) Atlantic coastal Brazil (Salvador, Vitoria, Rio de Janeiro, São Paulo); (3) Pacific Northern Andes of Latin America (Caracas, Cartagena, Guayaquil); and (4) Southeast Asia (Jakarta, Kuala Lumpur, Singapore, Manila). Though year-round warmth, the high even distribution of rainfall, the land use history of deforestation and now second-growth and urbanization of the coast, and the dependence of the uplands for water make the surface drinking watersheds of these regions very relevant beyond the temperate context of this book.

7.6.4 "Running Pure" and Forested Systems

As the case studies presented in this book show, forests provide the essential, and perhaps most effective, first barrier of water quality and quantity protection. Forests can act as natural filters and water retention systems while at the same time regulating and moderating flows. Forests protect soils and other factors such as stream water temperature and thereby provide ancillary benefits for aquatic life and other systems that rely on and develop in forested systems (de la Cretaz and Barten, 2007; Stanton et al., 2010).

A broad-scale survey and analysis of the importance of protected forests for maintaining drinking water quality for large cities is provided by the World Bank/World Wildlife Fund Alliance for Forest Conservation and Sustainable Use (Dudley and Stolton, 2003). This report, entitled "Running Pure: The Importance of Forest Protected Areas to Drinking Water," takes the largest 105 cities in the world and analyzes the degree to which their drinking water is protected and supplied by forests.

The conclusions from the report support many of the conclusions and recommendations that the authors of this book highlight (Dudley and Stolton, 2003). At least one-third of the world's largest cities obtain their drinking water from forested areas that are in some form protected (Dudley and Stolton, 2003). The drinking water supply systems for cities (populations greater than 100,000) identified in the report for temperate broadleaf and mixed biomes provide an initial set of candidates for comparison with our case study forested watershed systems (Table 7.4, Figure 7.4).

With the exception of the cities of New York and Boston, the case studies presented in this book are smaller than many of the systems that "Running Pure" examines. Northeast Asia (China), Western and Central Europe, and the Western and Eastern United States are home to an overwhelming majority of these larger population centers. In our analysis, we have chosen to focus on cities in temperate biomes outside of the United States, though, as previously mentioned, we believe there are parallels to be made for regions in certain tropical forest biomes with similar development histories. It is important to note, however, that the lessons learned in our case studies may most immediately have the greatest applicability in the United States itself. Outside the United States perhaps the most obvious similarity in resource issues are with Northeast Asia because of its temperate continental climate and long history of deforestation.

7.7 QUABBIN AND DAHUOFANG RESERVOIRS: EXAMPLE OF A COMPARATIVE ANALYSIS

Aside from the cities identified by "Running Pure," there are other places where there is wider applicability of the lessons learned in the Northeast. One such example

Table 7.4 Large Cities in Temperate Biomes outside the United States; All These Cities Have Forested Watersheds Supplying All (or Part) of Their Drinking Water Supply; the "Running Pure Report" Identified These Cities as Important

City, Country	Current Watershed Considerations	Literature
	Asia	
Seoul, South Korea	Comprehensive Han River Development Plan in place Nearly 40 years of water quality data Shrinking green space in watershed	(Chang, 2005) (Chang, 2008) (Hwang et al., 2007)
Busan, South Korea	Intense development pressure Largely gray infrastructure response	(Park and Lee, 2002) (Hwang et al., 2007)
Tokyo, Japan	Focus on gray infrastructure improvement and renovation Land acquisition in watershed (trial stage, 2009)	(BWTMG, 2009, 2010)
Beijing, China	Intense development pressure Intense agricultural development pressure Extreme seasonal variation in precipitation Large-scale afforestation to protect water quality	(Shuhuai et al., 2001) (Peisert and Sternfeld, 2005)
Shanghai, China	Rapid urbanization leading to water degradation in Hunagpu River Water emission trading (1987)	(Ren et al., 2003) (Bennett, 2009)
Tianjin, China	Drought and conflicts with Beijing over reservoir use Interbasin water transfer project	(Peisert and Sternfeld, 2005) (Jiaqi and Jun, 1999)
	Europe	
Bucharest, Romania	Drilled wells augmenting surface water supplies System privatized in 2000	(Chiru, 2006) (Ingram, 2003)
Istanbul, Turkey	Multiple use management in forest reservoir-based system Illegal human settlement issues	(Bekiroğlu and Eker, 2011)
Kharkiv, Ukraine	Reliant on surface water Pollution issues post soviet union	(Khmelko, 2012)
Kiev, Ukraine	Very little forested watershed Dnipro River severely polluted	(Dudley and Stolton, 2003; Pidlisnyuk et al., 2004)
Minsk, Belarus	Water pumped to Zaslavl Reservoir from adjacent watersheds	(Gol'dberg and Pluzhnikov, 1976) (Baranets, 2007)
Munich, Germany	Drinking water mainly from springs in forested area Payments to farmers in watershed to move to organic techniques, since 1992	(Grolleau and McCann, 2012)
Paris, France	Intense human pressure on watershed Combination of surface and ground water Focus on water treatment	(Barbier et al., 1992) (Even et al., 2007) (Billen et al., 2007)

(continued)

Table 7.4 Large Cities in Temperate Biomes outside the United States; All These Cities Have Forested Watersheds Supplying All (or Part) of Their Drinking Water Supply; the "Running Pure Report" Identified These Cities as Important (continued)

City, Country	Current Watershed Considerations	Literature
Prague, Czech Republic	Three major water treatment plants Surface water from rivers, reservoirs important Vltava River watershed 35% forested (1996)	(Möller et al., 2002) (Hejzlar et al., 1996)
Sofia, Bulgaria	Mountain watersheds serving lowlands Iskur Reservoir, interbasin transfers and conflict Long history of water projects	(Knight and Staneva, 1996a) (Knight and Staneva, 1996b) (Clark and Wang, 2003)
Stockholm, Sweden	Municipally owned water companies performing superbly Multiuse, active forest management in watershed	(Lobina and Hall, 2000) (Barucq et al., 2006) (Svensk Skogscertifiering AB, 1998)
Vienna, Austria	Forested watersheds managed specifically for drinking water	(Vacik and Lexer, 2001)
Warsaw, Poland	Main supply from three reservoirs, largely riverfed 60% of vistula watershed in agricultural production	(Denczew, 2001) (Andrulewicz, 2008)
	Oceania	
Brisbane, Australia	Forested watershed management historically very important Majority of current water passes through Mt. Crosby water treatment plan	(Dudley and Stolton, 2003) (Traves et al., 2008)
Melbourne, Australia	Forested watershed with almost 50% protected	(Dudley and Stolton, 2003)
Sydney, Australia	Sydney catchment authority 25% of watershed managed especially for water quality	(Dudley and Stolton, 2003)
	Americas	
Santiago, Chile	Water rights trading system and subsidies for the poor Vegetation in watershed severely impacted by development Municipal water company supplies majority of water Protected forests in upper catchment areas	(Briscoe, 1997) (Rosegrant et al., 2000) (Romero et al., 1999) (Dudley and Stolton, 2003)

Source: Dudley and Stolton, 2003.

Note: Moscow, Nizhnij Novgorod, and Samara, Russia; and Wuhan, China were identified by Dudley and Stolton (2003) but omitted from this table due to lack of access to data and literature.

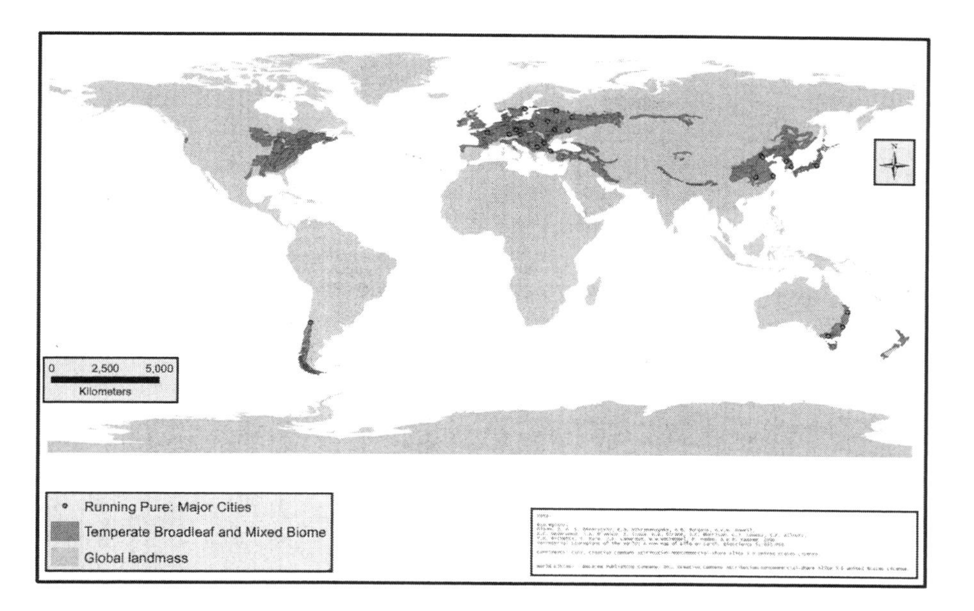

Figure 7.4 Major cities located in temperate broadleaf and mixed biomes that source drinking water from forested watersheds (Modified after Dudley and Stolton, 2003).

would be the city of Shenyang, China. Shenyang is located in the northeast of China, in Liaoning Province. It is temperate, continental, but also influenced by the monsoon. It tends to get less precipitation than the New England / New York City region, and the precipitation is concentrated in the summer months. Much of the vegetation has been cleared for agriculture (Boerma et al., 1995). The natural vegetation of the region includes similar genera to the northestern United States such as *Pinus, Quercus, Acer, Juglans,* and *Fraxinus* (Xiao et al., 2002).

Shenyang and the nearby city of Fushun rely heavily on the Dahuofang Reservoir for drinking water. This reservoir was constructed in 1958 and is comparable in size to the Quabbin Reservoir (see Table 7.5). The Dahuofang Reservoir serves significantly more people and also relies on a large-scale water transfer project that was

Table 7.5 A Comparison of the Physical Characteristics of the Quabbin and Dahuofang Reservoirs

	Quabbin Reservoir	Dahuofang Reservoir
Reservoir Capacity	1.56×10^9 m^3	1.28×10^9 m^3
Drainage Area	484 km^2	5,437 km^2
Surface area	99 km^2	114 km^2
Average Annual Precipitation	1178 mm	840 mm
Average Annual Evaporation	559 mm	940 mm
Year Completed	1939	1958

Sources: Kyker-Snowman et al., 2007; Shen et al., 2011.

completed in 2008 (Shen et al., 2011). Precipitation patterns differ between the two, with Liaoning Province much more drought prone than Massachusetts. There is also much more agricultural land and pressure to convert from the forest ecosystem in and around Dahuofang (Boerma et al., 1995).

While there are clearly many differences that complicate a comparison (land tenure, economic structure, regulatory framework), the case studies and the recommendations they lead to could provide valuable insights and points of comparison for watershed managers around the Dahuofang Reservoir. As the Quabbin watershed managers have worked to bring back forest cover and to diversify its structure and composition, so too could forest managers in Liaoning work to increase the biofilter capacities around Dahuofang (see Alcott et al., Chapter 3, this book). Both systems also have long-term monitoring systems in place that could yield valuable data and help to facilitate a comparison and a sharing of best practices (Kyker-Snowman et al., 2007; Shen et al., 2011). As the region around Dahuofang Reservoir develops, incentive and regulatory programs instituted by cities like New York and Boston will need to be implemented. An example might be to focus on source water protection in the multibarrier approach through (1) regulatory enforcement of best management practices of riparian zones, (2) payment schemes to farmers to improve management of animal wastes, and (3) eminent domain or incentive programs for land purchase by the water utility.

7.8 CONCLUSIONS

The forested watersheds that protect drinking water supplies in the northeastern United States are an esential component of the infrastructure that delivers this water to the end user. The case studies presented in this book highlight how watershed managers in this region are striving to best balance the quantities of green and gray infrastructure needed to meet the constantly changing demands that are placed upon them. By presenting these case studies together, we have shown that while different systems strike this balance in different ways, there are key similarities.

In this book, we have observed where the lessons learned from these case studies might have relevance and applicablity to other global regions. We began by identifying three lenses for analysis:

- Biophysical factors
- Forest history, land use, and infrastructure development
- Human institutions: legal, social, governmental, and regulatory environments

We then used these lenses for the Northeast and identified key factors that would either facilitate or complicate comparisons between regions. Source water protection efforts via land conservation and via payments for watershed services schemes gave us a way to bring many of the more complex human institutions into out analysis. Using PWS not only as a source protection tool and mechanism for internalizing the often externalized costs of water provision, but also as a way to analyze across watershed systems, prove to be promising areas for further research and analysis.

We then shifted to look at the biophyscial lens as way to most readily see where the lessons learned from our case studies might most immediately be helpful for watershed managers in other regions of the world. Using the work already done by Dudley and Stolton (2003), we highlighted areas where larger cities have developed in biomes similar to those of our case studies and with a similar reliance on upland, forested systems for their drinking water provision (Table 7.4). We then compared the Quabbin and Dahuofang Resrvoirs as an example to briefly illustrate how a comparison might play out and what could be gained from such an analysis. While we do not describe the details of each system we identify, we believe the protocol that we have used provides a useful way for comparing watershed across the globe. For this book the protocol can be used by the watershed manager and policymaker to identfy the common and different characteristics of their suface drinking water supply from those of northeastern United States. Such an analysis may provide some useful insights and lessons learned in our case studies to better balance the green and gray infrastructure components that serve to deliver drinking water to the end user in other regions.

REFERENCES

Aber, J., Cogbill, C., Colburn, E., D'Amato, A., Donahue, B., Driscoll, C., Ellison, A., Fahey, T., Foster, D., and B. Hall, (2010). *Wildlands and Woodlands: A Vision for the New England Landscape*. Harvard Forest, Harvard University, Petersham, MA.

Andrulewicz, E., (2008). Watershed Management System in Poland and Its Implications for Environmental Conditions of the Baltic Sea: An Example of the Vistula River Watershed. In: İ.E. Gönenç, A. Vadineanu, J.P. Wolflin, and R.C Russo (Eds.), *Sustainable Use and Development of Watersheds*,. Springer Netherlands, pp. 99–111.

Arscott, D.B., C.L. Dow, and B.W. Sweeney, (2006). Landscape Template of New York City's Drinking-Water-Supply Watersheds. *Journal of the North American Benthological Society* 25, 867–886.

Ausness, R., (1982). Water Rights Legislation in the East: A Program for Reform. Wm. & Mary L. Rev. 24, 547.

Bailey, R.G., P.E. Avers, T. King, and W.H. McNab, (1994). Ecological Subregions of the United States, section descriptions. USDA Forest Service, Ecosystem Management.

Baranets, G., 2007. Planning and Water Management for the Minsk Green Diameter. In: D.U. Vestbro (Ed.), *Rebuilding the City: Managing the Built Environment and Remediation of Brownfields*. Baltic University Press, Baltic University Urban Forum p. 10.

Barbier, J., O. Ricci, A. Montiel, and Y. Richard, (1992). Paris Improves Its Drinking Water Treatment Plants. *Water and Environment Journal* 6, 2–12.

Barry, R.G., and R.J. Chorley, (2003). *Atmosphere, Weather, and Climate*. Routledge, London, New York.

Barten, P.K., and C.E. Ernst, (2004). Land Conservation and Watershed Management for Source Protection. *Journal-American Water Works Association* 96, 121–135.

Barucq, C., J.P. Guillot, and F. Michel, (2006). Analysis of Drinking Water and Wastewater Services in Eight European Capitals: The Sustainable Development Perspective. Bureau d'Information et de Prévisions Economiques (BIPE), Issy-les-Moulineaux Cedex, France.

Bekiroğlu, S., and O. Eker, (2011). The Importance of Forests in a Sustainable Supply of Drinking Water: Istanbul Example. *African Journal of Agricultural Research* 6(7), 1794–1801.

Bennett, M., (2009). *Markets for Ecosystem Services in China. An Exploration of China's 'Eco-Compensation' and Other Market-Based Environmental Policies.* Forest Trends, Washington, DC.

Billen, G., J. Garnier, J.-M. Mouchel, and M. Silvestre, (2007). The Seine System: Introduction to a Multidisciplinary Approach of the Functioning of a Regional River System. *Science of the Total Environment* 375, 1–12.

Boerma, J.A.K., G. Luo, and B. Huang, (1995). People's Republic of China: Reference Soils of the Liaohe Plan, Liaoning Province. Soil Brief CN 11. Institute of Soil Science-Academica Sinica, Nanjing and International Soil Reference and Information Centre, Wageningen, The Netherlands, p. 16.

Brady, N.C., and R.R. Weil, (2008). *The Nature and Properties of Soils.* Pearson Prentice Hall, Upper Saddle River, NJ.

Briscoe, J., (1997). Managing Water as an Economic Good: Rules for Reformers. *Water: Economics, Management and Demand.* World Bank, 339–361.

Burns, D., T. Vitvar, J. McDonnell, J. Hassett, J. Duncan, and C. Kendall, (2005). Effects of Suburban Development on Runoff Generation in the Croton River Basin, New York, USA. *Journal of Hydrology* 311, 266–281.

Butler, B.J., (2008). Family Forest Owners of the United States, (2006). Gen. Tech. Rep. NRS-27. Newtown Square, PA: US Department of Agriculture, Forest Service, Northern Research Station 73.

BWTMG (Bureau of Waterworks Tokyo Metropolitan Government), (2009). Outline of the Tokyo Waterworks, Tokyo, Japan.

BWTMG, (2010). Tokyo Waterworks Management Plan 2010. Tokyo, Japan.

Chang, H., (2005). Spatial and Temporal Variations of Water Quality in the Han River and Its Tributaries, Seoul, Korea, 1993–2002. *Water, Air, & Soil Pollution* 161, 267–284.

Chang, H., (2008). Spatial Analysis of Water Quality Trends in the Han River Basin, South Korea. *Water Research* 42, 3285–3304.

Chapin, III, F.S., P.A. Matson, and P.M. Vitousek, (2011). Plant Carbon Budgets. In *Principles of Terrestrial Ecosystem Ecology.* New York: Springer, pp. 157–181.

Chazdon, R.L. (2003). Tropical Forest Recovery: Legacies of Human Impacts and Natural Disturbances. *Perspectives in Plant Ecology, Evolution and Systematics*, 6: 51–61.

Chiru, E., (2006). Water Supply of Bucharest—Past, Present, Future: A Study Case. In: P. Hlavinek, et al., (Eds.). *Integrated Urban Water Resources Management.* Springer, Netherlands, pp. 119–130.

Ciolkosz, E.J., W.J. Waltman, T.W. Simpson, and R.R. Dobos, (1989). Distribution and Genesis of Soils of the Northeastern United States. *Geomorphology* 2, 285–302.

Clark, W.A., and G.A.Wang, (2003). Conflicting Attitudes toward Inter-Basin Water Transfers in Bulgaria. *Water International* 28, 79–89.

Coles, J.F., T.F. Cuffney, G. McMahon, and K.M. Beaulieu, (2004). The Effects of Urbanization on the Biological, Physical, and Chemical Characteristics of Coastal New England Streams. US Department of the Interior, US Geological Survey.

Culliton, T., (1998). Population: Distribution, Density and Growth. National Oceanic and Atmospheric Administration (NOAA), *State of the Coast Report*, Silver Spring, MD.

de la Cretaz, A.L., and P.K. Barten, (2007). Land Use Effects on Streamflow and Water Quality in the Northeastern United States. Boca Raton, FL: CRC Press.

Denczew, E.S., (2001). Assessment of Infallibility of Water Supply in Warsaw. *Water International* 26, 448–450.

Dent, D., and S.J. Wright, (2009). The Future of Tropical Species in Secondary Forests: A Quantitative Review. *Biological Conservation* 142:2833–2843.

Dudley, N., and S. Stolton, (2003). Running Pure: The Importance of Forest Protected Areas to Drinking Water. A research report for the World Bank/WWF Alliance for Forest Conservation and Sustainable Use. Banco Mundial/WWF Alliance for Forest Conservation and Sustainable Use.

Ellis, E.C., and N. Ramankutty, (2008). Putting People in the Map: Anthropogenic Biomes of the World. *Frontiers in Ecology and the Environment* 6(8): 439–447.

EPA, (2006). The Multiple Barrier Approach to Public Health Protection. EPA 816-K-06-005.

Ernst, C., R. Gullick, and K. Nixon, (2004a). Conserving Forests to Protect Water. *Opflow* 30, 1–7.

Ernst, C., K. Hopper, and D. Summers, (2004b). Protecting the Source: Land conservation and the Future of America's Drinking Water. Trust for Public Land.

Evans, A., and R. Perschel, (2009). A Review of Forestry Mitigation and Adaptation Strategies in the Northeast U.S. *Climatic Change* 96, 167–183.

Even, S., G. Billen, N. Bacq, S. Théry, D. Ruelland, J. Garnier, P. Cugier, M. Poulin, S. Blanc, F. Lamy, and C. Paffoni, (2007). New Tools for Modelling Water Quality of Hydrosystems: An Application in the Seine River Basin the Frame of the Water Framework Directive. *Science of the Total Environment* 375, 274–291.

Foster, D.R., (1992). Land-Use History (1730–1990) and Vegetation Dynamics in Central New England, USA. *Journal of Ecology*, 753–771.

Gol'dberg, P., and V. Pluzhnikov, (1976). The Vileyka-Minsk Water System. *Power Technology and Engineering* (formerly *Hydrotechnical Construction*) 10, 1218–1222.

Griffith, G.E., J.M. Omernik, S.A. Bryce, J. Royte, W.D. Hoar, J.W. Homer, D. Keirstad, K.J. Metzlter, and J.G. Hellyer, (2009). Ecoregions of New England (color poster with map, descriptive text, summary tables, and photographs). Reston, VA: U.S. Geological Survey.

Grolleau, G., and L.M.J. McCann, (2012). Designing Watershed Programs to Pay Farmers for Water Quality Services: Case Studies of Munich and New York City. *Ecological Economics* 76, 87–94.

Hamilton, L.S., (2008). Forests and Water: A Thematic Study Prepared in the Framework of the Global Forest Resources Assessment 2005. Food and Agriculture Organization of the United Nations.

Hanson, C., J. Talberth, and L. Yonavjak, (2011). Forests for Water: Exploring Payments for Watershed Services in the U.S. South. World Resources Institute, Issue Brief 2.

Hejzlar, J., V. Vyhnalek, J. Kopacek, and J. Duras, (1996). Sources and Transport of Phosphorus in the Vltava River Basin (Czech Republic). *Water Science and Technology* 33, 137–144.

Hodgkins, G., R. Dudley, and T. Huntington, (2003). Changes in the Timing of High River Flows in New England over the 20th Century. *Journal of Hydrology* 278, 244–252.

Hodgman, T., J. Munger, J.S. Hall, and M.S. Ashton, (2012). Managing Afforestation and Reforestation for Carbon Sequestration: Considerations for Land Managers and Policymakers. In: M.S. Ashton, M.L. Tyrrell, D. Spalding, and B. Gentry (Eds.). *Managing Forest Carbon in a Changing Climate*. New York: Springer-Verlag, pp. 227–255.

Huntington, T.G., G.A. Hodgkins, B.D. Keim, and R.W. Dudley, (2010). Changes in the Proportion of Precipitation Occurring as Snow in New England (1949–2000).

Hwang, S.-J., S.-W. Lee, J.-Y. Son, G.-A. Park, and S-J.Kim, (2007). Moderating Effects of the Geometry of Reservoirs on the Relation between Urban Land Use and Water Quality. *Landscape and Urban Planning* 82, 175–183.

Ingram, G., (2003). Project Performance Assessment Report Romania: Bucharest Water Supply Project. Sector and Thematic Evaluation Group, Operations Evaluation Department. The World Bank. Washington, DC. Report No: 26002. http://lnweb90.worldbank.org/oed/oeddoclib.nsf/DocUNIDViewForJavaSearch/D774017DDF1C95A785256D42007B6222/$file/Romania_PPAR_26002.pdf

Irland, L.C., (1999). The Northeast's Changing Forests. Distributed by Harvard University Press for Harvard Forest, Petersham, MA.

Jiaqi, C., and X. Jun, (1999). Facing the Challenge: Barriers to Sustainable Water Resources Development in China. *Hydrological Sciences Journal* 44, 507–516.

Kargbo, D.M., R.G. Wilhelm, and D.J. Campbell, (2010). Natural Gas Plays in the Marcellus Shale: Challenges and Potential Opportunities. *Environ Sci Technol* 44, 5679–5684.

Keim, B., (1998). A Climate Primer for New England. New England's Changing Climate, Weather, and Air Quality. University of New Hampshire, Climate Change Research Center. Available from http://airmap. unh. edu/background/ClimatePrimer. html [accessed 12 February 2008].

Khmelko, I., (2012). Administrative Decentralization in Post-Communist Countries: The Case of Water Management in Ukraine. *Journal of Poltical Science, Government and Politics* 1(1). http://www.scientificjournals.org/journals2012/articles/1512.pdf

Kirshen, P., and N. Fennessey, (1995). Possible Climate-Change Impacts on Water Supply of Metropolitan Boston. *Journal of Water Resources Planning and Management* 121, 61–70.

Knight, C.G., and M.P. Staneva, (1996a). The Water Resources of Bulgaria: An Overview. *GeoJournal* 40, 347–362.

Knight, C.G., and M.P.Staneva, (1996b). Water Resources of Bulgaria: Introduction to a Special Issue of *GeoJournal*. *GeoJournal* 40, 343–345.

Kyker-Snowman, T., D. Clark, H. Eck, and J. French, (2007). Quabbin Reservoir Watershed System: Land Management Plan 2007–2017. Massachusetts Department of Conservation and Recreation, Office of Watershed Management, Belchertown, MA.

Lobina, E., and D. Hall, (2000). Public Sector Alternatives to Water Supply and Sewerage Privatization: Case Studies. *International Journal of Water Resources Development* 16, 35–55.

Lorimer, C.G., and A.S. White, (2003). Scale and Frequency of Natural Disturbances in the Northeastern US: Implications for Early Successional Forest Habitats and Regional Age Distributions. *Forest Ecology and Management* 185, 41–64.

McGranahan, G., D. Balk, and B. Anderson, (2007). The Rising Tide: Assessing the Risks of Climate Change and Human Settlements in Low Elevation Coastal Zones. *Environment and Urbanization* 19, 17–37.

McMahon, G., and T.F. Cuffney, (2000). Quantifying Urban Intensity in Drainage Basins for Assessing Stream Ecological Conditions. *Journal of the American Water Resources Association* 36, 1247–1261.

McWilliams, W.H., (2005). The Forests of Maine: 2003. US Dept. of Agriculture, Forest Service, Northeastern Research Station.

MDC/DWM (Metropolitan District Commission/Division of Watershed Management), (2001). Wachusett Reservoir Watershed: Land Management Plan 2001–2010. http://www.mass.gov/dcr/watersupply/watershed/documents/WachusettLMPES.pdf

Meck, S., (1996). Model Planning and Zoning Enabling Legislation: A Short History. *Modernizing State Planning Statutes: The Growing Smart Working Papers* 1, 1–17.

Möller, P., T. Paces, P. Dulski, and G. Morteani, (2002). Anthropogenic Gd in Surface Water, Drainage System, and the Water Supply of the City of Prague, Czech Republic. *Environ Sci Technol* 36, 2387–2394.

Nolon, J.R., P.E. Salkin, and M. Gitelman, (2008). *Cases and Materials on Land Use and Community Development.* Thomson/West, St. Paul, MN.

Oliver, C.D., and B.C. Larson, (1996). *Forest Stand Dynamics.* New York: John Wiley & Sons.

Olson, D.M., E. Dinerstein, E.D. Wikramanayake, N.D. Burgess, G.V.N. Powell, E.C. Underwood, J.A. D'amico, I. Itoua, H.E. Strand, and J.C. Morrison, (2001). Terrestrial Ecoregions of the World: A New Map of Life on Earth. *BioScience* 51, 933–938.

Paddock, L.R.C., (1990). Federal and State Roles in Environmental Enforcement: A Proposal for a More Effective and More Efficient Relationship, *The. Harv. Envtl. L. Rev.* 14, 7.

Park, S.S., and Y.S. Lee, (2002). A Water Quality Modeling Study of the Nakdong River, Korea. *Ecological Modelling* 152, 65–75.

Peisert, C., and E. Sternfeld, (2005). Quenching Beijing's Thirst: The Need for Integrated Management for the Endangered Miyun Reservoir. *China Environmen Series* 7, 33–45.

Pidlisnyuk, V., M. Borisyuk, and I. Pidlisnyuk, (2004). Sustainable Use of Water Resources: Perspective for Ukraine. Proceeding of the International Conference Integrated Management of Natural Resources in the Transboundary Dnister River Basin. Kyshyney, Moldova. 2004.

Porras, I., M. Grieg-Gran, and N. Neves, (2009). *All That Glitters: A Review of Payments for Watershed Services in Developing Countries.* International Institute for Environment and Development, London, UK.

Reich, P., and L. Frelich, (2002). Temperate Deciduous Forests. In: H.A. Mooney and J.G. Canadell (Eds.), *Encyclopedia of Global Environmental Change.* John Wiley &Sons, Ltd., Chichester.

Ren, W., Y. Zhong, J. Meligrana, B. Anderson, W.E. Watt, J. Chen, and H-L. Leung, (2003). Urbanization, Land Use, and Water Quality in Shanghai: 1947–1996. *Environment International* 29, 649–659.

Romero, H., M. Ihl, A. Rivera, P. Zalazar, and P. Azocar, (1999). Rapid Urban Growth, Land-Use Changes and Air Pollution in Santiago, Chile. *Atmospheric Environment* 33, 4039–4047.

Rosegrant, M.W., C. Ringler, D.C. McKinney, X. Cai, A. Keller, and G. Donoso, (2000). Integrated Economic-Hydrologic Water Modeling at the Basin Scale: The Maipo River Basin. *Agricultural Economics* 24, 33–46.

Shen, Y., J. Wang, B. Zheng, H. Zhen, Y. Feng, Z. Wang, and X.Yang, (2011). Modeling Study of Residence Time and Water Age in Dahuofang Reservoir in China. *Science China Physics, Mechanics & Astronomy* 54, 127–142.

Sherk, G.W., (1989). Eastern Water Law: Trends in State Legislation. *Va. Envtl. LJ* 9, 287.

Shuhuai, D., G. Zhihui, H.M. Gregersen, K.N. Brooks, and P.F. Ffolliott, (2001). Protecting Beijing's Municipal Water Supply through Watershed Managemen and Economic Assessment. *Journal of the American Water Resources Association* 37, 585–594.

Small, C., and J.E. Cohen, (2004). Continental Physiography, Climate, and the Global Distribution of Human Population. *Current Anthropology* 45, 269–277.

Small, C., and R.J. Nicholls, (2003). A Global Analysis of Human Settlement in Coastal Zones. *Journal of Coastal Research*, 584–599.

Stanton, T., M. Echavarria, K. Hamilton, and C. Ott, (2010). State of Watershed Payments: An Emerging Marketplace. *Ecosystem Marketplace*.

Svensk SkogsCertifiering AB, (1998). Draft Certification Summary on FSC-Evaluation of Stockholm Vatten AB. SSC Forestry, Uppsala, Sweden.

Tibbetts, J., (2002). Coastal Cities: Living on the Edge. *Environmental Health Perspectives* 110(11), A674–A681.

Traves, W., E. Gardner, B. Dennien, and D. Spiller, (2008). Towards Indirect Potable Reuse in South East Queensland. *Water Science & Technology* 58, 153–161.

Tyrrell, M.L., J. Ross, and M.J. Kelty, (2012). Carbon Dynamics in the Temperate Forest. In M.S. Ashton, M.L. Tyrrell, D. Spalding, and B. Gentry (Eds.), *Managing Forest Carbon in a Changing Climate*. New York: Springer-Verlag, pp. 77–107.

Vacik, H., and M.J. Lexer, (2001). Application of a Spatial Decision Support System in Managing the Protection Forests of Vienna for Sustained Yield of Water Resources. *Forest Ecology and Management* 143, 65–76.

van Velthuizen, H., B. Huddleston, G. Fischer, M. Salvatore, E. Ataman, F. Nachtergaele, M. Zanetti, M. Bloise, A. Antonicelli, J. Bel, and A. De Liddo, (2008). Mapping Biophysical Factors That Influence Agricultural Production and Rural Vulnerability. *Environment and Natural Resources Series* 11. Food and Agriculture Organization of the United Nations, Rome.

Vogel, R.M., N.M. Fennessey, and R.A. Bolognese, (1995). Storage-Reliability-Resilience-Yield Relations for Northeastern United States. *Journal of Water Resources Planning and Management* 121, 365–374.

Whitney, G.G., (1994). *From Coastal Wilderness to Fruited Plain: A History of Environmental Change in Temperate North America, 1500 to the Present.* New York: Cambridge University Press.

Wickham, J.D., T.G. Wade, and K.H. Riitters, (2011). An Environmental Assessment of United States Drinking Water Watersheds. *Landscape Ecology*, 1–12.

Wright, S.J. (2005). Tropical Forests in a Changing Environment. *Trends in Ecology and Evolution* 10: 553–560.

Xiao, X., S. Boles, J. Liu, D. Zhuang, and M. Liu, (2002). Characterization of Forest Types in Northeastern China, Using Multi-Temporal SPOT-4 VEGETATION Sensor Data. *Remote Sensing of Environment* 82, 335–348.

Zielinski, G.A., and B.D. Keim, (2003). New England Weather, New England Climate. New Lebanon, NH: University Press of New England.

Index

G

H

I

J

K

L

M

N

O

P

Q

R

S

V

W

X

Z